MW01274254

The Enlightenment of Thomas Beddoes

Thomas Beddoes (1760–1808) lived in "decidedly interesting times" in which established orders in politics and science were challenged by revolutionary new ideas. Enthusiastically participating in the heady atmosphere of Enlightenment debate, Beddoes' career suffered from his radical views on politics and science. Denied a professorship at Oxford, he set up a medical practice in Bristol in 1793. Six years later – with support from a range of leading industrialists and scientists including the Wedgwoods, Erasmus Darwin, James Watt, James Keir and others associated with the Lunar Society – he established a Pneumatic Institution for investigating the therapeutic effects of breathing different kinds of "air" on a wide spectrum of diseases.

The treatment of the poor, gratis, was an important part of the Pneumatic Institution and Beddoes, who had long concerned himself with their moral and material well-being, published numerous pamphlets and small books about their education, wretched material circumstances, proper nutrition, and the importance of affordable medical facilities. Beddoes' democratic political concerns reinforced his belief that chemistry and medicine should co-operate to ameliorate the conditions of the poor. But those concerns also polarized the medical profession and the wider community of academic chemists and physicians, many of whom became mistrustful of Beddoes' projects due to his radical politics.

Highlighting the breadth of Beddoes' concerns in politics, chemistry, medicine, geology, and education (including the use of toys and models), this book reveals how his reforming and radical zeal were exemplified in every aspect of his public and professional life, and made for a remarkably coherent program of change. He was frequently a contrarian, but not without cause, as becomes apparent once he is viewed in the round, as part of the response to the politics and social pressures of the late Enlightenment.

Trevor Levere is University Professor Emeritus in the Institute for the History and Philosophy of Science and Technology, University of Toronto, Canada.

Larry Stewart is a Professor in the Department of History, University of Saskatchewan, Canada.

Hugh Torrens is Emeritus Professor of History of Science and Technology at Keele University, UK.

Joseph Wachelder is an Associate Professor in the Department of History, Maastricht University, the Netherlands.

Science, Technology and Culture, 1700–1945

Science, Technology and Culture, 1700–1945 focuses on the social, cultural, industrial, and economic contexts of science and technology from the "scientific revolution" up to the Second World War. Publishing lively, original, innovative research across a broad spectrum of subjects and genres by an international list of authors, the series has a global compass that concerns the development of modern science in all regions of the world. Subjects may range from close studies of particular sciences and problems to cultural and social histories of science, technology, and biomedicine; accounts of scientific travel and exploration; transnational histories of scientific and technological change; monographs examining instruments, their makers and users; the material and visual cultures of science; contextual studies of institutions and of individual scientists, engineers and popularizers of science; and well-edited volumes of essays on themes in the field.

Also in the series

Francis Watkins and the Dollond Telescope Patent Controversy
Brian Gee, edited by Anita McConnell and A. D. Morrison-Low

From Local Patriotism to a Planetary Perspective
Martina Kölbl-Ebert

Science Policies and Twentieth-Century Dictatorships
Amparo Gómez, Antonio Fco. Canales, and Brian Balmer

Entrepreneurial Ventures in Chemistry: The Muspratts of Liverpool, 1793–1934
Peter Reed

Barcelona: An Urban History of Science and Modernity, 1888–1929
Oliver Hochadel and Agustí Nieto-Galan

Pursuing the Unity of Science: Ideology and Scientific Practice from the Great War to the Cold War
Harmke Kamminga and Geert Somsen

The Enlightenment of Thomas Beddoes: Science, Medicine, and Reform
Trevor Levere, Larry Stewart, and Hugh Torrens, with Joseph Wachelder

The Enlightenment of Thomas Beddoes

Science, medicine, and reform

Trevor Levere, Larry Stewart, and Hugh Torrens, with Joseph Wachelder

LONDON AND NEW YORK

First published 2017
by Routledge
2 Park Square, Milton Park, Abingdon, Oxon OX14 4RN

and by Routledge
711 Third Avenue, New York, NY 10017

Routledge is an imprint of the Taylor & Francis Group, an informa business

British Library Cataloguing in Publication Data
A catalogue record for this book is available from the British Library

Library of Congress Cataloging-in-Publication Data
Names: Levere, Trevor Harvey, editor. | Stewart, Larry, 1946– editor. | Torrens, H. S., editor. | Wachelder, Joseph, editor.
Title: The enlightenment of Thomas Beddoes : science, medicine and reform / [edited by] Trevor Levere, Larry Stewart, and Hugh Torrens with Joseph Wachelder.
Description: Milton Park, Abingdon, Oxon ; New York, NY : Routledge, 2017. | Series: Science, technology and culture | Includes index.
Identifiers: LCCN 2016021376 | ISBN 9781472488299 (hardback) | ISBN 9781315411934 (e-book)
Subjects: LCSH: Beddoes, Thomas, 1760–1808. | Physicians—England—Biography. | Medicine—England—History—18th century.
Classification: LCC R489.B4 E55 2017 | DDC 610.69/5—dc23
LC record available at https://lccn.loc.gov/2016021376

ISBN: 978-1-4724-8829-9 (hbk)
ISBN: 978-1-315-41193-4 (ebk)

Typeset in Sabon
by Apex CoVantage, LLC

Frontispiece: Thomas Beddoes, pencil drawing by Edward Bird. Wellcome Library, London.

To Jennifer, Shirley, and Uta

Contents

Acknowledgements

Many librarians and archivists have helped us, some over the decades in which we severally engaged with Beddoes, and we are much in their debt. We thank especially but not exclusively those in the Birmingham Central Library and Record Office, the Cornwall Record Office, Keele University Archives, the National Archives (Kew), the National Library of Scotland, the Wedgwood Museum, and the British Library. We are also indebted to Beddoes' earlier biographers, Roy Porter, Dorothy Stansfield, and Mike Jay. Trevor Levere thanks Robert Anderson, Rebecca Bowd, Heather Jackson, and David Knight, who read and commented on individual chapters, Kristen Schranz and Victor Boantza for impeccable research assistance, and the Social Sciences Research Council of Canada for their support. Hugh Torrens has incurred specific debts to Karen Bieterman (Schaumberg IL, USA), Alistair McLean (Sheffield), and Monica Price (Oxford), which are gratefully acknowledged. Jo Wachelder thanks Andrea Immel (Cotsen Children's Library (Princeton), Emma Jefkins (Science Museum London) and Tom Slootweg (Groningen University) for their help. We have exchanged, commented on, and edited one another's chapters, to our mutual benefit. And above all, we thank our spouses, who have had to live with Thomas Beddoes over the years, and to whom we dedicate this book.

Illustrations

[Unattributed photos are by Levere, from his library.]

Abbreviations

APS American Philosophical Society, Philadelphia
BCL Birmingham Central Library
BL British Library
Bodley The Bodleian Library, Oxford University
BWP Boulton and Watt Papers
C Chancery papers
CBL *A Catalogue of the very valuable and extensive library of Thomas Beddoes, M.D. of Clifton, near Bristol, lately deceased: containing a very capital collection of modern publications in all the departments of surgery and medicine; voyages and travels, antiquities, natural history and belles lettres: likewise all the late best German writers on the above subjects, which will be sold by auction by Leigh and S. Sotheby, Booksellers, at their House, No. 145, Strand, opposite Catherine Street, On Friday, November 10, 1809, and Nine following Days, (Sundays excepted) at 12 o'Clock*, Sotheby 60 (2), annotated auctioneer's copy in British Library (currently missing; there is a microfilm copy).
CED Desmond King-Hele, ed., *The Collected Letters of Erasmus Darwin* (Cambridge: Cambridge University Press, 2007)
CJB Robert G. W. Anderson and Jean Jones, eds., *The Correspondence of Joseph Black*, 2 vols. (Farnham, Surrey & Burlington, VT, 2012)
C. L. MS notes of Joseph Black, *Chemical Lectures* (ca. 1785), Royal College of Physicians of Edinburgh, MS 9.42, 9.43
Considerations Thomas Beddoes, M.D., and James Watt, Engineer, *Considerations on the medicinal use, and on the production of factitious airs Part I. By Thomas Beddoes, M.D, Part II. Description of an Air Apparatus; with hints respecting the use and properties of different elastic fluid. By James Watt, Esq.*; *Pt. III Considerations on the medicinal use, and on the production of factitious airs* [as observed or felt by several others] (London: J. Johnson, 1795); *Pts. IV & V, Medical Cases and Speculations; including [Part V] Supplement to the Description of a Pneumatic Apparatus for Preparing Factitious Airs; Containing a Description of a Simplified Apparatus, and of a Portable Apparatus. By James Watt, Engineer*, 3rd ed., corrected and enlarged (London: J. Johnson, 1796).

CRO	Cornwall Record Office, Truro
DSB	*Dictionary of Scientific Biography*, Charles Coulston Gillispie editor in chief (New York: Scribners, 18 vols. in 8, 1981)
EUL	Edinburgh University Library
FLS	Fellow of the Linnean Society
FRS	Fellow of the Royal Society of London
GRO	Gloucestershire Record Office
HO	Home Office papers
Jay	Mike Jay, *The Atmosphere of Heaven: The Unnatural Experiments of Dr. Beddoes and his Sons of Genius* (New Haven and London: Yale University Press, 2009)
JDB	Joseph DeBoffe
JHT	John Horne Tooke
JMA	John Murray Archives
JWP	James Watt Papers. The James Watt Papers were formerly in Doldowlod House, Wales, and are now in the Birmingham Central Library. They have been numbered and renumbered; they may be found with the help of archivists in BCL using either the old or the new set of numbers.
MBP	Matthew Boulton Papers
Memorial	Thomas Beddoes, *A Memorial Concerning the State of the Bodleian Library, and the Conduct of the Principal Librarian. Addressed to the Curators of that Library, by the Chemical Reader* (Oxford: 1787)
M.P.	Member of Parliament
MS	manuscript
MS DD DG	Davies Gilbert (Giddy) archive
MSS	manuscripts
NA	National Archives, Kew
NLS	National Library of Scotland
ODNB	*Oxford Dictionary of National Biography*, online edition; print edition, *Oxford Dictionary of National Biography*, 62 vols. (Oxford: Oxford University Press, 2014)
OED	*The Oxford English Dictionary*, 2nd edn., 20 vols. (Oxford: Oxford University Press, 1989)
Phil. Trans.	*Philosophical Transactions of the Royal Society of London*
Porter	Roy Porter, *Doctor of Society: Thomas Beddoes and the Sick Trade in Late- Enlightenment England* (London and New York: Routledge, 1992)
Stansfield	Dorothy Stansfield, *Thomas Beddoes M.D. 1760–1848* (Dordrecht: Reidel, 1984)
Stock (1811)	John Edmonds Stock, *Memoirs of the Life of Thomas Beddoes, M.D. with an Analytical Account of His Writings* (London: John Murray, 1811)

WC	Wedgwood correspondence
WMB	Wedgwood Museum, Barlaston, Staffs
W/M	Wedgwood-Moseley MSS

The Wedgwood papers were for some years held by Keele University Library; they have since been removed to the Wedgwood Museum.

Introduction

Trevor Levere and Larry Stewart

> [Thomas Beddoes is] a little fat Democrat of considerable abilities, of great name in the Scientific world as a naturalist and Chemist – good humored good natured – a man of honor & Virtue, enthusiastic & sanguine . . . His manners are not polite – but he is sincere & candid – . . . . The Doctor will settle at Clifton and if he will put off his political projects till he has accomplish'd his medical establishment he will succeed and make a fortune – But if he bloweth the trumpet of Sedition the Aristocracy will rather go to hell with Satan than with any democratic Devil.[1]

By the time that Edgeworth wrote his vignette, bracketed as it was by executions, France had ended the reign and life of Louis XVI in January 1793; he was soon to be followed to the scaffold by Marie Antoinette in October. Promise would descend into Terror in the unravelling of what had once appeared to have begun simply as a constitutional crisis, an echo of America and beloved of bourgeois reformers. In the English Midlands, across la Manche, James Watt and Josiah Wedgwood, with many others, watched with alarm as Britain and France were about to descend into war, their fears as much over the collapse of trade as political disorder. For his part, Beddoes would vigorously and publicly oppose British policy on numerous grounds.

France was the touchstone of much British debate, especially after 1789. As with the orthodox champion Edmund Burke from 1790, so too with the radical Beddoes there could be no misunderstanding of views on the importance of rights or of the collapse of constitutions. But soon, in Britain, any criticism of the government or the King would become illegal. In 1792, a propitious year in many respects, a proclamation by King George III had guarded his subjects against seditious writings, and against correspondence with democratic and antimonarchical individuals like Tom Paine and with organizations like corresponding societies agitating for reform. Ultimately, 1795 saw the publication of the two acts and trials for treason and sedition, which became standard instruments of Pittite repression in Britain. During the 1790s, forty-five individuals were tried for treason, forty-six for sedition, and twenty-two for political conspiracy. Of those found guilty, nineteen were imprisoned, thirteen were executed, and seven transported.[2] Beddoes, with considerable courage in the face of government fear and heavy-handedness, continued to oppose the policies of Pitt's government, and protested the repressive legislation of Pitt and Grenville that made free speech seditious; he was the first to call their legislation "gagging bills".[3]

It is in the rising turmoil, in the arc from Tom Paine to Edmund Burke, we will find the range of Beddoes' own unease. As he wrote to Davies Giddy: "The French Revolution is every day losing that amiable aspect which it wore in the beginning. It is no longer [a] revolution cemented by water instead of blood".[4] He was horrified at the excesses of the Paris mob, Revolutionary armed forces, and the dictatorship of the Committee of Public Safety during the Reign of Terror of 1793–1794.

Yet his political writings from 1795 to 1797 leave no room for ambiguity: *A Word in Defence of the Bill of Rights against Gagging-bills* (1795), *Where would be the harm of a Speedy Peace?* (1795), *An Essay on the Public Merits of Mr. Pitt* (1796), and *Alternatives compared; or, What shall the Rich do to be Safe?* (1797). Like many appalled by the Terror in France, he had seen great promise in the French Revolution. So, he was clearly a man of some honour and virtue, and a democrat, at a time when that word had the same resonance as "communist" was to have in McCarthy's America – or, some might say, "liberal" ever since. Government suspicions made Beddoes the subject of investigation by a spy from the Home Office who found, amidst the defence of security and fears of regicide, that he had been corrupting his students by advocating democracy. He clearly lived in decidedly interesting times, in which he was active and highly visible, to the detriment of his career and, as we shall see, to his plans for a medical reformation. But his career was no muddle of democratic airs or cauldron of chemical, medical, and geological innovation. His was a life of enlightenment and principle.

Beddoes had taken his B.A. at Oxford, studied anatomy under John Sheldon in London, the leading teacher of anatomy of his day, and then gone on to study medicine and chemistry at Edinburgh, where Joseph Black was Europe's foremost teacher of chemistry. Then, after visiting Lavoisier in Paris in 1787, he came back to Oxford to lecture in chemistry. He did this well enough for the Vice-Chancellor to want to appoint him to a chair in chemistry, only to be frustrated by the Home Office's fear of democrats. He had meanwhile given lectures in geology, and published three important papers on geology and chemistry in the *Philosophical Transactions of the Royal Society*, as well as translations of books on natural history and chemistry. He was early in adopting James Hutton's uniformitarian geology. He corresponded with Guyton de Morveau, whom he had met in Dijon, with Joseph Black in Edinburgh about a wide range of chemical topics, and with Erasmus Darwin in Derby about geology, mineralogy, and medicine. He was, in short, as his prospective father-in-law observed, a man "of considerable abilities, of great name in the Scientific world as a naturalist and Chemist".[5]

While at Oxford, Beddoes had spoken and written in favour of the French Revolution, against the abuse of power by Warren Hastings and the East India Company, and about the afflictions of Britain's poor. When the government's policy of repression cost him a professorship at Oxford, Beddoes fell back upon his medical qualification. He moved to Bristol, set up a medical practice, and – with support from the Watts, Wedgwoods, the chemical manufacturer James Keir, and others associated with the Lunar Society of Birmingham, as well as from his friend the ironmaster William Reynolds of Ketley, and Georgiana, Duchess of Devonshire – began a campaign for a Pneumatic Institution. This would be a practical site for investigating the therapeutic effects of breathing different kinds of air (often ill-defined gases produced in the laboratory or workshop) on a wide spectrum of diseases. He would thereby combine research and treatment, like many pneumatic enthusiasts often first testing the various gases on himself. His principal target was consumption, which we call tuberculosis, then a scourge at every level of society. He had high hopes for pneumatic medicine, the application of the latest discoveries of pneumatic chemistry to the treatment of respiratory diseases, and more. Erasmus Darwin cheered him on, and James Watt joined Beddoes in the development of apparatus for breathing the gases, essential to any pneumatic therapy.

The treatment of the poor, gratis, was an important part of the Pneumatic Institution, and Beddoes had long concerned himself with their moral and material well-being. He wrote pamphlets and small books about their education and their increasingly wretched material circumstances, proper nutrition, and the importance of available and affordable medical facilities. Chemistry and medicine were to co-operate in ameliorating the conditions of the poor, and Beddoes' democratic political concerns reinforced his social conscience. But those concerns also polarized the medical profession, and the wider community of academic chemists and physicians – all contaminated by political alarms. Joseph Banks would not support Beddoes' projects because he mistrusted his politics. James Watt, who otherwise encouraged Beddoes, wished that he would keep quiet about politics, which could only do harm to the acceptance of his science. This, in the end, Beddoes admitted. Edgeworth, again, had anticipated the reaction of the aristocracy to "any democratic Devil".[6] And the landed aristocracy along with the politically well-connected, Beddoes' targets but hardly his primary medical concern, were nonetheless those with power and patronage to offer. The influential Banks, at the Royal Society, dismissed anyone who might breathe democratic airs.

Beddoes, like others, recognized that prevention was the best guarantee of health. Besides writing his pamphlets and advising his patients, he sought to compile statistics of illness, a basic record for public health. He was not the first to do this, as bills of mortality had long existed, but was among the foremost English physicians of his own day in advocating and attempting to construct such records. Much needed to be done for the labouring masses, as midlands medics increasingly acknowledged. He was, in this and in much else, a vigorous advocate of reform, or, what comes to the same thing, a constant challenger of authority across the board. This, of course, hardly advanced his popularity in many quarters. He was also a friend of the eighteenth-century notion of improvement, in which, among many other things, science, technology, and medicine could be applied to increasing yields from fields and mines, and to bettering the condition of all classes of society, including agricultural labourers,[7] but especially industrial workers. Indeed, some may have been inclined to question the profit and promise in manufactures and the costs inflicted upon many in the rapidly growing industrial towns.

In this book, the breadth of his concerns in politics, education, chemistry, medicine, and geology will be the principal subjects. We hope to reveal how Beddoes' reforming and radical zeal were exemplified in every aspect of his public and professional life, and made for a remarkably coherent program of change. He was frequently a contrarian, but not without cause, and he needs to be seen whole and in the round – as part of the response to the politics and social pressures of the late Enlightenment.

Beddoes urged reform of the traditionally conservative – and in his view, reactionary – medical profession. He did so most pointedly in the last year of his life, in a small book addressed to Joseph Banks.[8] He observed that many doctors were ill-educated, pernicious empirical medics who injured the health of the populace, while physicians in universities were convinced that their program of education needed no improvement, although it was clear to Beddoes that a six-year medical syllabus was required, with the first four years to be spent at Edinburgh University. More political yet was his view that there were too few preventive institutions and there was a dearth of knowledge needed for identifying the needs of preventive medicine, while writers on the benefits of spas were generally duplicitous, self-serving quacks:

> I was going to shew, that Bath has no less capacity for preventing the public from feeling the extermination of quacks than Bristol. But the argument is not local. In all places, there may be found medical Jansenists to fill up the room of the suppressed Jesuits.

He did not help his case by referring to French proposals for medical reforms, in the midst of the ongoing war with France, which he had long opposed.[9] His stubborn campaign proved lengthy even as it was urgent. But hectoring the privileged profession was not to be his cause alone.

Beddoes wanted to improve the lot of the poor, at a time when events across the English Channel made the authorities and the aristocracy fearful of the power of the mob. There was a profound social awareness behind Beddoes' sense of rising afflictions for which there was no certain answer. Yet simply acknowledging the dire circumstances of the unwashed was enough to raise political suspicions. This was the alarm that since 1790 Burke promoted. This was the foundation for Pitt's gagging bills of 1795, which aroused fears that revolution in France might serve as a plot for incendiaries in Britain. There had been politically charged riots, the Birmingham one of 1791 prominent among them, but that one had been encouraged by local and national public figures as a way of suppressing dissent. Local government was particularly pleased about that riot, as even in the next year in Manchester where suspicions of the motives of magistrates arose. But other mobs were not to Westminster taste, worried as it was about bread riots and miners' protests, triggered by the ever-increasing price of bread accompanied by declining agricultural and industrial wages. But Beddoes did not want revolution at home, even if for some years he was sympathetic to a republic on the French model. As fears ballooned in government, in November 1792 he wrote to his friend and former student Davies Giddy, then county sheriff of Cornwall:

> France, you see, is for ever free & a REPUBLIC, the only form of govt. consistent with honesty & common sense. . . . But this Convention, pray heaven, it mend in its paces, as it grows older. . . . Shall the King be tried, condemned & executed? I vote for this measure. . . .
>
> The falsehoods, fabricated by ministerial hirelings, the common language held by the Aristocrates, the proclamation . . . have spread a firm persuasion among the people of the unrelenting wickedness of the French patriots & their favourers in England. This impression more than any thing else seems likely to produce popular violence in England. . . . . . . I know . . . no way of guarding the people agt. the terrible effect of absurd rumours <so good> as by opening the press, the fountain of truth; nor do I think it can be so widely diffused among farmers small traders & artisans as by means of a provincial newspaper. For such reasons I have for many months been extremely desirous to establish such a paper. It will be a losing project, at first; & when solicited my democratical acquaintance to join in bearg. the expence, they refused for want of discrimination in this instance. They said "you will inflame." It was in vain for me to reply that I shd. destroy the tendency to inflammation, & what was always a beneficial, wd. every day become a more urgent measure. . . .
>
> Much is to be apprehended if the mind of the [governing] class beforementioned be not elevated a little towards the level of reform. At present, it is certain that the moral is even worse than the physical condition of the labouring poor, peasants especially.[10]

A few days later, reading the signs from the north, he wrote more pointedly:

> Venit summa dies & ineluctabile tempus,[11] as I have preached for the last 3 months to our rich democrats & Aristocrates – but in vain – they

> begin now to tremble & sing low, the latter I mean. . .. The Nth. of England & Scotland all democratic – London rapidly democratizing – Vive l'égalité, vice G – S – the K – every[where] inscribed on its patriotic walls. No cause of apprehension but in the wretched state, moral & physical, which our happy C – n in Ch – & State has left the poor.[12]

As these letters demonstrate, he insisted on political reform, and an improvement in the circumstances of the poor, including the many labourers whom he followed with much concern. These goals were for him inextricably interlinked. As we shall see, they were more than sufficient for the government to see him as downright seditious. Beddoes was anxious to humanize the poor – a view that followed from a view opposing that of magistrates and champions of orthodoxy – but he intended that they would reject calls to a violent revolution. His understanding was not therefore fragmented, but all of a piece with enlightenment in mind. He wrote for the press, he wrote to his friends, he encouraged his students, he wrote about education, and he wrote booklets and broadsheets for the poor.[13] When the results of the research program at his Pneumatic Institution in Bristol destroyed his hopes for the cure of consumption and cancer by pneumatic medicine, he changed the name of the Institute in 1802 to the Preventive Institution, and again, in 1804, to the Medical Institution, for the Sick and Drooping Poor.[14]

Beddoes likewise advocated reform in chemistry. He became a supporter of the new French system, albeit not uncritically, at a time when chemistry was associated with radicalism – simply condemned as French innovation. Burke had slyly made the point – as though French chemistry in particular was inimical to good government. French chemistry carried with it French philosophy and the politics of the "calculators" he so despised. Once a supporter of liberty in America, Burke had since become the leading spokesman for opposition to France, and that included opposition to French chemistry, which its principal inventor, Lavoisier, had once described as revolutionary. For Burke, such chemists, in England or in France, had

> dispositions that make them worse than indifferent about those feelings and habitudes, which are the supports of the moral world. . .. These philosophers consider men in their experiments, no more than they do mice in an air pump, or in a recipient of mephitick gas.[15]

One of Burke's chief bugbears[16] was Joseph Priestley, dissenting clergyman, political reformer, chemist, and more besides. In a ground-breaking work on the chemistry of different gases, different kinds of air, Priestley had stated roundly that the progress of knowledge would

> be the means, under God, of extirpating *all* error and prejudice, and of putting an end to all undue and usurped authority in the business of *religion*, as well as of *science*; and all the efforts of the interested

> friends of corrupt establishments of all kinds will be ineffectual for their support in this enlightened age. . . . [T]he English hierarchy (if there be anything unsound in its constitution) has equal reason to tremble even at an air-pump, or an electrical machine.[17]

Metaphors be damned, battle was joined over experiments of many kinds – in laboratories, in politics, in the streets stormed by the unemployed. Guilt by association meant that English supporters of the new French chemistry were deemed as revolutionary fodder, and Priestley's reforming zeal made sure of that. No wonder that he was the principal target of the Birmingham riots in 1791. Those rioters would as happily been employed building a Bastille as bringing one down – and this was the difficulty any would-be reformer had then to face.

Priestley's publisher was Joseph Johnson, the leading radical printer and bookseller of the day. Beddoes, after an early spell with John Murray in Edinburgh, turned to Johnson in London, who became his main publisher. Few others would have dared print Priestley's or Beddoes' works. Johnson paid for his radicalism: he fell foul of the repressive government, and in 1798 was handed a prison sentence.

Even Beddoes' book collecting, which was remarkable, broke conventional boundaries. While lecturing at Oxford, Beddoes penned a trenchant criticism of Bodley's Librarian and the deficiencies of the Bodleian Library.[18] Beddoes argued vigorously that its foreign holdings were inadequate, especially where German books and periodicals were concerned. Beddoes' own library was extraordinary in its size and cultural reach; and approximately half of it was German. In time of war, Beddoes ignored national boundaries and found ways to acquire German and French books. Internationalism was not in vogue in England; but in building his remarkable library, Beddoes went beyond the boundaries.

But what attitudes did he import, beyond books? What radical legacies were there in new theories, what therapies might yet arise, what attempts to cure the incurable? What were the poisons equally of politics and industry that needed to be identified and debated – in which, at the same time, even solutions might lie? The Bastille had fallen, but at the same time, in Birmingham, James Keir had written about the need for the dissemination of chemical knowledge as a possible answer to the corruptions of the world. By then many aristocrats had escaped from riot in France, and tocsins had rung among the many boulevards and clubs of Paris. Within a year, the assault on the Bourbon monarchy alarmed princes and kings, and among the magistrates who demonstrated they could equally well raise a mob. Soon anyone associated with promoting revolution, like a Priestley or a Keir, even as champions of the benign English version a hundred years earlier, was treated as a pariah. In the gathering crisis after 1789, the force of books was shown by Edmund Burke and, in bitter answer, by Tom Paine. They drew their battle lines in Britain, but agreed on nothing about the afflictions of men or even what remedies to constitutions and corruption might be discovered and tried. For these reasons, as from Burke to Paine, from Priestley to Lavoisier, Beddoes also gathered energy to challenge authority. We may fairly owe much to Beddoes' earlier biographers, particularly to the late Roy Porter. Stock's life of Beddoes is cautious and underwhelming; Stansfield wrote the most comprehensive modern account of Beddoes, while never engaging adequately with his science; Porter treated Beddoes as an anti-Thatcher physician; Neil Vickers wrote about Beddoes as Coleridge's doctor; Mike Jay's book deals with Beddoes in political context, and then concentrates on his pneumatic medicine and the Pneumatic Institution. But they have not given sufficient attention to Beddoes as chemist, geologist, educator, author, and book collector, nor have they adequately integrated these several aspects of Beddoes' work into his reforming and democratic politics.[19] Seeking to restore Beddoes to his guiding concerns, we have written a book presenting these aspects in their broad and interlaced character. Beddoes' enterprise, for all its polymathy, was a remarkably coherent one, with politics and a social conscience as driving forces.

Notes

1 Richard Lovell Edgeworth's description of Beddoes, in a postscript to his daughter Maria's letter to Mrs. R. Clifton, 21 July 1793, quoted in Marilyn Butler, *Maria Edgeworth: A Literary Biography* (Oxford: Oxford University Press, 1972), 110.

2 Kenneth R. Johnston, *Unusual Suspects: Pitt's Reign of Alarm & the Lost Generation of the 1790s* (Oxford: Oxford University Press, 2013), 329–330.

3 T. Beddoes, "Postscript", in S. T. Coleridge, *The Watchman*, in *The Collected Works of Samuel Taylor Coleridge*, ed. Lewis Patton, general editor K. Coburn (London and Princeton: Routledge and Kegan Paul and Princeton University Press, 1970), 344. Beddoes first met Coleridge in Bristol at a public meeting to protest against the gagging bills. Beddoes became a regular contributor to Coleridge's radical periodical.

4 Beddoes to Giddy 11 May 1792 [?1793], CRO (see Abbreviations) MS DG 41/3.

5 See note 1.

6 See note 1.

7 T. Beddoes, *Good Advice for the Husbandman in Harvest and for All Those Who Labour Hard in Hot Berths; As Also for Others Who Will Follow It in Warm Weather* (Bristol: Mills & Co., 1808).

8 *A Letter to the Right Honourable Sir Joseph Banks . . . on the Causes and Removal of the Prevailing Discontents, Imperfections, and Abuses, in Medicine* (London: Printed for R. Phillips by T. Gillet, 1808).

9 Ibid., 6, 19, 45, 57–58, 68, quotation from 105, See also T. Beddoes, *Instructions on the Subject of Health* (London: Longman, 1807); *Where Would Be the Harm of a Speedy Peace?* (Bristol: N. Biggs, 1795).

10 Beddoes to Davies Giddy, 8 November 1792, CRO MS DD, DG 41/5.

11 Freely translated, "The last day is coming, and the inevitable doom." Virgil, *Aenied*, II 324.

12 Beddoes to Giddy, 19 November 1792, CRO DG 41/38.

13 T. Beddoes, *Extract of a Letter on Early Instruction, Particularly that of the Poor* (Madely: J. Edmunds, 1792); *The History of Isaac Jenkins, and of the Sickness*

of Sarah, His Wife, and Their Three Children (Madely, 1792); *A Guide for Self-Preservation, and Parental Affection or Plain Directions for Enabling People to Keep Themselves and Their Children Free from Several Common Disorders* (London: J. Murray and J. Johnson, 1793); *Reasons, for Believing the Friends of Liberty in France Not to Be the Authors or Abettors of the Crimes Committed in That Country* (London, 1793); *A Letter to the Right Hon. William Pitt, on the Means of Relieving the Present Scarcity, and Preventing the Diseases That Arise from Meagre Food* (London, 1796).

14 Robert Mitchell, "Introduction", in *Hygëia: Essays Moral and Medical*, reprint edition, ed. T. Beddoes (London: Thoemmes, 2004; reprint of the 1802 original), xi. T. Beddoes, *Rules of the Medical Institution, for the Benefit of the Sick and Drooping Poor* (Bristol: J. Mills, 1804). According to Stock (see Abbreviations), 415, "[A]n edition on larger paper was entitled Instruction for People of all Capacities respecting their own Health and that of their Children." We have not seen the latter edition.

15 Edmund Burke, *A Letter from the Right Honourable Edmund Burke to a Noble Lord, on the Attacks Made upon Him and His Pension, in the House of Lords, by the Duke of Bedford and the Earl of Lauderdale Early in the Present Sessions of Parliament* (London: Printed for J. Owen and C. Rivington, 1796), 62.

16 Other bugbears were Thomas Paine, author of *The Rights of Man* (1791), who argued that revolution was a legitimate response to government that failed to defend the natural rights of its citizens, and James Mackintosh, whose *Vindiciae Gallicae: A Defence of the French Revolution and Its English Admirers* (1791) was a rebuttal of Burke's attack on the French Revolution.

17 Joseph Priestley, *Experiments and Observations on Different Kinds of Air*, 3 vols. (London: J. Johnson, 1775–1777), I, xiv.

18 T. Beddoes, *A Memorial Concerning the State of the Bodleian Library, and the Conduct of the Principal Librarian: Addressed to the Curators of that Library, by the Chemical Reader* (Oxford, 1787).

19 Beddoes' biographies include Stock (see abbreviation list), Stansfield (see abbreviation list), Porter (see abbreviation list), Neil Vickers, *Coleridge and the Doctors (1795–1806)* (Oxford: Clarendon Press, 2004), and Jay (see abbreviation list).

1 Chemistry, consumption, and reform

Trevor Levere

Oxford University: student days

Industrialism figured largely in the landscape of Beddoes' Britain. He was born in Shifnal, Shropshire, the son of Ann (née Whitehall) and Richard, a tanner who was part of a wealthy and politically liberal commercial network.[1] Like other eighteenth-century trades, tanning was closely allied to chemistry, and regularly featured as such in dictionaries of chemistry and in lectures on applied chemistry, then known as the "chemistry of the arts". Beddoes' principal chemical discovery was Humphry Davy, who lectured on chemistry, included tanning, in his course of lectures at the Royal Institution of Great Britain in 1802. Indeed, his work on tanning seems to have been a significant part of the reasons for appointing him to the Royal Institution, and his Copley Medal, awarded to him by the Royal Society in 1805, was largely because of his lectures and paper on tanning.[2] Another West Country tanner was Tom Poole of Nether Stowey, a political radical, and friend of Davy, Coleridge, and the Wedgwoods; Davy dedicated his last, posthumously published work to Poole, "in remembrance of thirty years of continued and faithful friendship".[3] Richard Beddoes, well off and well connected, was wealthy enough to send his son to Oxford. Beddoes was a student at Pembroke College, Oxford, from 1776 to 1779. William Adams, the Master of Pembroke from 1775 to 1779, was keen on chemistry, "probably because of his close friendship with Samuel Johnson who was an enthusiastic amateur chemist".[4] Besides Beddoes, three other chemists – William Higgins, James Smithson, and Davies Gilbert – matriculated at Pembroke in these years.

Beddoes attended chemistry demonstrations in the basement of the Old Ashmolean, probably performed by John Smith, MD.[5] Beddoes appears, astonishingly, to have taught himself to read Dutch, French, German, Italian, Latin, Spanish, probably Portuguese, and possibly Swedish; he completed his acquisition of foreign languages while registered as an undergraduate student. In 1781, after taking his B.A., he moved to London to study anatomy under John Sheldon, who was soon to be recognized as the leading teacher of anatomy in London when he succeeded his former mentor William Hunter as Professor of Anatomy to the Royal Academy in 1783. In London, Beddoes would also have had opportunities to meet chemists and attend chemical lectures, but for all the probabilities, we have found no direct evidence of such activities.[6] Sheldon taught physiology as well as anatomy. Beddoes, who was familiar with Sheldon's principal published work,[7] translated a collection of Spallanzani's writings on digestion and other topics.[8] Digestion, fermentation, putrefaction, and the efficacy of manures in agriculture were widely discussed in contemporary writings and lectures. Chemistry, as taught at universities in England and Scotland and in much of Europe, was usually a part of medical instruction. That was pre-eminently the case at the University of Leiden, where Boerhaave's chemical lectures had helped to shape a generation of physicians and inform the medical schools at Edinburgh and Glasgow. The practitioners of the new chemistry as it was developing in France aimed to establish chemistry as an independent discipline, useful to pharmacists, industrialists, and physicians, but its own master. Nevertheless, it is often not productive to label eighteenth-century practitioners as chemists rather than physicians. Beddoes' career was to embrace medicine and chemistry, and as an advocate of pneumatic medicine, he exemplifies the intimate fusion of chemistry with medicine.

In 1783, Sheldon was excited by a non-medical application of chemistry when he learned of the first balloon ascent of Pilâtre de Rozier in Paris in November 1783 and promptly set about calculating the weight of the atmosphere, then "involved himself" in the making of a balloon, which was used in two attempts at flight in 1784; the second attempt was a disaster. Nothing daunted, Sheldon in October 1784 accompanied Blanchard in the first stage of a successful British voyage by balloon.[9] Beddoes, as Sheldon's student, could not have been ignorant of his teacher's enthusiasm for ballooning, and he was subsequently to work with and employ a future English balloonist, James Sadler of Oxford, who served as his instrument maker and laboratory assistant.[10] But by the time that Sheldon ascended with Blanchard, Beddoes was in Edinburgh.

Edinburgh University, 1784–1786

Beddoes was there at an exciting time, as a beneficiary of the late flowering of the Scottish Enlightenment. David Hume and Adam Smith are perhaps the best-known figures in that movement; in 1784 Smith was still lecturing in Glasgow, although Hume had died in 1776. The Scottish Enlightenment was practical as well as literary. David Hume asserted in 1742 that "*industry, knowledge*, and *humanity*, are linked together by an indissoluble chain, and are found, from experience as well as reason, to be peculiar to the more polished, and, what are commonly denominated, the more luxurious ages."[11] Forty years later, this was still true of Edinburgh, and for two years Beddoes benefited from that linkage. Roger Emerson has shown how fully medicine and chemistry were part of that practical culture.[12] Beddoes was able to benefit from the lectures of teachers with a European reputation, and especially from the chemical lectures and demonstrations of Joseph Black. In an undated letter, he wrote to Charles Brandon Trye about Black's lectures, and about his other teachers in Edinburgh. Trye had apprenticed as an apothecary in Worcester, then became a pupil of William Russell, senior surgeon to the Worcester Infirmary; he then moved to London to study under John Hunter, and was appointed house surgeon to the Westminster Hospital. Sheldon appointed Trye to assist him in his private medical school, which Trye did briefly until returning to Gloucester, where in 1783 he became house apothecary, and from 1784 served as surgeon to the charity.[13] Beddoes had met Trye while working under Sheldon, and was full of enthusiasm for Black's lectures:

> Black's course is the best I have ever heard or ever shall hear. But as you cannot see his exps. in a MS, you may again all the information this wd. give you from Bergman & the new edition of Macquer's dictionary when it is published. I ought perhaps to except the doctrine of latent heat, but of this I have a copy. . ..
>
> Cullen does nothing but read his textbook. . . . Gregory's chemical lectures are good. I shall have some notes of them. . ..
>
> I have borrowed MS lectures of the best Professors on the continent, & shall take notes from them. Richter of Göttingen is as highly spoken of as any man I ever heard of.[14]

Black was to be Beddoes' model, and so it will be worth our while to consider several aspects of his lectures. Black had made the first quantitative chemical experiments on a new gas or kind of air, which he called "fixed air" and which we know as carbon dioxide.[15] He was a skilled pneumatic chemist, using the simplest apparatus – even crude – to remarkably good effect.

He had developed a theory of latent heat, which was very much part of chemistry, heat then being considered as a chemical substance. He instructed those students who wished to inform themselves more fully to "read Dr. Boerhaave's treatise on fire, in the first part of his Chemistry; and Martin's Essay on heat and Thermometers. These have the greatest connection with the first part of our Course."[16] It was this view of heat, adopted by his most illustrious pupil, James Watt, which provided a foundation for Watt's improved steam engine.[17] Black was the leading chemical professor in Britain, and his reputation and lectures attracted students from all Europe and beyond,[18] including Russia, whither Catherine the Great tried to entice him. Black was not fond of travel, and declined to go.[19] His foreign students brought with them lecture notes from home, and as Beddoes told Trye, he was soon borrowing them.

Black declined to publish his lecture notes and it was left to his junior colleague and friend John Robison, long linked to James Watt, to edit and publish them after Black's death – like many sets of lecture notes used by accomplished lecturers, they were incomplete, allusive, and full of undated interpolations and corrections. Robison tells us that he used student notes to supplement Black's manuscript notes, and the published text, which includes passages by Robison, cannot be definitive.[20]

Numerous sets of lecture notes from Black's courses survive today, mainly written down by copyists, and although they, like Robison's text, lack the authority of Black's imprimatur, they give a vivid impression of the lectures. In what follows, we shall refer to a set of notes that are tentatively dated to 1785,[21] while Beddoes was a medical student. In these notes, chemistry is defined as

> the study of such phaenomena or properties of bodies as are discovered by variously mixing them together, & by Exposing them to different degrees of heat, alone, or in mixture, with a view to the enlargement of our knowledge in Nature, & to the improvement of the useful arts.

Heat indeed was the key to chemistry, both as a material participant, and as the principal cause of change:

> Heat is . . . the principle active matter in nature. . . . Nor are the effects of Mixture less various & extensive. The number & variety of the kinds of matter, produced by nature & art are infinite, & they all differ in the way of mixture.[22]

The multiplicity of chemical substances was such that in order to navigate the science, some classification or arrangement of substances was essential. "I do not find any [text] in which the arrangement is the best & the views are sufficiently extensive and therefore I use an arrangement of my own."[23] Given the general inadequacy of theory,

> the only thing we have left is simply to take a view of the object of Chemistry arrang'd in that order that will enable us to comprehend them most easily, so to distribute them into a number of Classes that will bear a resemblance in their natural qualities rejecting those that are dissimilar. . . . therefore I divide them into Salts, Earths, Inflammable Substances, Metals, and Waters.[24]

One "theory or opinion" that Black acknowledged in his lectures was that "the quality of Inflammability depends upon the presence of a particular principle or ingredient abounding in inflammable substances to which they give the name of the Phlogiston."[25] Through the 1770s, Black had taught that phlogiston existed and possessed absolute levity. As Carleton Perrin showed, Black had experimental evidence for his opinion.[26] The development of pneumatic chemistry gradually undermined the doctrine of levity. By the late 1770s, "Black was presenting the levity hypotheses in a lower key and with less personal identification,"[27] although he adhered to it until around 1780. Black's lectures by 1782–1783 indicate that Lavoisier's work had made an impression on him, and he relinquished the levity of phlogiston. By 1785, he acknowledged that the phlogiston theory was useful in explaining experiments and observations, but he gave it something less than a complete endorsement.[28] Cavendish's experiments on airs, with their implications for the composition of water, were carried out in the early 1780s, and published in 1784 and 1785. Priestley, visiting Paris, had shared his observations on different kinds of air with the French chemists, including his account of vital or eminently respirable or dephlogisticated air, beginning in the early 1770s; and Lavoisier had begun his assault on phlogiston in the early 1780s, although his reflections on phlogiston were not published until 1786.[29] Black was extremely well connected with chemists internationally, but chose in his lectures to be cautious, and not to subscribe immediately to claims against the phlogiston theory. As Perrin has shown, his "[s]tudents preceded their teacher in adopting Lavoisier's theory in Edinburgh."[30]

Black's caution is well represented in the notes of his lectures in 1785. These notes were written approximately two years before the publication of the new French nomenclature with its wholesale rejection of the phlogiston theory and concomitant new arrangement of substances, four years before the publication of Lavoisier's textbook and manifesto of the new chemistry, published in the same year in English translation.[31] When the French theory was published, Black was sympathetic but still cautious, declining to use the new nomenclature because that would have entailed an endorsement of the new theory.

Black was an elegant demonstrator in his own lectures. His public experiments always succeeded. He warned against the excesses of theory, and told his students that "Chemistry . . . is an experimental Science, more perhaps than any other, & pursues its enquiries & makes its discoveries chiefly in the way of experiment."[32] He stressed the usefulness of chemistry, a science "dedicated to the improvement of the essential & ornamental arts of life".[33] "Improvement" was perhaps the defining word for Enlightenment Scots, as it was for many Englishmen, including industrial leaders in Birmingham and the Midlands, James Watt among them. Black, with "improvement" as his watchword, was concerned to distinguish between the chemical arts or trades, and the business of the chemical "Philosopher, who studies and acquires knowledge for the improvement of his mind, or to improve the branch he applies to for the production of new Arts or Manufactures, or the improvement of those already established".[34] His students were to be chemical philosophers. Black wanted chemistry to be a science akin to Newtonian natural philosophy (although he was privately aware that chemistry had not yet reached that point).[35] We shall see just how far Beddoes chose, and how far he was able, to emulate his teacher.

Meanwhile, still at Edinburgh, Beddoes was translating chemical works for John Murray, chief among them *A Dissertation on Elective Attractions* by the Swedish chemist Torbern Bergmann.[36] Bergmann's work on affinities necessarily addressed the problem of the arrangement of chemical elements and compounds. The project involved a vast number of experiments, and he realized that the project was beyond one man. He corresponded with Louis Bernard Guyton de Morveau in France, encouraging him to see the project through. Bergmann died in 1785, the year in which Beddoes' translation appeared.[37] Beddoes became a member of the university's Chemical Society,[38] founded and encouraged by Black. The University of Edinburgh Library has a bound manuscript volume that includes a list of the members, and two papers by Beddoes, apparently presented at a single meeting, "An Attempt to Point out some of the Consequences which Flow from Mr. Cavendish's Discovery of the Component Parts of Water", and "A Conjecture concerning the use of Manure".[39]

Cavendish's "Discovery" was published in two papers in the *Philosophical Transactions of the Royal Society of London* in 1784 and 1785,[40] so Beddoes, in looking at the question of the "Component Parts of Water", was addressing a current and hot topic. He first looked at the debate between phlogistonists and supporters of Lavoisier's new antiphlogistic chemistry, and remarked that

> I think it is common to both parties to assert more than they can prove . . . If two different hypotheses can be adjusted to any set of appearances, both become uncertain. . . . Perhaps we may conclude without error that they are both false, or at least imperfect.[41]

What was not in doubt for Beddoes was the accuracy of Cavendish's experiments, which, almost uniquely among "modern philosophers", stood up robustly and accurately to the test of repetition. To those who were reluctant to accept the implication of Cavendish's experiments, that water was a compound and not a simple body, "I can only say, that if they are sure that God has bestowed upon them faculties by which they can discover the property of natural bodies without the aid of experience, they do right in persisting."[42] Since Cavendish had produced water by the combustion of inflammable air (hydrogen) with vital or dephlogisticated air (oxygen), it is tempting to see these substances as the components of water, and water as a compound; but contemporary debate was both more subtle and more complicated, and it is arguable that Cavendish, at least initially, saw the production of water as a condensation rather than as a chemical combination.[43] If inflammable air was phlogiston itself, and dephlogisticated air was water deprived of phlogiston, such an argument was at least worth considering.

Beddoes believed that one the consequence of Cavendish's experiments was a reprieve for phlogiston:

> Many Operations can now be reconciled with the existence of Phlogiston which admitted of no Explanation before. Indeed it seems to me that but for this discovery phlogiston must have been totally abandoned. When phlogiston is burnt in vital Air, & the whole of it disappears, what account could have been given of this, different from Mr. Lavoisiers, if we did not know that the production of water will account both for the diminution of the Air, and the increase of weight of the consumed body.[44]

Debates about phlogiston and the composition of water and of different kinds of air (gases) were very much at the forefront of chemical debate in the 1780s. In England, Joseph Priestley and Richard Kirwan were the leading phlogistonists, Black had met Kirwan and knew of Priestley through his published work and through Watt,[45] and it is not surprising to find Beddoes commenting on the debate. It is, however, somewhat less obvious that Beddoes followed his discussion of the consequences that followed from Cavendish's work on the formation of water by "A Conjecture concerning the use of Manure" – but it makes sense. First, many chemists including Black, William Cullen, and Kirwan were contributing to the ideology of improvement by providing a chemical account of manures and fertilizers and manures, with advice to farmers and landowners for obtaining greater yields.[46] Second, the work of Priestley, Senebier, and Ingenhousz had led to an understanding of the symbiosis between animal and vegetable respiration;[47] in the processes of plant growth and respiration, and under the stimulation of light, vegetables were shown to produce vital or dephlogisticated air. According to Cavendish's theory, since water appeared to be produced from inflammable air (perhaps itself phlogiston) and dephlogisticated air (perhaps water deprived of phlogiston), it was reasonable to conjecture if not conclude that this air came from water. Beddoes, commenting on the enrichment of atmospheric air by plants disengaging dephlogisticated air, thought it probable that nature had even more powerful stimulants waiting to be discovered. Here Beddoes had touched for the first time on the possibilities of pneumatic medicine, the use of different kinds of air for therapeutic purposes: "And I doubt not that in time a rational System of Vegetable Medicine may be constructed if the subject be properly prosecuted."[48] Beddoes considered that the role of plants in pneumatic chemistry meant that vegetable medicine and pneumatic medicine would, when developed, prove to be closely allied. His expansive optimism and willingness to engage in speculation were not well attuned to Black's empirical caution. Here was a trait in which Beddoes would persevere, occasionally to good effect. Beddoes was braver than most in advancing controversial views, in science, medicine, and politics.

Beddoes later told Joseph Banks that his residence in Edinburgh occupied "three winters and one summer, [during which I] was perpetually at the lectures of the professors and in the societies of the students".[49] In those years, he had the great benefit of Black's lectures, with their emphasis on careful and precise experiment rather than theory. If Beddoes had absorbed Black's ethos of precise quantitative experiment, he would have been more cautious and less prominent in his own day. His breadth of experimentation is closer in style to Priestley's.[50] But he did absorb Black's teaching about the utility of chemistry to medicine and agriculture, and about the problem of chemical classification; when he came to deliver his own lectures, first in Oxford and then in Bristol, Black was his principal model.[51] Meanwhile, still a student, he also became a friend of a young visitor to Edinburgh, the physician and chemist Christoph Girtanner, who, born in Switzerland, obtained his doctorate from Göttingen, where he passed most of his career. We are told that Girtanner and Beddoes corresponded for years.[52] It may well have been through Girtanner that Beddoes' works were acquired by the Georg-August University of Göttingen soon after publication,[53] and his theories were briskly debated, especially those relating to Brunonian medical theory and to pneumatic medicine. John Brown was lecturing at Edinburgh while Beddoes studied there, and as Beddoes wrote to Trye, "We are all mad after theory & between the Brunonians & Cullenians Truth & Nature are too much forgotten."[54] The debate about Brunonianism was even more intense in Göttingen, so that in 1782 the cavalry was actually called out to put down a riot between Brunonians and their critics.[55] The central doctrine in Brown's turgid and convoluted lectures was that

> The Principle and Power in the Body by which [they] Act is called Excitability and the Common Effect of the Exciting Powers are sense and motion. . . . The Seat of Excitability is in the Nervous System and Muscular Fibres.[56]

Brown's "excitability" was analogous to "irritability" in the system of Albrecht von Haller.[57] Diseases were produced by a surfeit or a deficiency of excitability. Beddoes was to blow hot and cold about this theory, but it was in Edinburgh that he was exposed to it intensely and, as Vickers has shown, it became an issue in pneumatic medicine, where irritability or excitability were linked to breathing oxygen.[58]

Oxford University: the chemical reader

In December of 1786, Beddoes returned to Oxford to obtain his MD. In March 1787 he repeated the experiments on the production of artificial cold first carried out by Richard Walker, apothecary to the Radcliffe Infirmary.[59] In the following May, he published his memorial on the state of the Bodleian Library and the culpable negligence of Bodley's Librarian, the Rev. John

Price.[60] Beddoes remained in Oxford until the late summer of 1787, when he visited France, partly in preparation for his first course of lectures: he had been appointed Reader in Chemistry at Oxford University (or lecturer, or professor – the title he used was decidedly fluid, and his income came from the students, not from the university). He went to Dijon, where Guyton de Morveau had his laboratory in the Academy. Guyton had been offering a course in chemistry at the Academy since 1776. He was a successful lecturer, and his course attracted larger audiences year by year. Two full-time assistants prepared the demonstrations. The contents of the course were published in 1777–1778.[61] Guyton's textbook and lectures were "perhaps the first to be based on a theory of chemical reaction", and for a decade he made use of the phlogiston theory. He also used traditional chemical nomenclature until 1782, when he began to use his own new system.[62] Guyton had been experiencing difficulties at the Academy, which had been accumulating substantial debts; some of the literary members of the Academy blamed him because of the costs of the laboratory and his chemical course.[63] In the late winter or early spring of 1787, Guyton went to Paris, where he spent several months. He worked in Lavoisier's laboratory, became converted to the new antiphlogistic or oxygen-based chemistry, and collaborated with Lavoisier, Fourcroy, and Berthollet in their revolutionary revision of chemical nomenclature.[64] It made a lot of sense for Beddoes to visit Guyton and discuss the new French chemistry, and to inspect Guyton's laboratory in preparation for setting up his own in Oxford, and for planning his lectures. He also visited the hospital, learning about Guyton's disinfector and the latest ways of controlling contagion. His visit to Dijon coincided with a holiday visit to Guyton by Lavoisier, Berthollet, and Fourcroy, accompanied by their wives, and by mathematicians Gaspard Monge and Alexandre-Théophile Vandermonde, which enabled him to meet the leading French chemists and advocates of the new antiphlogistic chemistry.[65]

After Dijon, Beddoes went to Paris,[66] where he spent several weeks. He again met leading chemists, listened to their arguments and discussed the new French chemical nomenclature, and possibly witnessed some of their key and controversial experiments. When he visited Lavoisier, he gave him a copy of James Hutton on rain,[67] and it would have been strange if he had not attended one of the Lavoisier's frequent dinners for scientific visitors.[68] On his return to Oxford, fresh from his discussions with French chemists, Beddoes had "almost accepted the antiphlogistic theory".[69]

He spent the Michaelmas (fall) term of 1787 preparing his lectures, and on 6 November wrote from London to his mentor Joseph Black.[70] It is an important letter for understanding Beddoes' chemical views:

> Dear Sir
>
> The want of some tolerable manual of chemistry & the increasing ardour for the pursuit of that science at Oxford & I believe every where else induced me to promise to attempt something of the kind myself.

> I did not fail to perceive the various difficulties of such an undertaking at the time & they have become more numerous the more I have reflected on the subject. I have collected all the modern elementary books,[71] which are not a few, especially the German, but I find no assistance for the method which constitutes the chief difficulty of such a work. We seem to me to have arrived at that period of chemical knowledge when the necessity of a new arrangement appears evident without being possessed of sufficient information to discern clearly what that arrangement ought to be.

This line of argument was certain to appeal to Black, who had complained in his own lectures that "I do not find any [textbook] in which the arrangement is the best & the views are sufficiently extensive and therefore I use an arrangement of my own."[72] Beddoes continued with a brief statement of the importance of pneumatic chemistry for chemical theory, beginning with Black's discovery of fixed air, and passing, by way of Priestley's numerous additions to the list of different kinds of air, to the latest discoveries of Cavendish and the French chemists:

> Before your discovery of fixed air the acids were entitled to the first or at least to the second place, on account of the simplicity of their composition & because they were the principal agents in the operations of chemistry. In consequence of the direction which that discovery gave to the pursuits of Nat. Philosophers many aeriform substances wch. had before eluded the senses were submitted to expt. & though many curious facts were ascertained they remained in great measure solitary, their connection with the phenomena that had been formerly observed was unperceived & the lecturer or the writer was left to introduce them where he cd. most conveniently without disarranging his ancient system. By degrees it was perceived that the new [airs] were of too much importance to act the part of significant accessories; their claims to higher stations began to introduce disorder into the old arrangement & many I believe, considering them as unwelcome intruders, began to despair altogether of the theory of Chemistry. But the light which has been afforded by the recent discoveries of Mr Cavendish, Mr Lavoisier, Berthollet & some others, seems to me to suggest better hopes, except to those who have had the folly or the misfortune to fix their opinions inalterably.[73]

Fixing his own opinions inalterably was never Beddoes' style. The latest discoveries led him to reject phlogiston, and to accept Lavoisier's argument that gases were compounds of ponderable substances with the principle of heat. In the following year, however, he changed his mind:

> Dr Priestley seems totally to have overthrown the antiphlogistic theory – I am anxious to hear what the French chemists have to say on the other

> side – I have seen some of their private objections to Dr Priestley's inferences, but they are totally insignificant.[74]

By 1791, he inclined once more to Lavoisier, having acquired the necessary apparatus to work his way through the Frenchman's experiments. But, like Black, who introduced Lavoisier's chemistry piecemeal into his lectures in the 1790s, Beddoes only partly accepted the French chemistry. From the beginning, he had rejected the notion of caloric as material heat. In the 1790s and beyond, he found only more reasons to confirm him in that view.

Although he did not accept Lavoisier's caloric in 1787, he did agree that Black's own work on latent heat furnished a particularly cogent example of the chemical combination of fire or heat with ponderable substances, associated with changes of physical as well as chemical state:

> These discoveries lead to three important changes, to the rejection of phlogiston, 2 & consequently no longer to consider inflammable bodies as containing one common principle (which, with other considerations too long to be mentioned here leads me to <reject> destroy the class) & 3rdly to place the elastic fluids or the greatest part of them before the acids immediately after the doctrine of heat. I wd not make a class of elastic fluids any more than of inflammables. That state, as you have taught us, depends upon the combination of fire & the qualities of the body which was before solid or liquid may be \very/ different in other respects though they may agree in having <that> such an attraction for fire as to enable them to retain enough to keep their particles at a certain distance from each other, which is all that is essential to the aeriform state.[75]

Beddoes knew that his mentor was very cautious when it came to new theories:

> I do not expect that you & our other philosophers will approve of these changes immediately on hearing them mentioned, perhaps not after considering the arguments I have to advance in favour of them. I shall indeed be very far from satisfying myself. I only hope to propose a plan which by gradual improvements may comprehend the new discoveries. You will allow, I dare say, that we have none at present which is capable of doing this. For the rest I wd. adopt the order of your lectures entirely, as after much comparison & reflection I do not think it possible to have contrived a way of presenting the facts relating to chemistry that were known at that time more clearly & scientifically. But till the science shall be rendered perfect, successive changes of method will be requisite, for as new [airs] or new qualities of [airs] already known are detected, they will require to be arranged somewhere; by which means the former order will be more or less disturbed in proportion to the quantity & importance of the new matter to be introduced.[76]

The theory of chemistry was embedded in its language, as Condillac and after him Lavoisier and his collaborators were well aware: "We think only through the medium of words. – Languages are true analytical methods. . .. The art of reasoning is nothing more than a language well arranged."[77] Black would not use the new French nomenclature. Beddoes was less concerned with the theory it entailed than he was with some infelicitous translations of particular terms:

> There remains still a difficulty which at present appears to me more insuperable than that of arrangement, I mean, the system of denomination, I know not whether you have seen the *nouvelle Nomenclature chymique,* which is the production of the ablest heads in France. I think they have been very far from successful in all cases or, if they have, that it will be impossible to introduce a corresponding nomenclature into our language. As an example of the first objection, I may mention "azote", the denomn of phlogistd air, which is not as far as we know more pernicious to life than other gases, whatever the French chemists may pretend, & which therefore is not properly distinguished by this appellation. Again, "Potasse" is the new name for veg. alkali, but how can one adopt it in English, when it is already appropriated to another substance. The distinction of the difft states of the same and by difft terminations will also be excessively hard & difficult in English. I intend to print what I have to say upon these subjects along with the first part of my syllabus, which I hope will be ready by the beginning of March. With your permission I will publish it in the form of a letter to you. In the mean time I wish I cd induce you to reflect a little on the subject & to favour me with a few lines upon it.[78]
>
> I am just returned from the continent, where I have picked up some news;[79] but I find that I have not room enough left to set down any of it.

Beddoes has left only fragments of lectures that were presumably for his course on chemistry,[80] one discussing heat (including latent heat) and its effects: "Two hypotheses respecting the cause of the sensation we denominate *heat* – the English or mechanical doctrine – the foreign or chemical – Neither yet applied in a satisfactory manner to the principal phaenomena – language confused".[81] No sets of notes by students are known. Beddoes later stated that he used Isaac Milner's Cambridge lecture syllabus.[82] How he managed to reconcile this with his demonstrations of Lavoisier's key experiments is hard to imagine. Milner accepted the phlogiston theory and the Aristotelian four elements (air, fire, earth, and water); he described a fair amount of traditional apparatus, but none of the apparatus needed for pneumatic chemistry, of which Beddoes claimed to have a full set; and he discussed respiration and atmospheric air within the framework of phlogiston theory. Beddoes' correspondence with Black provides the fullest picture we have of his lectures, although on occasion it needs to be taken with a grain of salt.

Black replied very promptly and encouragingly to Beddoes' first letter from Oxford. He recognized that the researches of French chemists were forcing a revision of the table of simple substances, and regretted that the business of lecturing left him no time for detailed consideration and correspondence about the right new arrangement of chemistry:

> I have the pleasure to receive your letter of the 6th Curr^{t} and very much approve of your trying to <*form a new*> Plan <*for the*> a new arrangement <*of*> & classification of the objects of Chemistry. There is no doubt that many of the Sub^{ces}. which not long since were considered as simple have lately been shown to be compounded & that others are now viewed in a very diff^{r} light from that in which they formerly appeared, as you have been on the spot where these improvements have been the most ardently promoted & discussed & \as you/ are otherwise very well prepared for such an attempt I <*hope?*> am happy to hear that you think of engageing in it & have no objection against your addressing it to me but on the contrary shall think it an honour.[83]

Beddoes wrote again to Black on 23 February 1788.[84] He fulsomely and oleaginously reiterated his intention to dedicate his textbook (which he never wrote) to Black: "For as I am in no haste to be rich, I choose rather to pay the debts of gratitude than to carry offerings to the vanity of the great."[85] Then, with a mixture of accurate statement and hyperbole, he announced:

> I have begun my lectures & although my numbers cannot be put in competition with one at Edin^{r} yet the desire of knowing something of Chemistry seems to be spreading, some people have perceived that it is neither a petty branch of medicine nor one of the black arts, as they are termed, but simply an inviting & important part of Nat. Phil. I think my numbers will be greater this than the last course, though I had then the largest class that has ever been seen at Oxford, at least within the memory of man, in any Department of knowledge.[86]

If Beddoes had large audiences, that was fortunate for him. He received no stipend from the university, and so his earnings depended entirely on the payment of course fees from students attending his lectures. But his lecture demonstrations were less than perfect. Some chemists seem to have the chemical equivalent of a gardener's green thumb, which makes everything prosper. Henry Cavendish was one such chemist. Beddoes, however, was neither an experienced nor a highly skilled experimentalist. He appealed to Black:

> What I find most difficult is to repeat some of those apparently simple exp^{s}. which in your hands are so striking and so instructive. I have not yet learned how to show the gradual approach towards saturation by

> throwing slowly a powdered salt into water. What salt do you use? & how do you perform the exp^{t}? How do you contrive to make that capital exp^{t} which shews the burning of iron in deph^{d} air? I mean to attempt it, but am told that the vessel has been frequently in other hands burst with great violence? do you put sand at the bottom? I know the form of the vessel &c. What salt do you use to shew the effects of agitation upon mixture?[87]

Black visited Oxford later that year,[88] greatly impressing the Dean of Christ Church and other dignitaries; but, although Beddoes had been made an honorary member of the Common Room at Christ Church, there is no evidence that Black visited him,[89] although Beddoes continued to lecture with a special dispensation from John Cooke, Vice-Chancellor of the University.[90]

Looking back on his Oxford lectures in the summer of 1793, Beddoes despondently admitted to Thomas Wedgwood, his friend and patient, son of the potter Josiah and one of the first to explore photochemistry, that "I know very well that some of my chemical lectures at Oxford were dull – The subject is often so to those who look only to be entertained by shewy experiments."[91] This is a remarkable comment in an age where public spectacle was crucial to the promotion of science. A second-hand report of Beddoes' lectures, by one who had been a student when Beddoes was Reader, but who never attended them, none the less fits Beddoes' depressed account:

> He was a man of science and of genius, but by no means successful as a lecturer. His figure and delivery were ungraceful, his language inflated and ambitious, and he was so singularly awkward in the mechanical part of his experiments that they generally failed, and he was then compelled to proceed in his discourse on the hypothesis that the result had been the reverse of that which the eyes of his audience would have led them to believe.[92]

A theoretical framework was difficult to establish while contradictory reports of experiments were circulating; there was no textbook that fitted Beddoes' approach to chemistry, which grasped at too many different analogies to be coherent; and he needed to build up a set of demonstration apparatus. By April 1791 he felt that at any rate he had achieved the latter goal. The completion of this collection went along with the end of his vacillation between Priestley and Lavoisier, between dephlogisticated air and oxygen. Indeed, acquiring and building the apparatus needed in order to repeat Lavoisier's experiments was in itself a major step towards reaching a decision about the rival theories. Apparatus is designed to function within a theoretical framework, and the results obtained with that apparatus tend accordingly to validate the theory;[93] there is a kind of circularity tying theory to apparatus via experiment, and sometimes the demonstration of new theories depends upon the use of new apparatus. That was what Lavoisier himself believed when he wrote his *Traité* of 1789, giving a third of the book to the description of apparatus and instructions for using them:

> perhaps, if, in the different papers that I have given to the Academy, I had dwelt at greater length to the details of manipulation, I would have made myself more easily understood, and the science would have made more rapid progress. . . . It will easily be perceived that this third part could not have been extracted from any existing work, and that, in its principal articles, I could only be helped by my own experience.[94]

In 1789, Josiah Wedgwood was supplying Beddoes with porcelain tubes for the laboratory, and Beddoes acquired a burning glass from the internationally renowned William Parker of Fleet Street, the leading London supplier of scientific glassware, especially lenses.[95] That was an instrument equally useful for experiments in Priestley's phlogistic scheme and in Lavoisier's anti-phlogistic one. In 1791, a year in which he ordered chemical apparatus of the same size and composition as "Dr. Priestley's last tubes and retorts",[96] he told Black that he had

> a very valuable assortment of chemical apparatus – a gazometer very much improved upon Mr Lavoisier's[97] \&c/ so that I am able to shew any & every expt. in his book – It has been constructed by a pastry cook in this place, a perfect prodigy in mechanics, who has invented & executed an improved barometer of which the column of [mercury] is not altered by temperature; and by the help of which I can measure the height of a room as accurately as by a rule; an air pump which exhausts perfectly – & of course is constructed on principles totally new – a balance which I have seen turn with [1/100th] of a grain, \when/ loaded with a pound at each arm – I have all these instruments in the Elaby. & several more of less importance – He is besides now taking out a patent for some new machinery which I believe will supersede all the water-wheels, steam engines &c now in use.[98]

The "pastry cook" was James Sadler, son of an Oxford cook and confectioner. Sadler was by turns balloonist, chemist, mechanic, and engineer. His career was later frustrated by patent disputes about steam engines with Boulton and Watt.[99] Even the excellence of his chemical apparatus is doubtful. Stock in 1811 wrote that Sadler "so multiplied machinery, that those parts of it which he had contrived, have been said to resemble . . . the complicate[d] engine which, in one of Hogarth's prints is applied to the purpose of drawing a cork."[100] Beddoes' enthusiasms were always as immoderate as his depressions: 1/100th of a grain equals about $3 \times 10 - 7$ lbs., which means that Beddoes was claiming a sensitivity of one part in 300,000, which was very nearly equal to the best mechanical balances made in the eighteenth century. These were Lavoisier's great balances, made by Fortin and Mégnié, and the comparable instruments made by Jesse Ramsden, the finest instrument maker in Britain, and by John Harrison, who in addition to his balance built a chronometer that ultimately, after much controversy, satisfied the Royal Navy's needs for determining longitude. Each of these could weigh to about one part in 400,000.[101] No other instruments came near these three triumphs of precision craftsmanship. To put this in perspective, Joseph Black achieved his results with a balance sensitive to about one part in 200. Beddoes' claim here is simply not credible, nor is his account of a new barometer that could measure the height of a room. His gasometer, however, was effective. Beddoes sketched his own proposed instrument (Figure 1.1) in his correspondence with Watt, who incorporated it into the breathing apparatus that he made for pneumatic medical purposes.[102] This gasometer used a weight over a pulley to control the pressure on the gas in the gasholder, on the same principle as those advertised by Dumoutiez in Paris from 1795.[103]

If one leaned towards simplicity, improvements over Lavoisier's design (Figure 1.2) were possible, and several chemists and instrument makers devised such simplifications and improvements. As for Sadler's steam engines, quite apart from the legal squabbles with Boulton and Watt, they were less successful than he and Beddoes believed. The Shropshire ironmaster William Reynolds,[104] a friend of Beddoes, had professional dealings with Boulton and Watt, making several engines for them. His company was sufficiently independent to give Sadler's machine a fair trial. The results were disappointing: "We have made no further progress in the trial of Sadlers Engine we are convinced it will answer no good purpose for large Engines." The report on the engine allowed that it could be useful for light work, for example in driving a lathe, but it would not serve for pumping out mines or driving large machinery.[105]

For all the problems with his engines, Sadler seems to have constructed a set of apparatus for Beddoes that enabled him to perform Lavoisier's key experiments, and to become persuaded that Priestley and the phlogistonists were wrong, and that on the whole the French chemists were mostly right. Beddoes was never fully persuaded that Lavoisier's list of elements was altogether right. But by now, he was fully convinced that the phlogiston theory was wrong. Lavoisier's *Traité* of 1789 made the definitive case, and in 1791 Beddoes congratulated Black on giving up phlogiston: "I am glad to see your solemn renunciation of the old chemical theory in the Annales de Chymie."[106] In the same year, he wrote to Joseph Banks that the "[e]laboratory at Oxford has been for a considerable time past undergoing a thorough repair, which I hope will render it one of the best in Europe; in the mean time, I have been prevented from making any experiments".[107]

The lack of experiment was a reflection of Beddoes' newly dejected state. He complained to Joseph Black that "Science & Chemistry in particular will

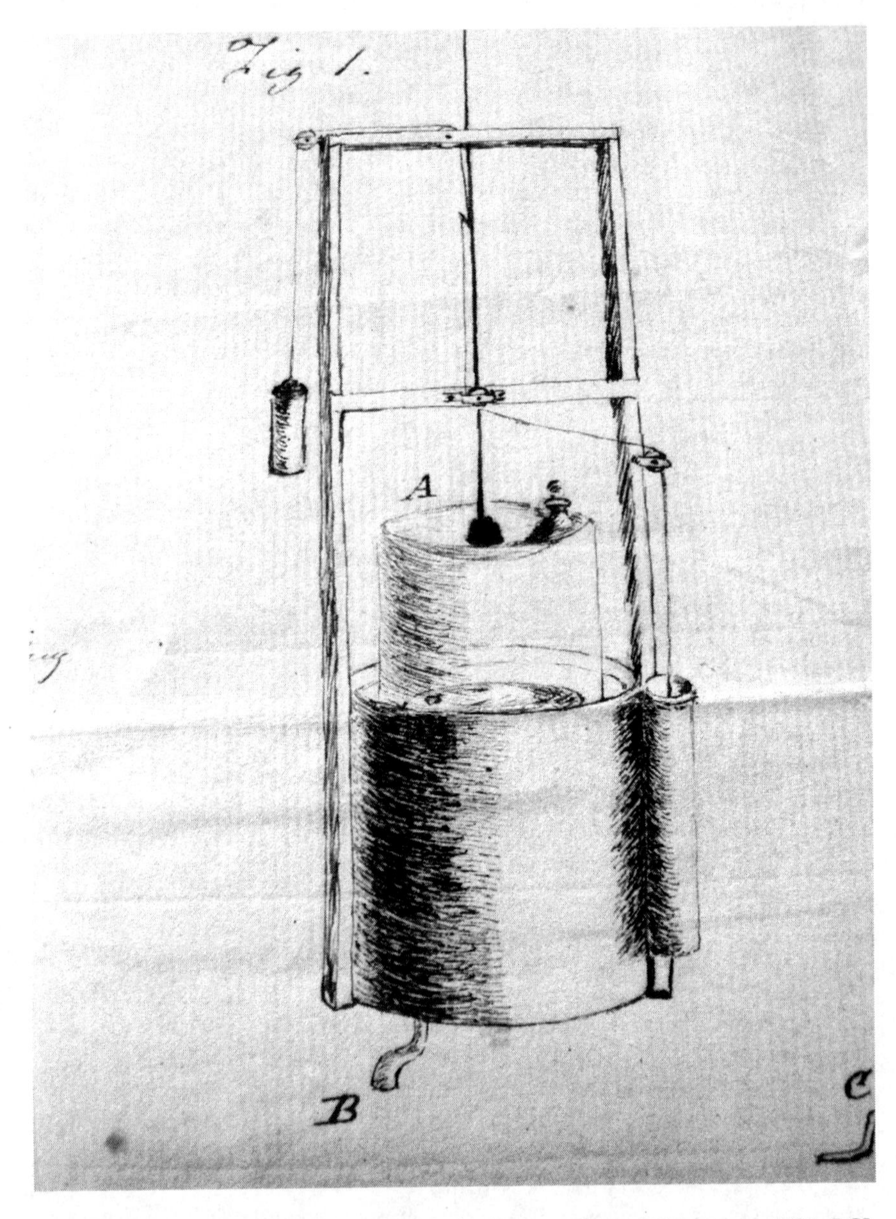

Figure 1.1 Beddoes' gasometer: sketch in his letter to Watt, 27 February 1795, BCL JWP 4/65 no. 18, MS 3219/4/27:13. Birmingham Central Library.

Figure 1.2 Lavoisier's gasometer: *Traité élémentaire de chimie* (1789). Wellcome Library, London.

never flourish in the shadow of ecclesiastical and scholastic institutions," and he told William Withering of the Lunar Society of Birmingham that the

> spirit of Chemistry has almost evaporated at Oxford, as indeed I always expected it would. The young men are generally speaking regardless of every thing. . . . Accordingly my class was chiefly composed of the senior members of the University . . . the stock of curiosity seems nearly exhausted.[108]

What had happened to cause Beddoes to lose his self-confidence and his students? The French Revolution had happened, with all the alarms it conjured in its wake in the minds of the English establishment. Beddoes at first supported the Revolution without equivocation or concealment. He was an old acquaintance of Sir James Mackintosh, M.P., whose *Vindiciae Gallicae* of 1791 he saw as an able answer to Burke's *Reflections on the Revolution in France* of 1790. What could one do in Oxford with an avowed democrat ? One could shun his lectures, and this the young men of Oxford did, although according to Davies Giddy, Beddoes' former student and thereafter lifelong friend (and a future President of the Royal Society of London), the Dean of Christ Church had attended his lectures.[109]

In January 1792 Beddoes wrote to Giddy:

> [Y]ou will readily believe that anticipated disapprobation falls light. You know as well as I do, the orbit in which Oxford minds move. I suppose one might trace a chain of ideas from the French Revolution to doubts concerning the extensive usefulness of hospitals; & one might venture to foretell that neither the one nor the other would be well received in the house adjacent to the Divinity-school or the tower of St. Angels. Establishments would be equally respected in both.[110]

In July 1792, Beddoes proposed a petition, never sent, in which the dangers to apprehended from popular unrest were seen as far less alarming than "the bigotry to be apprehended against the protesting populace".[111] At about the same time, he issued his *Letter to a Lady*, in which he represented the oppression of the poor as likely to lead to violence; injustice invited rebellion.[112]

By this time, Beddoes was an unsettling man to have in Oxford; nor was Oxford any longer congenial to him. He decided to resign, and in June went to the Vice-Chancellor, John Cooke. Beddoes later described the interview to Giddy:

> I said I had no desire to read any more lectures; but that as I had given no chemical lectures that year, & as I thought it unjustifiable to keep possession of the Elab[orator]y with[ou]t reading lectures [a most uncharacteristic sentiment in eighteenth-century Oxford!] & as I cd. not immediately with convenience remove my specimens &c, I would not formerly resign till *this* summer [1793]. I added that I wd. offer to read *this* spring & if I had a tolerable class, wd. proceed – that course however shd. certainly be the last.[113]

Cooke, however, asked Beddoes to draft a memorandum making the case for a chair in Chemistry. Beddoes drafted the memorandum, stressing the cost of apparatus and the expense of running the laboratory, as well as the desirability of parity with Cambridge. Cooke sent the memorandum to the Prime Minister, Lord North, along with a proposal that Beddoes be the first holder of the new chair:

> [He] (tho' as he says for a short time) has so unequivocal a claim in his Line to our utmost assistance and has given such ample satisfaction by Courses of Chemical Lectures to a larger class of Pupils than ever was

> before collected in that branch of useful Science, that I trust your Lordship's known candour and goodness will pardon this last attempt in your Vice-Chancellor to render some material & lasting Service to your University.[114]

Beddoes' politics, the ruthlessness of Henry Dundas (the Home Secretary), and widespread fear of French democracy ensured that these hopes were frustrated. Dundas was informed that, in spite of Beddoes' thorough knowledge of chemistry and his well-attended lectures, "in his political character he is a most violent *Democrate* and that he takes great pains to seduce Young Men to the same political principles with himself."[115] There would be no chair in Chemistry for Oxford, certainly not one with royal approval.

Even without Beddoes, chemistry (especially but not exclusively French chemistry) was coming to have revolutionary associations. Joseph Priestley was the principal target for political opposition to chemistry. He had been sympathetic to the American colonies. Later, he was equally sympathetic to the French Revolution, and to what was soon seen by some in England as the disease of democracy. The fall of the Bastille in 1789 nearly coincided with the publication of Lavoisier's manifesto for the chemical revolution, his *Traité élémentaire de chimie*.[116] But the association between chemistry and explosive revolt was more than a coincidence of dates. Priestley had earlier thrown down a challenge, in the preface to his three-volume work on different kinds of air. The rapid progress of knowledge witnessed over the past century

> will, I doubt not, be the means, under God, of extirpating *all* error and prejudice, and of putting an end to all undue and usurped authority in the business of *religion*, as well as of *science*. . .. And the English hierarchy (if there be anything unsound in its constitution) has equal reason to tremble even at an air-pump, or an electrical machine.[117]

The English hierarchy took him at his word and became highly suspicious of his activities, as suggested in the cartoon of "Dr. Phlogiston" (Figure 1.3).

In 1791, riots broke out in Birmingham following a dinner commemorating the Glorious Revolution of 1688. Coincidentally or by design, this fell on the second anniversary of the fall of the Bastille. "The King's head on a platter" was the revolution toast (A Birmingham Toast) supposedly given by Priestley in a cartoon published later that month (Figure 1.4). Priestley was in fact not present. His laboratory, house, and chapel were nevertheless all wrecked by the rioters. Priestley withdrew first to London, and then ultimately to the safety of Northumberland, Pennsylvania, where he spent the remainder of his life.

Following the Birmingham riots, government spies kept tabs on Priestley, reporting back to the Home Office. In 1792, a Home Office list of seditious

Figure 1.3 "Doctor Phlogiston, the priestley politician or The political priest", Annabal Scratch, fecit. ([England]: Published as the Act directs by Bentley & Co., 1 July 1791). Wellcome Library, London.

Figure 1.4 "A Birmingham toast, as given on the 14th of July by the – revolution society", by James Gillray ([London]: Pubd. by S.W. Fores, N. 3 Piccadilly, 23 July 1791). Library of Congress.

persons included Beddoes and Priestley.[118] The link between revolution and revolt was further underlined when Priestley's American friend Benjamin Franklin was made an honorary French citizen. It was not too far a stretch to link Priestley and Franklin with Thomas Paine, Thomas Cooper, and the young James Watt junior.

Edmund Burke was the most articulate British opponent of the French Revolution, and he saw the spectre of Priestley threatening stability at home:[119] "Churches, play-houses, coffee-houses, all alike are destined to be mingled, and equalized, and blended into one common rubbish; and well-sifted, and lixiviated, to crystalize into true, democratick, explosive, insurrectionary nitre."[120] And in case anyone missed the reference to Priestley, Burke underlined it: "The wild *gas*, the fixed air, is plainly broke loose."[121] Joseph Priestley and his fellows were, in Burke's view, deeply destabilizing the kingdom, from their dissenting pulpits and that other pulpit, the laboratory, as well as in their political agitation, and in their coffee-house machinations. Priestley used to confer in coffee houses with Benjamin Franklin and Richard Price – a dangerous trio. They aroused sometimes venomous criticism.

The association in the minds of critics of French democracy between French politics and French chemistry was to linger throughout the decade, fading but slowly. Beddoes, preparing to leave Oxford in April 1793, commented that he had experienced a surprising amount of civility from his colleagues. He wrote to Giddy, "It can be no wonder that men's exterior should become more smooth as their alarms subside; yet it feels to me as a mark of the increased liberality of the age."[122] Beddoes' optimism was sorely tested by subsequent events in France.[123]

Beddoes, going from Oxford to Bristol, would move from giving lectures on chemistry to practicing medicine as a physician, not uniquely but very visibly. He would combine medicine and chemistry in general with pneumatic chemistry in particular, in the form of pneumatic medicine.

The birth of a pneumatic physician

While still a student in Edinburgh, he had remarked on the importance of gas chemistry for plant and animal respiration. He had attended Black's lectures, with their discussion of fixed air and the numerous new kinds of air discovered by Joseph Priestley and others, and of the combustion of inflammable and dephlogisticated air to produce water. Priestley referred to vital air – or eminently respirable air or dephlogisticated air – our oxygen. Beddoes, like Black, gave lectures on chemistry, devoting some of them to the new chemistry of gases or airs; he used his gasometer in lecture demonstrations. Beddoes wrote to Black from Oxford in 1788 that of

> the Chemical news wch I have lately heard, the most important is, that Dr Priestley is repeating Mr Cavendish's exps on water with airs as dry as possible & instead of water, he gets an acid from the combination of the solid basis of dephlogd. [air] with that of inflammable air. Dr. Withering has the acid to examine, but has not found out what it is. It is not the nitrous, or seems not to be it.[124]

Dr. William Withering, like Priestley, was a member of the Lunar Society of Birmingham. Priestley's first publication on the subject of air was a small pamphlet on the method of impregnating water with fixed air,[125] to produce our soda water, seen then as having therapeutic value. Another member of the Lunar Society was Dr. Erasmus Darwin, who was probably both the richest and the bulkiest physician of his day. Darwin and Beddoes had been corresponding at least since 1787, when they discussed the chemistry of airs.[126] Beddoes' interest in this subject dates from his student days, and he always saw the study of airs as important for the practice of medicine. He was in correspondence with chemists and physicians on the subject of airs even before he began to lecture at Oxford.

In 1788 Edmund Goodwyn, who had been a medical student at Edinburgh at the same time as Beddoes, published a book on the connection of life with respiration.[127] Beddoes, who corresponded with Goodwyn, read this eagerly and with admiration, and in 1788 took the opportunity of expressing his admiration publicly. Frustrated by obstructive regulations in the Bodleian Library, he had been unable to obtain the treatises of John Mayow; but he was able to borrow a copy privately, and was mightily impressed by it. He translated and edited it, published it, and dedicated it to Goodwyn.

> Should I ask you, who of all your acquaintance is the person least likely to be overtaken by surprise, you would, I think, name a certain Northern Professor [viz. Black], to whom you and I may have our obligations. . .. Yet, at sight of the annexed representations of Mayow's pneumatic apparatus, this sedate philosopher lifted up his hands in complete astonishment.[128]

The reason for surprise was that Mayow's apparatus (Figure 1.5) was remarkably similar to that used by pneumatic chemists a century later, as these illustrations show.

His Fig. 1, with its burning glass and pneumatic trough, is similar to apparatus used by Lavoisier in his early experiments (see Figure 1.6); his Fig. 4 is like Stephen Hales' pneumatic trough (see Figure 1.7), and his Fig. 2 anticipates Priestley's experiments using mice to determine the goodness of air for respiration (see Figure 1.8).

Beddoes was not troubled by anti-Whig historiography, and announced that

> Mayow . . . silently and unperceived in the obscurity of the last century, discovered, if not the whole sum and substance, yet certainly many of those splendid truths which adorn the writings of Priestley, Scheele, Lavoisier, Crawford, Goodwyn, and other philosophers of this day.[129]

He wrote to Erasmus Darwin, an author, an Englishman and a physician, that another

> Author, an Englishman, and a Physician, clearly and distinctly discovered, a little after the middle of the last century, several capital branches of the pneumatic theory of chemistry; and he employs, not vague terms, and clear experiments, and an apparatus such as will astonish every intelligent person who sees it for the first time. I have employed a few intervals of leisure, to brush the dust off the memory of this, perhaps, next to him who awes all men to a respectful distance, the greatest of our natural philosophers.[130]

Beddoes probably met Erasmus' son Robert in Edinburgh, where Robert Darwin studied medicine before going to Leyden and obtaining his M.D. in 1785. John Wedgwood, Josiah's oldest son, wrote to his father in October 1785 from Edinburgh, where Robert Darwin introduced him and his brother Josiah junior to Joseph Black, to whom they presented a "thermometer", which was probably Wedgwood's pyrometer.[131] Erasmus and Beddoes

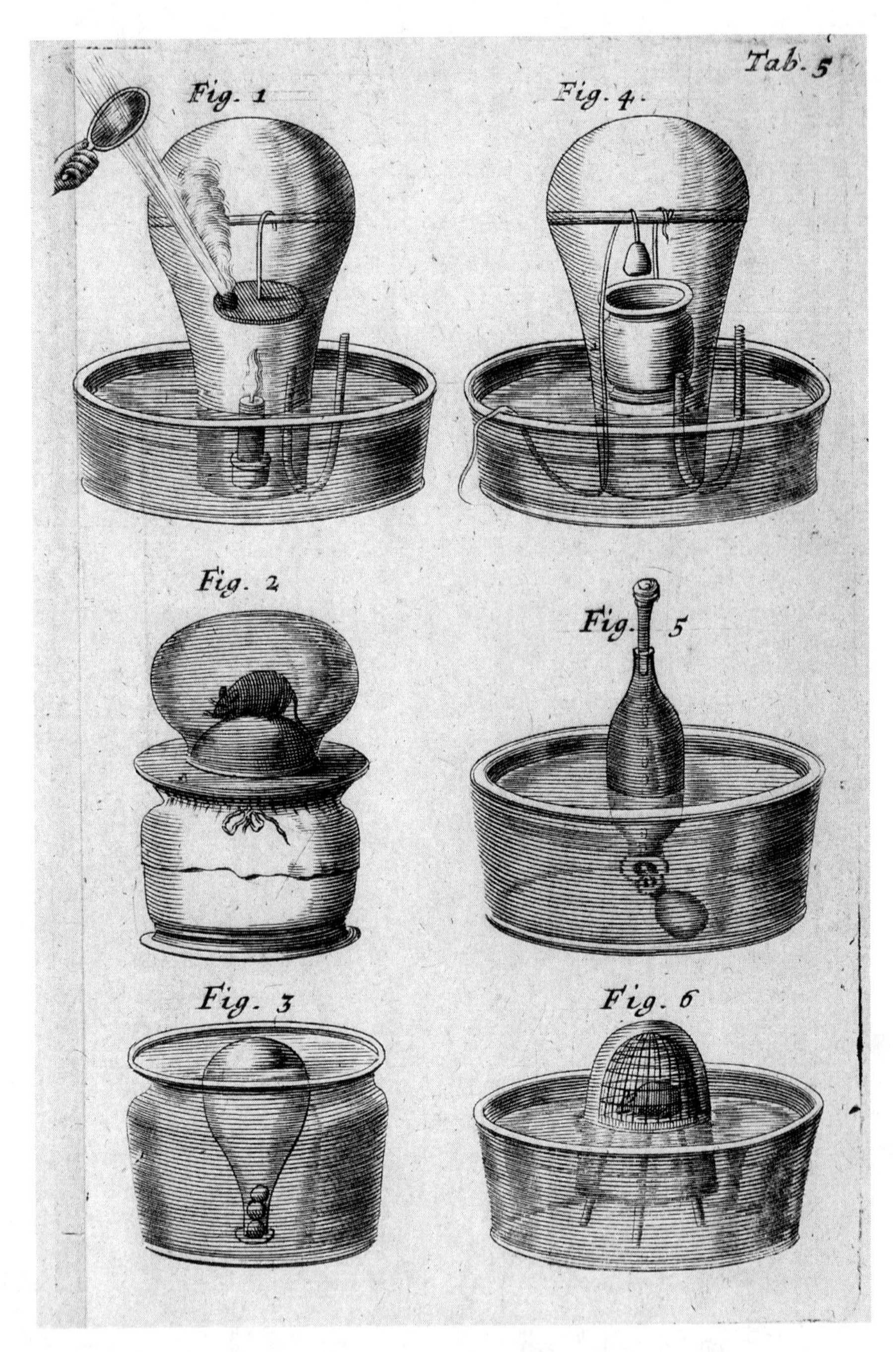

Figure 1.5 Mayow's pneumatic apparatus: [John Mayow], *Tractatus quinque medico-physici* (Oxford: 1674), plate 5. Wellcome Library, London.

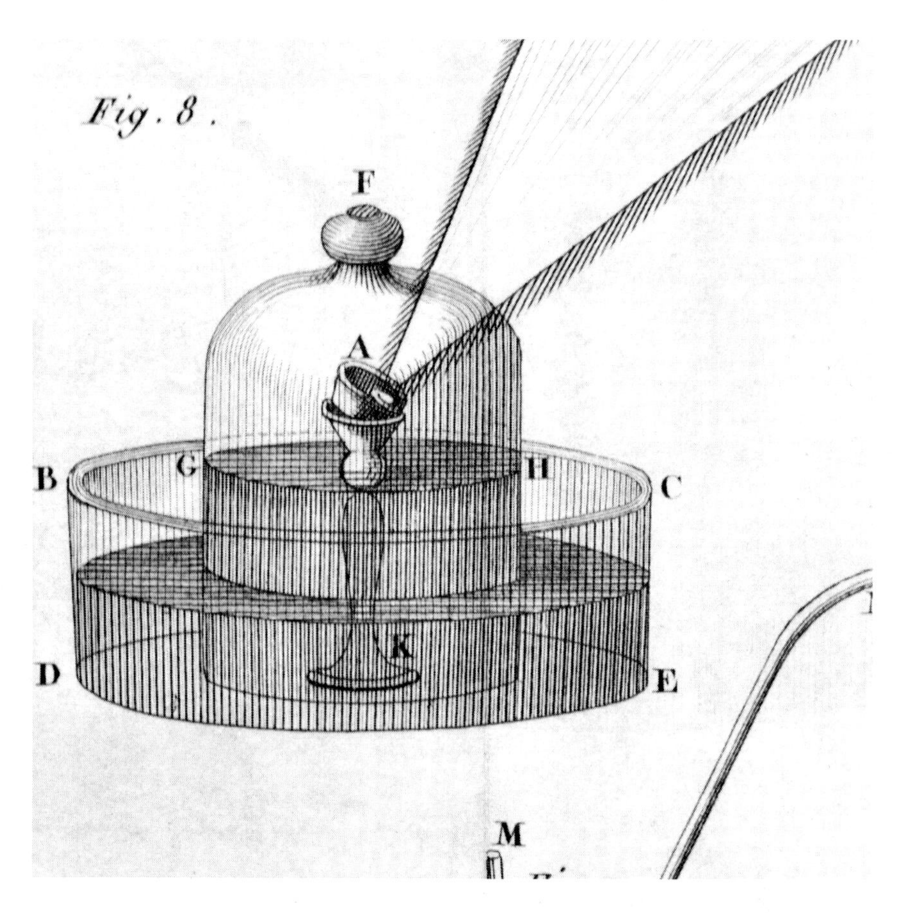

Figure 1.6 Lavoisier's pneumatic trough: Lavoisier, *Opuscules physiques et chimiques*, 2nd edn. (Paris, 1801), plate 2 fig. 8.

met when Beddoes was briefly home in Shifnal in Shropshire, before removing to Oxford and preparing his chemical lectures. Erasmus may have been combining the collection of mineral specimens with a visit to his son Robert, who had been a physician in Shrewsbury, the county town of Shropshire, since the fall of 1786. Beddoes and Darwin talked about mineral specimens ("fossils"), a theme followed up in correspondence.[132] Towards the end of 1787, Darwin wrote to Beddoes, asking him where he wanted his specimens to be sent,

> or whether you may not probably visit Derbyshire next summer, when I could supply you more amply, and more to your intention, and where I should be happy to see you, and glad if this motive may attract you

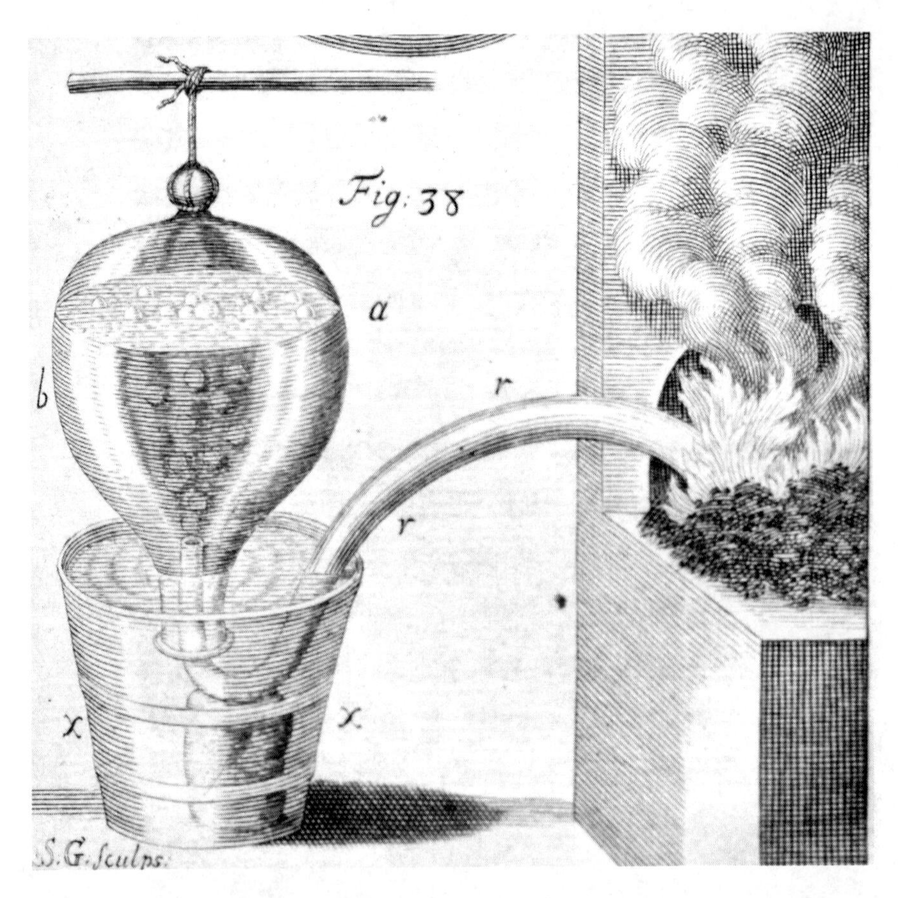

Figure 1.7 Stephen Hales' pneumatic trough: Hales, *Statical Essays: containing Vegetable Staticks . . .*, 3rd edn., I (London: for W. Innys and R. Maney, 1738), plate 17 fig. 38.

> into our hemisphere. I shall be likely, perhaps, to send you fossils which you do not care for, and not send you such as you want. I should therefore advise you to come over next summer, and I will load you home again; otherwise I will send you about one hundred weight when you please.[133]

Beddoes spent the summer of 1787 on the continent. Later that year, Darwin urged him again to visit Derby in the following summer. They had both theorised about coal strata, and "If the two boxes of coal-borings would be interesting to you to see, I will either send them to you or shew them to you in the summer."[134]

Darwin repeated the invitation, which may have been one of the motives that took Beddoes to Birmingham in 1791, since Derby was a mere 37 miles

Figure 1.8 Priestley's pneumatic trough and eudiometers: Joseph Priestley, *Experiments and Observations on Different Kinds of Air*, 3 vols. (London: J. Johnson, 1775–77) I, frontispiece. Wellcome Library, London.

from Birmingham, and could easily be included in a geologizing trip. In Birmingham, he decided to visit James Keir,[135] who had translated Macquer's *Dictionnaire de Chimie* into English, adding numerous notes, as well as tables and plates.[136] Keir had been born in Edinburgh, where he attended the university to study medicine, including chemistry. He gave up medicine for the army, and, finding little intellectual stimulus there, resigned his commission as a captain and moved to Birmingham, where he could pursue chemical manufactures and enjoy the conversation of his friend Erasmus Darwin and other members of the Lunar Society.

Beddoes arrived unannounced at Keir's door:

> Fortunately he was at home. As our opinions in chemistry were different & in politics the same, only that I have scoured more of the rust of prejudice off my mind, & as he is the intimate friend of Darwin [. . .] we shd. have been unlucky indeed if we had wanted conversation during the two days I passed with him.[137]

They talked about chemistry, medicine, and politics. Politics in Birmingham revolved around responses to the French Revolution, and the Church and King riots on the occasion of the second anniversary of the fall of the Bastille.[138]

The mob had been allowed to riot for four days before the dragoons were called in, so that the rioters were unopposed in their destruction of Joseph Priestley's house, library, and laboratory. Keir had saved his own house by "taking measures for a vigorous defence"[139] – but he remained a friend to democracy.

Keir tried a series of different manufactories and different partnerships. By the 1790s, he was running what may have been the largest soap factory in the world, and also producing red and white lead.[140] Keir showed Beddoes his factory, in which the production of caustic soda for soap production was the key.[141] "His whole secret", Beddoes reported to Davies Giddy, "consists in the decomposition of neutral salts". As Beddoes was about to leave, Keir urged him "as far as civility wd. permit to assist him in his great chemical work, offering me any articles I might choose".[142] Keir had published the first part of his own dictionary of chemistry two years previously,[143] using his personal copy of his translation of Macquer as part of the foundation,[144] and editing, removing, or adding material in the margins. Black had used Keir's translation as one of the key books for students taking his classes. But chemistry had made huge steps since 1779, when the third and final volume of that translation had been published. It was time for a new dictionary, and Keir set about writing one. He knew that Macquer was preparing a new dictionary of chemistry for publication in the *Encyclopédie méthodique*, and that would have made it unnecessary for Keir to publish his own dictionary,

> if I had not at the same time been told that Mr. *Morveau*, having enlisted himself under the banners of the antiphlogistic sect, had adopted their new *nomenclature*, which being adapted to their peculiar opinions, cannot be considered as the general language of chemistry.[145]

Keir made his own additions to Macquer (Figure 1.9), and also made use of Leonhardi's "addition to his German translation of Mr. Macquer's work"; he said that he found himself writing about 80% of the new dictionary.[146] He scrupulously indicated his borrowings from Leonhardi and Macquer. In the first part of the new dictionary, Keir got through the A's as far as the article on Affinity. In the article on ACIDS, he acknowledged recent discoveries indicating that

> each acid consists at least of two parts, *pure air* united with its peculiar *basis*. . . . Whether they accede or not to Mr. *Lavoisier's* doctrine respecting the non-existence of phlogiston, most chemists of this day,

ACID

Nitrous acid is one of the most powerful menstruums in chemistry: not that it is the strongest acid; for in strength it is inferior to vitriolic acid, and even in certain circumstances to marine acid; but on account of the facility, of the quickness, and of the activity with which it dissolves most substances.

The bodies upon which it acts most forcibly are, phlogiston, alkalis fixed and volatile, metallic substances and earths, particularly of the calcareous and absorbent kinds.

Nothing can equal the impetuosity with which nitrous acid joins itself to phlogiston. It is so great, that probably nitrous acid has a stronger affinity with phlogiston than the vitriolic has: and it is probable too, that this proceeds from the phlogiston being one of the principles of nitrous acid.

The phenomena exhibited by nitrous acid with all matters containing phlogiston, are different according to the state of these matters, and of the acid itself.

When the phlogiston of the substances applied to nitrous acid is in small quantity, and inveloped in much uninflammable matter, and when the nitrous acid is diluted in much water, from which it cannot disengage itself in the act of combination, then it dissolves these substances always with more facility, quickness, and activity than the other acids, all circumstances being supposed alike; but it dissolves them without being itself decomposed, and forms along with them new combinations. But when the substances to which the nitrous acid is applied, contain much phlogiston, as sulphur, oils, charcoal, and many metallic matters; and when the nitrous acid is dephlegmated as much as is possible; or when it can, in the very act of combination, become dry, and receive at the same instant the heat of ignition, whether this heat be applied to it, or it be generated by the violence of the re-action; then does the nitrous acid, in this state of heat and dryness, combine intimately with the phlogiston, and form with it a kind of sulphur or nitrous phosphorus, which is instantaneously inflamed and decomposed in such a manner, that not only the phlogiston, but also the nitrous acid itself, is entirely burnt and destroyed. And this combustion is effected differently from all others, without the necessity of access of air. Hence the inflammations, the detonations, the explosions, which always happen when all these circumstances concur. *See* DETONATION *of* NITRE *with sulphur*, &c. GUNPOWDER, FULMINATING POWDER, FULMINATING GOLD, INFLAMMATION *of* OILS.

VOL. I. C Nitrous

Figure 1.9 Keir's copy of his translation of Macquer's dictionary, with his marginalia for his own forthcoming dictionary of chemistry: [P.J. Macquer], *A Dictionary of Chemistry*, trans. Keir (1777).

agree to the admission of air as a necessary constituent part of all acids, which his experiments seem to have demonstrated.[147]

Beddoes was not prepared to write an equal share of the remaining volumes: "Besides he was the most able & conspicuous defender of the old system: to me the truth seemed to be on the opposite side."[148] Beddoes, for all his reservations, sided with Lavoisier against the phlogistonists, of whom Priestley, Kirwan, and Keir were the principal British chemists. They were in a dwindling minority. Keir struggled for a while to complete the work, reporting to Josiah Wedgwood that "My dictionary goes on very slowly, as I have been much engaged some time past, but it still goes on & a little of the 2nd part is printed."[149] Keir must have been demoralised about his new dictionary, because the first part is all that ever appeared, and it did so in 1789, the year in which Lavoisier published his *Traité*. Beddoes for once had made a wise decision.[150]

Keir's chemistry was literary, but also experimental, industrial, and political. In the preface, Keir observed that

> the age in which we live, seems to me . . . the most distinguished for the sudden and extensive impulse which the human mind has received, and which has extended its active influence to every object of human pursuits, political, commercial, and philosophical. The diffusion of a general knowledge, and of a taste for science, over all classes of men, . . . seems to be the characteristic feature of the present age.[151]

If knowledge was power, then power derived from the education of the people. The message was not as threatening as Priestley's warning that tyrants had reason to tremble at an air pump,[152] but its implications were the same. Keir was writing shortly before the fall of the Bastille; and when he observed that "[T]he pneumatic chemistry has been particularly productive of new *theory*," and that it was brilliantly advanced in France, with its combination of passion for novelty and philosophical ardour,[153] he was innocently providing advance ammunition for Burke's literary and political attack on the French Revolution.[154] But Beddoes was quite clear that Keir was no firebrand. He was rather "a very accurate & impartial observer, unless you shd. suppose his station at the revolution dinner might bias his judgment. For a reformer he is very temperate – one of your half-way politicians".[155]

Keir continued to engage in chemical research. In October he wrote to Josiah Wedgwood, expressing his "great pleasure to find there is any one in this country who has courage & ability to attempt such delicate & difficult experiments as those you intend", and telling him about his "experiments on metallic solutions by acids, from which I find the state of phlogistication of acids to be a matter of great moment".[156]

Beddoes' friendship with Keir continued, and in 1793 Beddoes bound himself to Keir and the ironmaster William Reynolds for the enormous sum of 10,000 pounds sterling. This was by far the largest contribution to Beddoes' career, and thus to the program of research and practice in pneumatic medicine on which Beddoes embarked immediately on his resignation from Oxford in 1793.[157] But the individual with whom Beddoes worked most closely in the early 1790s, and who was a constant support in pneumatic medicine and pneumatic chemistry, was the engineer James Watt.[158]

In the summer of 1793, Watt wrote to Black:

> We have had no philosophical news since the affair of the frogs electricity except that Doctor Beddoes is applying the antiphlogistic Chemistry to Medicine Azote & other poisonous airs to cure Consumptions & oxigene for spasmodic asthmas he is at Bristol wells for the greater practice.[159]

In 1793, Beddoes published *A Letter to Erasmus Darwin, M.D. on a New Method of Treating Pulmonary Consumption*.[160] Darwin read the

Letter before it was published and responded with generous enthusiasm, urging Beddoes to extend his researches to different kinds and proportions of gases. Beddoes duly published Darwin's endorsement, which included new observations:

> Your treatise on Consumption I have read with great pleasure, and am glad to find that you are about to combat this giant-malady, which has hitherto baffled the prowess of all ages, and which in this country destroys whole families, and, like war, cuts off the young in their prime of life, sparing old age and infirmity.
>
> The few observations I have made on this disease, since you request them, are at your service; as I wish to contribute even a mite to your great design. I hope you will be led to try a variety of experiments with mixtures of airs; your very ingenious reasonings from the scarlet colour of the blood in consumptive patients, and from the inflammatory size or coagulable part of it, and from the delay of the progress of consumption in pregnant women, indicate indeed hyperoxygenation to be the cause of this fatal disease. But by instituting other experiments with different kinds or proportions of airs, you will, at least, if your first experiments should not succeed to your utmost wish, hold out hopes to those unfortunate young men and women; who, if they knew the general fatality of their disease under the present modes of treatment, would despond at the commencement of it, or wish to try some new kind of medicine.[161]

Whether it succeeded or not, pneumatic medicine was promising, and needed extended trials. It was clear that gases played a role in disease as well as health; the chemistry of animal respiration was reasonably well understood, and pneumatic medicine was a logical path to explore. By 1793 Beddoes explained his own path to pneumatic medicine, giving himself a retroactive academic promotion:

> Early in the course of my medical studies I met with the writings of Mayow; and not long afterwards I became intimate with Dr. Edmund Goodwyn, to whom we owe a most masterly experimental investigation of the function of respiration. These circumstances led me to reflect with peculiar earnestness upon the action of the atmospheric air on the blood during its passage through the lungs; and at Edinburgh I made or witnessed many experiments upon animals, tending to illustrate this important subject. Being afterwards appointed to the Chair of Chemistry in the University of Oxford, I found myself obliged to acquire as minute a knowledge of the properties of elastic fluids as possible.
>
> From the moment I became acquainted with the effects of pregnancy in suspending the progress of Chemistry, I conceived hopes that by combining this fact with the discoveries daily making in pneumatic

> chemistry, a successful method of treating this disease, in some of its stages at least, might be devised. During the same interval the idea of turning these discoveries to the benefit of consumptive patients occurred to others: And several attempts, not absolutely unattended with success, were made to palliate their symptoms. But they were random trials, guided by no fixed principle; and, if we except those of Dr. Fourcroy, it was impossible that they should have been successful even if the means, properly employed, had been adequate.[162]

Beddoes experimented with pure air (oxygen), inflammable air (hydrogen), fixed air (carbon dioxide), and nitrous air (nitrous oxide); pure air was most successful. He used his friends, colleagues, and patients as guinea pigs, and he also did a lot of self-experimentation: "I was rather fat, but during this process I fell away rapidly, my waistcoats becoming much too large for me." He suffered nose bleeds and a tendency to cough, indicating a tendency to haemoptysis, and these effects "so strongly corroborate my theory of Consumption, that I am almost afraid lest they should be considered as imaginary of fictitious". Fortunately, there were others who could corroborate his symptoms.[163] He went on to list the diseases that "could safely and probably usefully be treated with oxygene": typhus, hysteria, anasarca and hydrothorax, diabetes, damaged liver, ulcers, palsy, and schirrus. Then, grasping at straws, he went on: "In Hydrophobia and some other desperate diseases, no one, I suppose, would object to the trial of various mixtures of air, even though we are totally deserted by theory."[164] But there were chemical reasons for pursuing pneumatic medicine:

> There are many diseases in which neither patients nor practitioners have much reason to be satisfied with the state of medicine; and multitudes will, no doubt, concur with me to put it upon a better footing. Many circumstances, indeed, seem to indicate that a great revolution in this art is at hand. We owe to PNEUMATIC CHEMISTRY the command of the elements which compose animal substances; now it is difficult not to believe that much depends on the due proportion of these ingredients; and it is the business of PNEUMATIC MEDICINE to apply them with caution and intelligence to the restoration and preservation of health.[165]

At this date, he did not describe an apparatus for administering airs, although no later than 1795 he had a model that anyone on the spot could inspect. One visitor was Georgiana, Duchess of Devonshire. In 1795 she wrote to her brother that she had known Beddoes since 1792,

> and followed his discoveries in pneumatical chemistry and his application of them to health, and in my own mind I have not the least doubt that in many cases they would cure disorders and in almost all give great relief. His proposals are very fair and candid, and he is full of genius

> and good sense in everything but the one subject of politics, in which he has neither judgement, taste or temper.[166]

Now, in 1795, after Beddoes extended his trials with Watt, she paid a visit to Bristol. With Beddoes practicing pneumatic medicine in Hope Square, she closely examined his apparatus. Beddoes was impressed by her knowledge of modern chemistry. She came back for a second and longer visit. During her visit, he first had the idea of an experimental hospital for the administration of pneumatic remedies. Immediately after her departure, he wrote to Darwin.

> It appeared to me that it would be more practicable to determine the effects of elastic fluids in one year, if we had from six to twelve patients in a house with the apparatus, than in twelve years of private practice. Six or seven hundred pounds would provide so many with air and food, and support all expences. Bad as the times are could not one find benevolent people enough to assist in the execution of so grand a design?[167]

He mentioned his idea to the Duchess, who asked him to write giving her details of the project.[168] Meanwhile, he expected that he would soon have an improved apparatus, and would publish a description with engravings in a few months in a new publication.[169]

The new publication was the fruit of a collaboration with James Watt, an engineer first but also a chemist.[170] There was a strong connection between Watt's practical ventures and his theoretical ideas. Watt's work directed at the improvement of the steam engine led him to conceive a chemistry of heat that went beyond Black, arguing that different airs were modifications of a single air, and that the amount of heat chemically combined was a major determinant of the unique properties of each different kind of air. That indicated to Watt that airs were interconvertible, and he also considered airs as solutions. His understanding of pneumatic chemistry bore directly on his approach to pneumatic medicine.[171]

As we know, Watt had an overwhelming personal reason to follow Beddoes' researches and to collaborate with him in exploring ways to treat consumption. His beloved daughter Jessy (Janet) was suffering from the disease, which killed her in 1794. Watt, grief-stricken and well aware of the wide spread of pneumatic interests, threw himself vigorously into Beddoes' project.[172] His close friend Erasmus Darwin had written poignantly sharing Watt's grief; a month later, he wrote to him again:

> My dear friend,
>
> It gave me great satisfaction both on your account & on that of the public, that you are employing your mind on the subject of medicinal airs, of which indeed Dr. Beddoes had before inform'd me. You will do me a great favour in sending me an apparatus, or a description of

one, & an account how easily to obtain the gases; I have now an asthmatic patient would readily try oxygen gas mix'd with atmospheric air.
In your letter to me, you speak of obtaining inflammable air from Zinc. Do you hold the theory, that the inflammable air (Hydrogen) is extracted from the Zinc, as your expression seems [to] countenance, or do you suppose that it is from the decomposition of water according to the idea of poor Lavoisier?[173]

Beddoes and James Watt

Watt was no follower of unfortunate Lavoisier's chemistry ("poor" indeed, as Darwin wrote, meaning unfortunate, since he had recently been guillotined at the height of the Terror in France). Whereas Beddoes had been an early convert to the new French chemistry, albeit not without some reservations, Watt remained a supporter of the phlogiston theory. In spite of Richard Kirwan's surrender to the new chemistry in 1791, Watt took comfort in the constancy of Joseph Priestley, his fellow Lunatick (member of the Lunar Society), and a convinced phlogistonist until his death in exile in America in 1804. In spite of this theoretical divergence, Watt and Beddoes were able to collaborate, first in the design of a new breathing apparatus for the administration of therapeutic airs, and then in the project of Beddoes' Pneumatic Institution. Watt is generally and rightly seen as an engineer, as he proudly subscribed himself. But as David Miller has shown, the chemistry of steam, the material nature of heat, and the chemistry of airs were all part of Watt's own chemistry. In the 1790s, Watt believed that different airs were modifications of a single air, and only the quantity of heat combined with that air accounted for the differences in different kinds of air.

Both Watt and Beddoes, then, adhered to theories in chemistry, but both of them saw fit to assert that they were influenced only by facts. On that basis, he and Beddoes could work together. Beddoes' experiments were often hard if not impossible to replicate, which would have strained Watt's patience. Even more of a problem was suggested by Beddoes' assertion that "sensory experience [could be] identified with the unmediated knowledge of intuition", and that there was a "need to engage the fancy in scientific education".[174] Beddoes' flights of theory were also an embarrassment for Watt.[175] Nonetheless, there was enough empirical evidence, whether confirmed by replication or not, and enough urgency in the face of Jessy's illness, for Watt to work with Beddoes, even though there were many others, including Dr. R.J. Thornton and Beddoes' friend Edmund Goodwyn, then working on aerial remedies.[176] The collaboration of Beddoes and Watt led to the publication in 1794 of their *Considerations on the Medicinal Use, and on the Production of Factitious Airs*.[177] Beddoes wrote most of the chemical and medical parts of the work, although Watt contributed to these; Watt was responsible for the part dealing with the construction and use of a breathing apparatus for pneumatic therapies, although Beddoes advised Watt on parts of it, including a device for the production of airs (Figure 1.10), and a gasometer (Figure 1.1), having already designed and used his own breathing apparatus.[178]

Beddoes hoped that the apparatus "would soon come to be ranked among the ordinary articles of household furniture";[179] accompanied by instructions for use, it would, as Erasmus Darwin had indicated, "put the experiments into many other hands."[180] Once Watt had the production of the apparatus in hand, Darwin asked him to send one, along with directions for producing the airs; he had an asthmatic patient who would benefit from pneumatic therapy. Once it arrived, Darwin was enthusiastic about the "magnificent apparatus", parts of which he described as "very ingenious, and truly *Wattean*".[181]

The apparatus (Figure 1.11) consisted of: (1) an alembic (for which a fire-tube was often substituted), in which the airs or gases were generated; (2) a "refrigeratory" for cooling and washing the airs; (3) "hydraulic bellows" (generally called a "gasometer") for receiving and measuring the air as it came from the refrigeratory; and (4) an air holder to receive the cooled and washed airs.

Figure 1.10 Beddoes, sketch of an alternative apparatus for the production of gases: Beddoes in letter to Watt 12 December 1794, BCL MS 3219/4/27:8. Birmingham Central Library.

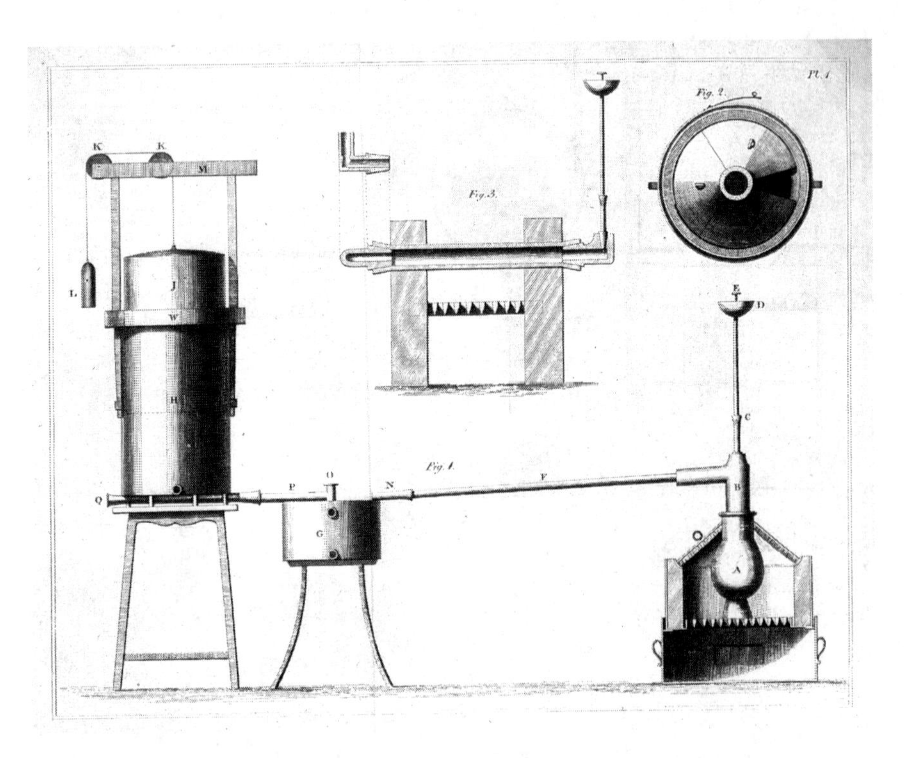

Figure 1.11 [Beddoes and] Watt, Alembic and hydraulic bellows or gasometer: Beddoes and Watt, *Considerations* 2nd ed. (1795), plate 1. Wellcome Library, London.

When the patient was present, it was often convenient to dispense with the air holder and instead to use oiled silk or linen bags, "or such other vessels as shall be thought convenient for mixing it with the proper proportion of common air, and also for the patient to inhale from".[182] Watt described the apparatus at length and also the ways of procuring the different airs. He was even-handed in referring readers to Priestley and to Lavoisier. He recommended heating manganese (our manganese dioxide) as the best way to obtain oxygen or dephlogisticated air; the quality of the manganese was important. In one of Watt's early trials, using manganese supplied by Beddoes, he found that the air was

> mostly fixt air, secondly dephld. thirdly fixt air again, by taste & smell for I did not apply other tests – lastly by the addition of water I got some of the most pungent fixt air I ever smelt. . . . I durst not try to breath either it or the others. I . . . breathed a little of the dephld. mixt with common air, but found some fixt air in it which as usual hurt my lungs, and means must be found to clear it easily of this [part].[183]

Shortly thereafter, Watt was generating more oxygen from manganese, and his son Gregory

> attempted to smell this air as it issued from the hole in the refrigeratory while he was pressing down the bellows, a little after he found himself on the floor looking at the ceiling, & wondered where he was, his awaking he says was rather pleasant, no vertigo or sickness ensured.[184]

Watt concluded that he needed more directions from Beddoes, as well as the best source of manganese, and he also decided that he would have to revise his contributions. Self-experimentation was the order of the day, and it could be thoroughly dangerous. Watt, for example, had tried the effects of hydrocarbonate (carbon monoxide), and found it "extremely deleterious".[185] No matter how many careful instructions were issued, unskilled use of the pneumatic apparatus could be very dangerous.

Dangerous in a different fashion, in the eyes of many, were French theories in science and politics. Beddoes and Watt feigned agnosticism about chemical theory, and this enabled Beddoes to begin his account with a brief résumé of the chemistry of the atmosphere, which "consists of two kinds of air, quite distinct in many properties. One is the kind called VITAL, DEPHLOGISTICATED, or OXYGENE AIR, and by a variety of names besides. The other has been named AZOTIC, PHLOGISTICATED, FOUL, or BAD AIR.".[186] Maintaining their putative even-handedness about theory enabled Beddoes to refer to the writings of Priestley, Scheele, Cavendish, and Lavoisier for ways of analysing air, but then he slipped into the new nomenclature: "These authors explain much of the nature of oxygene and azotic air."[187] Several chemists, physicians, and natural philosophers – including Goodwyn, Priestley, Senebier, Lavoisier, and Ingenhousz – had studied the chemistry of respiration.[188] To present this in the language of the new French chemistry implied, for many, acceptance of the new French theories of chemistry. That was why Black, for one, was cautious in his lectures when it came to French chemical nomenclature, even though he adopted many of its ideas.[189]

Edmund Burke had described the French Revolution as "a revolution of doctrine and theoretic dogma", a "philosophic revolution". French philosophers, including natural philosophers, men of science, were "only men of theory", and "they despise experience as the wisdom of unlettered men."[190] Making theories was a French disease, and Watt, chagrined by his son James's temporary adherence to the Jacobins, would have none of it.

> My Son James's conduct has given me much uneasiness, though I have nothing to accuse him of except being a violent Jacobin, that is bad enough in my eyes, who abhor democracy, as much as I do Tyranny, being in fact another sort of it.[191]

So there was no avowed truckling with theory. When in the same letter Watt wrote to Black about Beddoes' new medical practice in Bristol, he knew what underlay that practice: "Doctor Beddoes is applying the antiphlogistic Chemistry to Medicine, Azote & other poisonous airs to Cure Consumptions & oxygene for spasmodic asthmas."[192] Watt was no friend to antiphlogistic chemistry, and he for some years he had had his own theories about the nature of different kinds of air: they were all modifications of a single air, produced by different combinations of that air with heat. As David Philip Miller has shown,

> Watt's chemistry of airs, as he mobilised it in the mid-1790s, was still grounded in his early ideas of the interconvertibility of airs as modified by the subsequent debates on phlogiston theory, but . . . new ideas about airs as solutions had also been added. . . . [H]is approach to pneumatic medicine was, for all his vaunted even-handedness about facts in public, based on a theoretical position at odds with the new chemistry.[193]

Watt made no secret in his correspondence of his hostile view of French chemistry and French theories. Beddoes was surely responding defensively to Watt's phlogistic stance in his letter to Watt in March 1795:

> I do not believe in my own theories – for instance, I do not believe in the hyperoxygenation of the system in consumption – My first speculations were merely attempts to put facts together and to deduce such conclusions as might be put to the test of exp[eriment].[194]

Such a defensive posture would have been useful, if only Beddoes' reputation as a democrat had not given it the lie. One of Watt's correspondents was James Lind, a Scottish-born physician, a fellow of the Royal Society of London and of the Royal Society of Edinburgh, and, from around 1793, resident in Windsor and physician-in-ordinary to the royal household.[195] Lind and Watt were both in touch with Jean André de Luc, a Swiss meteorologist and geologist, whose theory of the formation of rain furnished him with evidence against the new French chemistry. Lind told Watt,

> He is much pleased to be informed by you of your having made the different Airs, without either Water, or Acids, He being no friend to Theories, either Philosophical or Political that are founded on falsehood, and are propagated by force, and such he thinks French Chemistry and Politics to be, Indeed there seems to have been a wonderful coincidence in sentiments, of some of the modern Chemists in this country that I have declined taking a part in the Pneumatical practice of Medicine from a detestation of having any connexion with such a set of miscreants, notwithstanding I am induced to believe from several reasons that in many Diseases, the practice wilſ be of <real> use.[196]

Watt, unsurprisingly in the light of such remarks, was increasingly frustrated by the effect of Beddoes' politics on the reception of pneumatic medicine:

> Doctors in London in General condemn the practice [of pneumatic medicine] in toto & some other people are sure it must be bad 1st because you believe in Lavoisier's theory, 2d because you have the character of a Jacobin 3dly because they have found out from some expressions in your tracts on air that you are a Materialist 4thly because in trying to do good some animals may be suffocated & some men get some new and unheard of diseases 5th 6thly 7thly for various other equally good reasons best understood by themselves. To this I have answered *fas est et ab hoste doceri*[197] that I never am too scrupulous in inquiring into the theories either in religion politicks or chemistry of those who are able & willing to do me or society any good, that I am no Jacobin nor Materialist, nor believe in Lavoisier's theory; but that I have an unaccountable tendency to believe in facts which pass under my own observation that though I think part of your theory false yet I see no reason to doubt that the lungs can absorb oxygene or carbone & that I am sure the vital & carbonic airs are powerful stimulants of the whole nervous system. I wish however you would observe the above maxim, ab hoste doceri, your republicanism may do more hurt by preventing the Pneumatic practice than it ever can do good, leave chemical or medical theory as much out of play as possible but ply them hard with facts – these are understood by every body & all the Doctors in London cannot overturn them. If they write against you answer them with more facts.[198]

More pithily, Watt expostulated later that year: "Why will you waste your time writing against P[itt] & C[o.]? You will do more harm to Pneumatics than you can possibly do good to the nation – amend your ways."[199] Beddoes admitted the charge.[200] A year later, Watt wrote despairingly of Beddoes as political critic:

> Beddoes is publishing some select cases with observations in his manner; some of them are interesting.
>
> In his politicks he is incorrigible, at least by me who have given him up in that line, and I am sorry to see that he is publishing another hit at the Doctors which I agree with you can do nothing but procure him enemies, who will be absurd enough to make war upon his doctrines in revenge.[201]

The immediate trigger for Watt's complaint was Beddoes' latest squib, *A Letter to the Right Hon. William Pitt, on the Means of Relieving the Present Scarcity, and Preventing the Diseases that Arise from Meagre Food* (London, 1796). Beddoes did not take Watt's advice. In 1795, he had published *Where would be the harm of a Speedy Peace?* (Bristol) and *A Word in Defence of the Bill of Rights, Against Gagging Bills* (Bristol); in 1796, he published his *Essay on the Public Merits of Mr. Pitt* (London); and in 1797, *Alternatives Compared; Or, What Shall the Rich do to be Safe?* (London). Opposition to the war with France, support for democracy and freedom of speech, warnings of retribution to the rich if they did not mend their ways, and a constant opposition to William Pitt's policies all combined to make Beddoes appear as a firebrand at best, seditious at worst. That was not calculated to win over the establishment, which included the Royal Society of London, the Royal College of Physicians, and the moneyed gentry. It is not coincidental that Beddoes, although a contributor of several papers to the *Philosophical Transactions of the Royal Society of London*, was never elected FRS.

Pneumatic chemistry and Beddoes' pneumatic institution

Beddoes was constantly arguing for the application of science– chemical, medical, and mechanical – to the purposes of everyday life. The poet, philosopher, and political journalist Samuel Taylor Coleridge reviewed Beddoes' *Letter to . . . Pitt*:

> To announce a work from the pen of Dr. Beddoes is to inform the benevolent in every city and parish, that they are appointed agents to some new and practicable scheme for increasing the comforts or alleviating the miseries of their fellow-creatures.[202]

In his *Letter*, Beddoes argued that science could and should be systematically applied to the relief of hunger among the poor. The necessary investigations could be

> undertaken on two conditions: if a moderate contribution could be collected from the rich, and if a few persons acquainted with chemistry and medicine would unite with a few others of leisure and activity to conduct a train of simple experiments.[203]

Could not the vegetable foods – including potatoes, which cattle converted into meat – be transformed by the art of the chemist to provide nutrition for mankind? Barley, wasted as Beddoes saw it on the production of strong beer, could be saved and used as food for the poor. When Pitt embarked on his war with France, there was widespread, even overwhelming public support for the enterprise; Beddoes rhetorically asked Pitt,

> with what enthusiasm did every man of genius spring forward to assist you? . . . And if the recent glories of your administration should become the subject of an epic poem, one book should be devoted to the enumeration of those heroes of chemic and mechanic science that thronged to your standard.[204]

In fact, of course, it was in France that science was co-opted in the service of the state, in stark contrast to Britain. Pitt did not respond to Beddoes' unsubtle irony. When Beddoes published his *Letter to Erasmus Darwin* (1793), he had made extravagant claims for the role of chemistry in health and nutrition.[205] But, as we know, this was a notion that some increasingly shared in industrial towns. Soon thereafter, he published a two-page broadsheet, *Reasons for believing the friends of liberty in France not to be the authors or abettors of the crimes committed in that country*, which included the assertion that the French, "the most injured and most enlightened people upon Earth, . . . within these few years (since Despotism has been overawed by Liberty) have improved Science more than all other Nations put together."[206]

The opposition struck back with a anonymous pamphlet, probably written by George Canning, entitled *The Golden Age, a Poetical Epistle from Erasmus D – n M.D. to Thomas Beddoes M.D.*[207] The pamphlet opened with a quotation, slightly altered, from Beddoes' *Reasons for believing the friends of liberty in France not to be the authors and abettors of the crimes committed in that country*, 9 October [1793]:[208] "The French, that most injured and most enlightened people upon earth, within these few years (since Despotism has been overawed by Liberty)[209] have improved Science more than all other Nations put together." For the followers of Burke, who by now loathed all things French, such statements were self-parodying, inviting ridicule or abuse. The same author or authors also reprinted, this time accurately, one of Beddoes' overly optimistic hopes: "May we not, by regulating the vegetable functions, teach our Woods and Hedges to supply us with Butter and Tallow?"[210]

The real Erasmus Darwin, M.D., unlike the spurious Erasmus D – n M.D., had written *The Botanic Garden: A Poem, in Two Parts* (London: for J. Johnson, 1791), which, besides its botanical and erotic content, had significant political content, including opposition to slavery and leanings to democracy, which were to earn him political enemies. But here in *The Golden* Age, written by the Anti-Jacobins[211] in the style of Darwinian verse, was ridicule directed at Beddoes:

> Boast of proud Shropshire, Oxford's lasting shame,
> Whom none but Coxcombs scorn, but Fools defame,
> Eternal war with dullness born to wage,
> Thou Paracelsus of this wondrous age;
> BEDDOES, the philosophic Chymist's Guide,
> The Bigot's Scourge, of Democrats the Pride;
> Accept this lay; and to thy Brother, Friend,
> Or name more dear, a Sans Culotte attend.[212]

Calling Beddoes a democrat, let alone a sans-culotte, came close to accusing him of sedition, or at least disloyalty in time of war. Louis XVI's execution on 21 January sent shudders throughout the monarchies of Europe. Ten days later, France declared war on Great Britain and the Dutch Republic. France declared war on Spain in March. The Revolutionary Tribunal was established on 10 March to deal with counter-revolution. The Committee of Public Safety was created in April. In July, Robespierre was elected to that Committee, and government by Terror began in September. Beddoes' timing, publishing a defence of France in wartime at the start of the Terror, was as risky as it was provocative.

No wonder that Watt despaired of Beddoes' politics, which undoubtedly did get in the way of fundraising for his project in pneumatic medicine and for an institution where pneumatic therapies could be tested. Beddoes had written to Erasmus Darwin about his idea of a Pneumatic Institution and discussed it with Georgiana, Duchess of Devonshire. He had also written about it to his friend Davies Giddy, who had been High Sheriff of Cornwall in 1792–1793, and would go on to become Deputy Lieutenant in 1795. Giddy was worried that Beddoes' plan would appear empirical, without any apparent connection to scientific theory. Beddoes replied that that objection had occurred to him, "& of course I took much pains to guard agt. it".[213] A four-page advertisement, in the form of *A Proposal for the Improvement of Medicine*, was printed in the summer of 1794, and in the copy in the Wedgwood Archives there is also an advance notice of the pneumatic apparatus that Watt had designed, with advice from Beddoes. Would-be subscribers to the proposed Medical Pneumatic Institution were invited to send donations to the one of the four banking houses acting as trustees for the Institution:

> I flatter myself that in a work entitled, *Observations on Consumption, Fever, and other diseases*, in my *Letter* to Dr. Darwin, and in a late collection of Letters from different correspondents on the subject of pneumatic medicine, it is abundantly proved, that the application of elastic fluids application of elastic fluids to the cure of diseases, is both practicable and promising. This method of treatment has been very lately adopted abroad, and appears, as far as it has been tried, to have exceeded rather than disappointed expectation. A series of experiments upon animals, of which an account shall soon be given to the public, will, I think, confirm the hopes entertained by many friends of humanity, concerning the medical efficacy of elastic fluids. Although, however, I might be allowed to suppose that enough has been done to encourage further enquiry, I am sensible that facts are wanting fully to establish general conclusions. To what precise extent, therefore, the new mode of practice may be advantageous, remains to be decided by cautious experiment.
>
> This object, I conceive, may be much more effectually accomplished in two years by means of a small appropriate Institution, than in twenty years of private practice; and persons of high respectability, both belonging to the medical profession, and others, have expressed their

> wishes that some attempt might be made to carry such a design into execution. . . . ,
>
> Such an Institution should be conducted with a view to the attainment of two objects 1. to ascertain the effects of these powerful agents in various diseases, and 2. to discover the best method of procuring and applying them. . . .
>
> A dwelling house, capable of receiving 12 patients, may, as it appears to me, be made fully to answer the purpose; since in many cases the airs may be administered without keeping the patient constantly in the house. In two or three years, such an establishment ought to render itself useless, by so far simplifying methods and ascertaining facts, that every practitioner of medicine, at least, may both know how to procure and how to apply the different elastic fluids, supposing they should be found serviceable in any species of disease. . ..
>
> For the whole, three or four thousand pounds would probably suffice; but the plan might be contracted or enlarged, according to the amount of the contributions. . . .
>
> It is, I believe, in the highest degree improbable that such an establishment should be totally unproductive of benefit. But even in the worst event, to have the merit of the project decided by a proper trial, will afford a sort of melancholy satisfaction to persons labouring under diseases at present invariably fatal, and to their friends.[214]

The Institution would need apparatus to generate and administer the airs, and a superintendent "to direct the chemical processes, and to administer airs and medicines under the direction of the physician".[215]

The members of the Lunar Society were co-opted in the campaign for funds. Particularly generous support came from the Wedgwoods. Joseph Banks, firmly declining to have anything to do with Beddoes' proposals, opined that Beddoes' experiments were more likely to lead to "a waste of human health" than to any improvement in the art of medicine."[216] The medical community was essentially polarized, with most of the London medical establishment opposing Beddoes, while extensive support came from Edinburgh and the Midlands of England. If one can judge by the spread of the apparatus, support was even broader. By 1795, Beddoes reported that "a ward in the Birm: hospital is to be appropriated this spring to pneumatic medicine."[217] Indeed, the campaign to expand the use of pneumatic medicine and Watt's apparatus in hospitals and infirmaries was extensive, partly driven by proponents of social reform. In 1796, Watt produced simplified apparatus consisting of one chamber with a furnace for the production of the airs, connected by a tube to a second chamber in which the airs could be stored.[218] 1796 was a splendid year for Beddoes: "The resort hither is beyond all example, and the harvest for Death and the Doctors has been plentiful accordingly."[219]

Bristol was not just a good place for medical practice; Beddoes was also keen on public lectures on medical and other topics, chemistry prominent among them. In 1796, he was promoting lectures on anatomy and physiology, along with the utility of physical education, by two surgeons at the Bristol Infirmary, F.C. Bowles and Richard Smith, while also preparing for chemical lectures. He wrote asking Giddy to obtain specimens from copper furnaces: "[I]f I receive them in March it will be time enough . . . They are for the chemical lectures . . . I shall have two good operators – & trust they will get up an infinity of lectures."[220] But at that point, he did not have good assistants, and he was to be disappointed for over a year in his search for good chemists.[221] In December 1797, he told James Watt junior that the lecture hall was far from crowded for the chemistry lectures.[222]

But if Beddoes' public lectures were a limited success, the campaign for the Pneumatic Institution was going well. Donations kept coming, and by 1798 Beddoes was able to buy a house in Dowry Square, Clifton, near Bristol, and to announce the opening of the Pneumatic Institution:

> Patients will be treated gratis.
>
> The application of persons in confirmed Consumption is principally wished at present; and though the disease has heretofore be deemed hopeless, *it is confidently expected that a considerable portion of such cases will be permanently cured*.
>
> It has been perfectly ascertained by experience, that none of the methods to be pursued are hazardous or painful.
>
> Attendance will be given from Eleven till One o'clock, by THOMAS BEDDOES, or HUMPHRY DAVY.[223]

The apparatus first used in the Institution was Watt's breathing apparatus, and it is probable that Watt supervised its installation. Subsequently, Clayfield, probably with input from Davy, designed and used a new breathing apparatus, and joined James Sadler in a balloon ascent in 1810.[224]

Humphry Davy, Thomas Beddoes, and nitric oxide

The story of Davy's appointment as the chemical operator, soon to become the chemical superintendent at the Pneumatic Institution, is as remarkable as it is familiar.[225] Gregory, James Watt's son by his second wife, went to Cornwall in the winter of 1797–1798 in hopes that the milder climate there would aid his recovery from consumption. He lodged at the house of Mrs. Davy, a woodcarver's widow and her two sons, and, very probably because of a shared enthusiasm for chemistry, soon became friendly with the older boy, Humphry, who, just one year Gregory's junior, was then an apprentice to an apothecary in Penzance. Davies Giddy also learned of Davy's fascination with chemistry, and took an interest in him. In 1798, Beddoes was geologizing in Cornwall, heard about Davy from both Giddy and Gregory Watt, read Dav's account of his experiments on heat and light, and was much impressed. Beddoes was soon to hire Davy, but until the last minute, Davy was not under consideration, except perhaps as a very junior assistant. In October 1797, Beddoes was sounding weary of the search for a chemical operator: "After so many applications I am disappointed in not hearing of an operator – I expect a young man lately from Edinr. who wd make a good one, if he be willing & able bodied – which last I doubt."[226]

At one stage,[227] Beddoes was negotiating with "Scherer," a German chemist whose pride appears to have been wounded by the offer of a subordinate position.[228] If it was A. N. Scherer, a measure of pride would not have been unreasonable, since he had just started editing and producing one of the few chemical journals of the day. But a more likely candidate is Johann Andreas Scherer, who wrote on chemical nomenclature, Mayow,[229] and eudiometry, and would have been a good fit with (if not a rival to) Beddoes.[230] Then Beddoes started to negotiate with a young Irishman, "Mr. [Cadwallader] Boyd, at Dublin", recommended to him as being as good as William Higgins, who in Oxford had been operator to the Professor of Chemistry, William Austin; Higgins may also have helped Beddoes when he became lecturing at Oxford.[231] Beddoes told Boyd that Dr. Clement Archer, a Bath physician, had led him "to entertain the highest opinion of your Chimical abilities. I had intended to write to Mr. Klaproth but what Mr A. said prevented me".[232] Soon afterwards, Beddoes wrote to James Watt junior, enclosing a copy of the letter he had sent to Boyd, which spells out just what Beddoes expected of his chemical operator:

> It goes for nothing to be a speculatist: he must be a practical chemist & it is essential that he shd. not have lain principally in the line of pharmacy or any single line[.] [W]ith the more complicated exps. with gases he shd. be familiar; & also with those processes of fusion & calcination which require the reverb[er]atory air and muffle furnaces.[233] to sum up all he should be conversant with whatever illustrates the philosophy of chemistry as we find it in Lavoisier. . . . [3 words illegible] He shd. be master of the blow pipe – at least understand how to bend & manage the glass tubes, and <understand> the construction of furnaces. I should not think my self bound to terms with a person who did not answer to the above description liberally interpreted . . . I should esteem it a singular mark of Candour if not having been conversant in the practice of experiment belonging to phi[loso]phical Chemistry, you should decline the proposal . . . PS you would have excellent assistance I expect a young man who has applied close to chemistry (I mean worked hard) *for 4 or 5 years* he alone wd. be adequate in case of need to the situation of assistant.[234]

Boyd's reply to that daunting letter, like most of the letters to Beddoes, appears not to have survived. None the less, it appears that Boyd was appointed as chemical operator at the Pneumatic Institution,[235] since in the following month, Beddoes wrote to James Watt junior that "Clayfield & I like the Irish operator very well so far."[236] Dr. William Clayfield, a young physician in Bristol, joined Beddoes in chemical experimentation, inhaled nitrous oxide, and devised an improved breathing apparatus (Figure 1.12); he later became the proprietor of a commercial alkali works.[237]

Having fixed on Boyd as the chemical operator or assistant, Beddoes turned to Davy as the superintendent of the chemical laboratory:

> I have been corresponding lately with Humphry Davy of Penzance . . . I think him most admirably qualified to be the superintendent. I have read the acct. of some exps. of his; & he appears to me to have uncommon talents for philosophical investigations. He has besides entered with ardour into the career of chemical philosophy. Giddy entertains the same high opinion of his talents.[238]

It took some diplomacy to release Davy from his apprenticeship, and although Davy was looking forward to the opportunities offered by the invitation from Beddoes, it was not until October that Davy formally took up his post in Bristol.[239] Beddoes, choosing Davy, had made an inspired appointment, although Davy was for a while seduced by the temptations of hypothesizing, and burned his fingers by publishing speculations along with experiments; Beddoes, himself prone to hypothesizing, was not at all put off, and in 1799 published Davy's essay on heat and light as the long first part of a volume of contributions to physical and medical knowledge.[240]

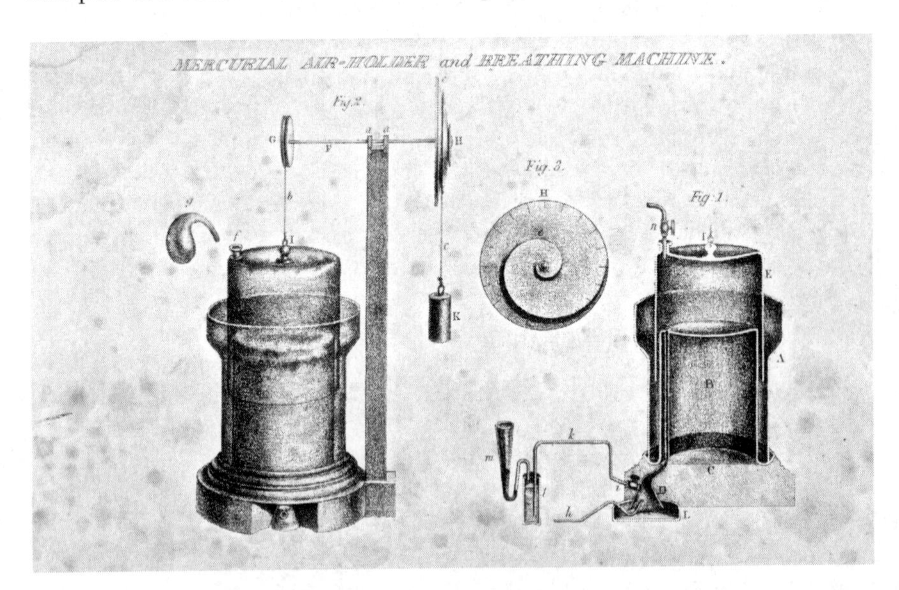

Figure 1.12 Clayfield and Davy, mercurial air-holder and breathing machine: Frontispiece to *The Collected Works of Sir Humphry Davy* 3.

As early as January 1798, Davy, who had already assembled and experimented with chemical apparatus, including an air pump, had been looking forward to the creation of a "superb laboratory" and giving advice to his Penzance friend Henry Penneck,[241] to whom he sent apparatus. He told him, "Chemical Implements are not kept ready for sale in the Bristol glass houses or I should have procured them long before. I was obliged to superintend the execution of them."[242] He listed the apparatus sent, including a blow pipe, common retorts, one tubulated retort, two cylinders, a large glass receiver, a double-necked receiver, curved tubes, glass tubes, matrasses,[243] and flint:

> With this apps: a few crucibles a small furnace & an Argand lamp[244] you may make all the expts essential to Philosophic Chemistry. . .. I hope to commence experiments on pneumatic Medicine in about a month.
>
> We had intended to lecture on Chemistry in March but as the Theatre is not yet erected we have not been able yet to procure a convenient room. It is therefore uncertain whether we shall lecture in the spring or defer our course till Autumn, when the Theatre will be erected & we shall be able to give our first course with every advantage. Whenever we lecture we are certain of a numerous class. . . . Beddoes will offer a chemical course, probably lecture on Medicine & Physiology. . .. I consider Dr. Beddoes as the most truly liberal candid & philosophic physician of the age, as he has sufficient [illegible] to give up his own theories, as well as those he has adopted whenever they appear contrary to facts. . . . If you have . . . made any trials with the nitrous acid . . . I shall be proud to insert them in the West country Physical & medical Contributions, which will I hope be out in three weeks.[245]

Davy stated ambitiously that he intended by the year's end "to publish a much larger work in the Laws of Corpuscular Motion & the connexion of chemistry with the Laws of life".[246] Beddoes thought that was a reasonable goal for a brilliant neophyte.

The anticipated delay in offering chemical lectures was an embarrassment for Beddoes:

> The public here is in great consternation on acct. of the difficulty of obtaining a room for the chemical course. I think the whole difficulty wd. subside if I cd. get a large Black's furnace – Can you get me one made – not luted – or tell me where to procure the drawing.[247]

In March, Beddoes told James Watt junior that he was immersed in preparing the lecture space: "We are going on / masons, carpenters, [joiners] &c &c – preparing these lectures which you projected."[248] He asked Watt to order him "a chemical lamp with a larger burner than usual" along with "6 well made Argands – Single I think will do as well. . .. I find the contracted Argands chimneys will admit the wick to burn much higher as Morveau says".[249] As he had done at Oxford, he planned to use Isaac Milner's syllabus, "a very good syllabus for its period". Since, as we have seen, Milner accepted the phlogiston theory, and listed the elements of bodies as air, earth, fire, and water, this was unduly generous of Beddoes, who, typically, indicated that he would not be tied down in advance of the course:

> My intention is to give a syllabus by spurts – If they have the heads of each lecture before it is given I do not think that not having the whole at first will be a disadvantage, & I shall better mature it.[250]

He wrote to several friends asking for mineral samples.[251] Although recognizing that his request was like asking for a loan of the Ark of the Covenant, he asked for, and received on loan, a sample of Herculaneum glass from Matthew Boulton, in order to enliven his history of chemistry.[252]

Beddoes published his "Specimen of an Arrangement of BODIES according to their PRINCIPLES" (Figure 1.13) in a volume that he edited and published in 1799.[253]

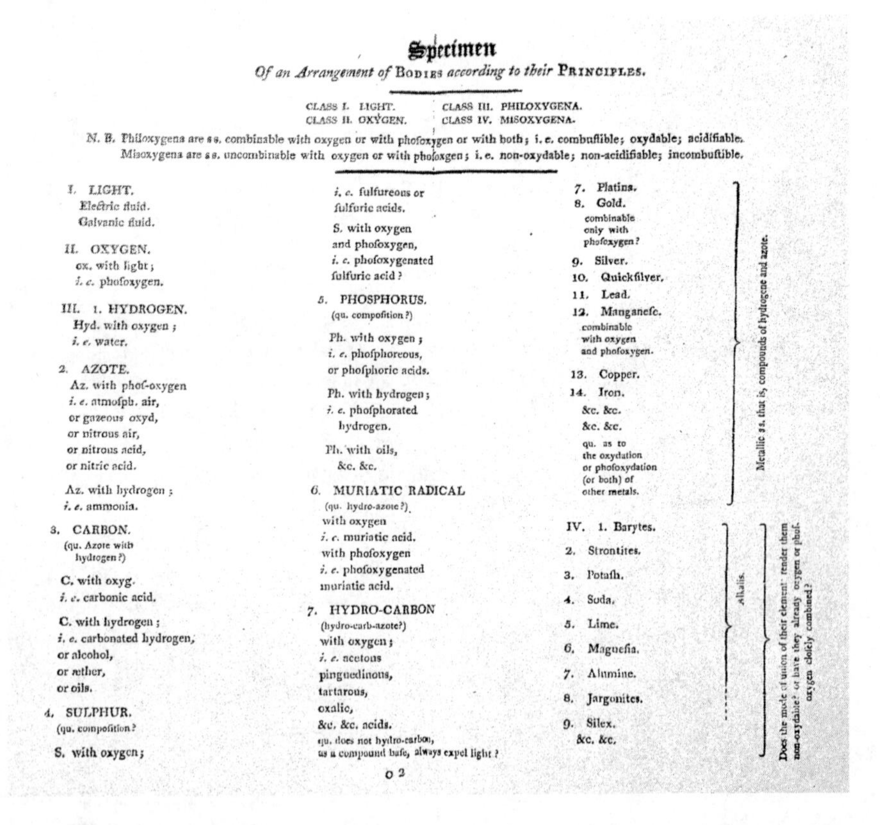

Specimen

Of an Arrangement of BODIES *according to their* PRINCIPLES.

CLASS I. LIGHT. CLASS III. PHILOXYGENA.
CLASS II. OXYGEN. CLASS IV. MISOXYGENA.

N. B. Philoxygena are ss. combinable with oxygen or with phosoxygen or with both; i. e. combustible; oxydable; acidifiable.
Misoxygena are ss. uncombinable with oxygen or with phosoxgen; i. e. non-oxydable; non-acidifiable; incombustible.

I. LIGHT.
Electric fluid.
Galvanic fluid.

II. OXYGEN.
ox. with light;
i. e. phosoxygen.

III. 1. HYDROGEN.
Hyd. with oxygen;
i. e. water.

2. AZOTE.
Az. with phos-oxygen
i. e. atmosph. air,
or gazeous oxyd,
or nitrous air,
or nitrous acid,
or nitric acid.

Az. with hydrogen;
i. e. ammonia.

3. CARBON.
(qu. Azote with hydrogen?)

C. with oxyg.
i. e. carbonic acid.

C. with hydrogen;
i. e. carbonated hydrogen,
or alcohol,
or æther,
or oils.

4. SULPHUR.
(qu. composition?

S. with oxygen;
i. e. sulfureous or
sulfuric acids.

S. with oxygen
and phosoxygen,
i. e. phosoxygenated
sulfuric acid?

5. PHOSPHORUS.
(qu. composition?)

Ph. with oxygen;
i. e. phosphoreous,
or phosphoric acids.

Ph. with hydrogen;
i. e. phosphorated
hydrogen.

Ph. with oils,
&c. &c.

6. MURIATIC RADICAL
(qu. hydro-azote?)
with oxygen
i. e. muriatic acid.
with phosoxygen
i. e. phosoxygenated
muriatic acid.

7. HYDRO-CARBON
(hydro-carb-azote?)
with oxygen;
i. e. acetous
pinguedinous,
tartarous,
oxalic,
&c. &c. acids.
qu. does not hydro-carbon,
as a compound base, always expel light?

7. Platina.
8. Gold.
combinable only with phosoxygen?
9. Silver.
10. Quicksilver.
11. Lead.
12. Manganese.
combinable with oxygen and phosoxygen.
13. Copper.
14. Iron.
&c. &c.
&c. &c.
qu. as to the oxydation or phosoxydation (or both) of other metals.

Metallic ss. that is, compounds of hydrogene and azote.

IV. 1. Barytes.
2. Strontites.
3. Potash.
4. Soda.
5. Lime.
6. Magnesia.
7. Alumine.
8. Jargonites.
9. Silex.
&c. &c.

Alkalis.

Does the mode of union of their elements render them non-oxydable? or have they already oxygen or phosoxygen closely combined?

O 2

Figure 1.13 Beddoes, Specimen of an Arrangement of BODIES according to their PRINCIPLES: *Contributions* (1799), facing 211.

It had been printed for his 1798 course in Bristol, modified only to reflect Humphry Davy's work on the chemistry of light and the nature of heat. Caloric was omitted, mainly because of

> the strange abuse of the doctrine of *latent heat* by Mr. Lavoisier. This I consider as a compleat *reductio ad absurdum* of the hypothesis of caloric, and a humiliating example of the frequent inability observable in men of the most energetic understanding, to push their skepticism far enough.[254]

There was no way that nitre, for example, could contain enough caloric to produce the gases and rise of temperature involved in the firing of gunpowder; and Davy's experiments on melting ice by friction, like Count Rumford's experiments on producing heat by boring a cannon barrel, argued for heat as produced by motion, and against Lavoisier's matter of light. But Beddoes had "expunged the matter of heat *or* caloric" from his chemical system long before Davy carried out his experiments, and before he knew of Rumford's "ingenious labours"; and Davy, while still an apprentice to an apothecary, had conducted his experiments in ignorance of Rumford's key paper.[255] As for latent heat, Beddoes saw that as his mentor Joseph Black's discovery, and although Lavoisier made use of the concept in the ice calorimeter that he and Pierre-Simon Laplace used in their researches on caloric, he chose not to account for it in his theory.

There were four fundamental classes of chemical substances in Beddoes' scheme: (1) Light, including the electric and galvanic fluids;[256] (2) oxygen, and its combination with light, Humphry Davy's "phosoxygen";[257] followed by (3) hydrogen, azote (nitrogen), carbon, sulphur, phosphorus, the muriatic radical,[258] hydro-carbon (the base of organic acids when combined with oxygen), and (4) metals including gold, silver, mercury, lead, and iron (which Beddoes viewed as compounded of hydrogen and azote, and so not ultimately elemental), and the alkalis, soda, potash, lime, and so on. Beddoes later modified and extended this view of the compound nature of metals so that by 1808 (the year he died) he was arguing that all nature consisted of two elements – only now these were hydrogen and oxygen.[259]

Davy, just twenty years old when he joined the Pneumatic Institution, was to provide inspiration for Beddoes in several realms of chemistry: the chemistry of light, pneumatic chemistry, and, later on, galvanic chemistry. In September 1798, Beddoes wrote enthusiastically to James Watt junior: "Davy is here; & I do not recollect to have conversed with a person of so great talents for experimental investigation – Papers of his on the chemistry of light & heat are in the press."[260]

"Phosoxygen", our oxygen gas, was for Davy a compound of oxygen with light, just as for Lavoisier it had been a combination of oxygen with caloric. Breathing oxygen gas helped with some medical conditions, but it was nitrous oxide that made the greatest stir at the Pneumatic Institution.

Joseph Priestley had discovered oxygen (or rather dephlogisticated air) and nitrous oxide (which he called "dephlogisticated nitrous air"), along with more kinds of air (gases) than anyone before him,[261] and the young Davy was a great admirer of his work. Priestley was soon to reciprocate that admiration.[262] Nitrogen was the principal constituent of atmospheric air (80% against oxygen's 20%), but it would not support life, which is why the French nomenclature called it "azote". This idea had been taken up and extended by Samuel Latham Mitchill, Professor of Chemistry, Natural History, and Agriculture at New York; he had been the first American advocate of Lavoisier's chemistry and was a former pupil of Joseph Black. He identified nitrous oxide as the principle of contagion. Different

> contagions and poisons may consist of the same materials, varying but in their proportions, or in some unimportant circumstance, and . . . the virus of syphilis, small-pox, and measles, and of the spider, rattlesnake, and other venomous creatures, as being all of animal production, may consist in the main of azote and oxygene, combined perhaps with some other ingredient; and there is high probability that marsh miasmata will be found little else than a similar compound. The ichor of cancer and other corroding ulcers is very probably pretty much the same thing.[263]

Beddoes at first accepted these views. Meanwhile, Priestley had found that nitrous air supported combustion. Davy directed his attention to nitrous oxide in 1798, anxious to test Mitchill's theory; he breathed in nitrous oxide mixed with atmospheric air, without experiencing any dangerous symptoms. On Mitchill's hypothesis, Davy should have succumbed to a plague of contagions, but he remained healthy. Mitchill's theory was soon overthrown. Davy regarded nitrous oxide as a beneficent stimulant, increasing the circulation of the blood, and accordingly likely to be helpful in treating diseases of deficient sensibility, such as paralysis. But Davy then retreated from regarding nitrous oxide as a universal stimulant. Beddoes, however, viewed it with enthusiasm. He believed that it overcame paralysis, and "I see every day consumptive patients on the verge of consumption recovering – Mr. Watt tells me they have not such success at Birm^{m}." Indeed, for Beddoes, nitrous oxide was "the most beneficial discovery ever made".[264]

Davy produced the gas by distilling ammonium nitrate at a controlled heat. Concerned that nitrous oxide might be a depressant, he first tested the pure gas on animals, without ill effect, and then on 11 April 1799 breathed it himself. Davy found that the gas was exhilarating, without debilitating after-effects.[265] Beddoes, observing the trial, described the scene as

> the most extraordinary I had ever witnessed, except [in the case of an epileptic patient]. . .. I saw and heard shouting, leaping, running, and other gestures, which may be supposed to be exhibited by a person who gives full loose to feelings, excited by . . . joyful . . . news.

Beddoes tried nitrous oxide on himself, and seemed to himself "to be bathed all over with a bucket of good humour, and a placid feeling pervades his whole frame".[266]

Davy and Beddoes tried the gas on friends, relatives, visitors, and patients, and Beddoes took part in the experiment. Josiah and Thomas Wedgwood both experienced unpleasant sensations, but Coleridge was among the majority who sampled the gas and enjoyed its effect. James Watt wrote to Joseph Black that nitrous oxide

> seems to have a great power on the nervous system, in some patients it gives great hilarity & disposition to motion, without leaving lassitude – But in those disposed to be Hysterical it excites the fit. Some paralyticks have been cured by it & others relieved; but powerful as it is upon the nerves it seems to have little action upon the arterial system as it does not affect the pulse. It is not <*acid &*> irritating and does not seem to hurt the lungs of those who have breathed it, though it is breathed pure.[267]

Beddoes was sanguine. The project of pneumatic medicine was off to a fine start, convincing him that chemistry was the key to understanding the body.

> Our observations . . . present themselves as a pledge, that by ascertaining the action of the elements entering into his composition, Man may, some time, come to rule over the causes of pain or pleasure, with a dominion as absolute as that which at present he exercises over domestic animals and the other instruments of his convenience. . .. To me it seems that there is no possibility of attaining a scientific knowledge of the interior conditions, upon which the sensible actions of the living system depend, without the most extensive application of chemistry to physiology. When a complete chemical acquaintance with the functions of the lungs, the stomach, and the skin, shall be obtained, (which attainments are very possible) the state of the body will become a matter of calculation; and so will the means of correcting it, when it deviates from that condition which is most desirable.

But he cautioned his readers that it was essential to use pure gases – impurities could lead to serious or even fatal consequences.[268] Purity was a large issue in pneumatic laboratories as many, like Watt, could attest.

Coleridge remarked that Beddoes was forever chasing theories: "Beddoes hunting a Pig with a buttered Tail – his whole Life and outcry of Eureka's and all eurekas Lies."[269] Certainly Beddoes was often sanguine, indeed overmuch so. But gradually his optimism about pneumatic medicine was undermined by the brutal reality of consumption: tuberculosis and cancer did not yield to any gases, however agreeable the short-term effects of breathing nitrous oxide or oxygen. James Watt, whose collaboration with Beddoes had morphed into the program of the Pneumatic Institution, remained convinced that some pneumatic medicine was helpful. He chided Beddoes: "You also seem to be among the sceptics in pneumatic medicine, you seem to have discarded both H. carbonate & oxygene which have done good here."[270] Watt was desperately seeking a cure for his son's consumption, which was ultimately to kill Gregory in 1804. But Beddoes by then had shifted his emphasis to preventive medicine and the compilation of medical statistics, and here his enterprise was well conceived. He had changed the name of his institution to the Institution for the Sick and Drooping Poor, and had moved quarters to Broad Quay. He intended to reinstate pneumatic medical trials once he had gained the confidence of his poor patients and had "established the preventive scheme – I have no confidence in my old speculations nor tenderness for them – but I hold it as a fact that the gasses [*sic*] have salutary powers – and that in a high degree". Beddoes found it hard to give up. He did gain the confidence of the poor; around 1801, he told Tom Wedgwood:

> I do go on well with the poor, especially with their young families – They get more regular every week.[271] . . . What I want is to have 4 or 5 young medical men – 3 perhaps wd do – I wd parcel out all Bristol between them & make them call; occasionally to see the situation of the family & to ensure general regularity of attendance. The young men [can] be had cheap. I think £550 a year wd confer the full benefit of preventive & to a great extent of curative medicine to all Bristol & the counties within attending distance.[272]

In spite of these hopes and partial successes, Beddoes, as we know, was losing confidence in pneumatic chemistry applied to medicine. He concentrated instead on providing a health system for the poor and their families, first at the Pneumatic Institution, and latterly at the Institution for the Sick and Drooping Poor. At the end of his life, his confidence failed. In his last letter to Davy, written days before his death, he lamented that, "like one who has scattered abroad the *avena fatua* of knowledge from which neither branch, nor blossom nor fruit has resulted, I require the consolation of a friend."[273]

His radical friends did not consider him a failure. Coleridge in 1796 had begun his review of Beddoes' *Letter to . . . Pitt* with a precise identification of his mixture of science, politics, and philanthropy:

> To announce a work from the pen of Dr. Beddoes is to inform the benevolent in every city and parish, that they are appointed agents to some new and practicable scheme for increasing the comforts or alleviating the miseries of their fellow-creatures. . .. In a strain of keenest irony the Doctor notices the singular fact that, while the French have pressed into their service all the inventive powers of the chemist and

> mechanic, the sons of science in Britain (almost without an exception) are know to regard the system and measures of the Minister with contempt or abhorrence; nor does he omit to glance on the recent practice of electing Members of the Royal Society from the colour of their political opinions.[274]

When Beddoes died, Coleridge wrote to Davy "that more Hope had been taken out of my life by this than by any former intent. For Beddoes was good and beneficent to all men".[275] The enemies of liberalism must have heaved a hearty sigh of relief.

Notes

1 Michael Neve, "Beddoes, Thomas (1760–1808)", ODNB (see abbreviation list).
2 William Nicholson, art. "Tanning", in *A Dictionary of Practical and Theoretical Chemistry, with Its Application to the Arts and Manufactures, and to the Explanation of the Phaenomena of Nature* (London: Richard Phillips, 1808); Humphry Davy, II, "An Account of Some Experiments and Observations on the Constituent Parts of Some Astringent Vegetables; and on their Operation in Tanning", 246–287, and "Observations on the Process of Tanning", 287–296, in *The Collected Works of Sir Humphry Davy, Bart . . .*, ed. John Davy, 9 vols. (London: Smith, Elder & Co., 1839). David Knight, *Humphry Davy: Science & Power* (Oxford: Blackwell, 1992), 45–46.
3 Humphry Davy, *Consolations in Travel, or the Last Days of a Philosopher*, ed. John Davy (London: John Murray, 1830).
4 Frederick Kurzer, "Chemistry in the Life of Dr. Samuel Johnson", *Bulletin for the History of Chemistry*, **29** (2004), 65–88. Robert J. P. Williams, Allan Chapman and John S. Rowlinson, eds., *Chemistry at Oxford: A History from 1600 to 2005* (Cambridge: Royal Society of Chemistry Publishing, 2009), 65. In 1771 Johnson attended John Warltire's scientific lectures at Ashborne, and "was much stimulated by its metallurgical and mineralogical content". (Hugh Torrens, "John Warltire", ODNB.) He reported chemical and ballooning intelligence to Hester Thrale, and also went to see some experiments made by "a physician at Salisbury, on the new kinds of air": Johnson to Thrale, 22 September 1783, in *The Letters of Samuel Johnson*, ed. Bruce Redford, 5 vols. (Princeton: Princeton University Press, 1992–94), IV, 203–205, 218 and nn. 3 and 4.
5 R. T. Gunther, *Early Science in Oxford* (Oxford: Oxford University Press, 1923), I, 58. Smith died in 1797.
6 There is, for example, no record of Beddoes attending a meeting of the Chapter Coffee House Society: Trevor Levere and Gerard L'E. Turner, *Discussing Chemistry and Steam: The Minutes of a Coffee House Philosophical Society 1780–1787* (Oxford: Oxford University Press, 2002).
7 John Sheldon, *The History of the Absorbent System, Part the First: Containing the Chylography, or Description of the Human Lacteal Vessels, with the Different Methods of Discovering, Injecting, and Preparing Them, and the Instruments Used for These Purposes* (London: Printed for the author, 1784).
8 Lazzaro Spallanzani, trans. Beddoes, *Dissertations Relative to the Natural History of Animals and Vegetables . . .* (London: John Murray, 1784), trans. from *Opuscoli di fisica animale e vegetabile, aggiuntevi alcune lettere relative ad essi opuscoli dal Signor Bonnet e da altri scritte all'autore*, 3 vols. (Venezia: Presso Giammaria Bassaglia, 1782), with other texts. Beddoes' work for John Murray is discussed in Chapter 4.

9 Stansfield (see abbreviation list), 18.

10 See notes 99, 100, and 224. It seems probable that Sheldon and Sadler knew one another, but we have found no evidence to prove this. Sadler was the son of a cook and confectioner. He released a 36-foot hydrogen balloon in Oxford on 9 February 1784, and on 4 October of that year made the first ascent of any English aeronaut in a 170-foot hot-air balloon, rising 3,600 feet above Oxford and landing 6 miles away. He had designed and built everything himself. He experimented with his own steam engine, and was one of the first to use coal gas for lighting. See H. S. Torrens, "Sadler, James (*bap.* 1753, *d.* 1828)", ODNB.

11 David Hume, *Essays and Treatises on Several Subjects . . . in Four Volumes . . ., Containing Essays, Moral, Political, and Literary* (London: A. Millar, 1760), II, Pt. II, essay ii, 30.

12 Roger L. Emerson, *Essays on David Hume, Medical Men and the Scottish Enlightenment: 'Industry, Knowledge and Humanity'* (Farnham, Surrey, and Burlington, VT: Ashgate, 2009).

13 D'A. Power revised by Michael Bevan, art. "Trye, Charles Brandon (1757–1811), surgeon", ODNB. On connections with Worcester, see Chapter 3.

14 Beddoes to Charles Brandon Trye, n.d. [1784 or 1785], Gloucestershire Record Office, MS D303 C1/61, published in T. H. Levere and P. B. Wood, "Thomas Beddoes and the Edinburgh Medical School: A Letter to Charles Brandon Trye, ca. 1785", *University of Edinburgh Journal*, 22 (1986), 36–39.

15 Joseph Black, *Dissertatio medica inauguralis, de humore acido a cibis orto, et magnesia alba . . .* (Edinburgh: G. Hamilton et J. Balfour, 1754).

16 *Chemical Lectures* (ca. 1785), Royal College of Physicians of Edinburgh, MS 42, M9.43, 11–12. Henceforth C. L. George Martine, *Essays and Observations on the Construction and Graduation of Thermometers, and on the Heating and Cooling of Bodies*, 3rd edn. (Edinburgh: Alexander Donaldson, 1780), vol. 1, 11–12.

17 David Philip Miller, *James Watt, Chemist: Understanding the Origins of the Steam Age* (London: Pickering and Chatto, 2009).

18 Joseph Black to James Watt, 15 March 1780, BCL JWP (see abbreviation list) 4/44/55, published in CJB (see abbreviation list) I, 414–415.

19 See R. M. G. Anderson, "Joseph Black", ODNB.

20 Robison wrote to James Watt about his preparation of the published edition of Black's lectures, which he described as "a most oppressive and dispiriting job": see Robison to Watt 18 January 1802 and 19 April 1803, in *Partners in Science: Letters of James Watt and Joseph Black*, eds. Eric Robinson and Douglas McKie (Cambridge, MA: Harvard University Press, 1970), 373–377. Watt, along with David Brewster, contributed to Robison, *A System of Mechanical Philosophy* (Edinburgh: J. Murray, 1822). Robison dedicated to Watt his edition of Black's lectures, *Lectures on the Elements of Chemistry, Delivered in the University of Edinburgh, by the Late Joseph Black, M.D., Professor of Chemistry in That University . . .*, 2 vols. (London: Printed by Mundell and son for Longman and Rees, and Edinburgh: William Creech, 1803).

21 C. L. (see abbreviation list), 11.

22 C. L., 50, 51.

23 C. L., 7.

24 C. L., 399.

25 C. L., vol. 2, 260.

26 Carleton E. Perrin, "Joseph Black and the Absolute Levity of Phlogiston", *Annals of Science*, 40 (1983), 109–137.

27 Ibid., 130.

28 C. L., 260–261.

29 Henry Cavendish, "Experiments on Air", *Phil. Trans.*, **74** (1784), 119–153 and **75** (1785), 372–384. A comprehensive account of Cavendish's life and work, including his experiments on air and water, is given in Christa Jungnickel and Russell McCormmach, *Cavendish: The Experimental Life*, 2nd edn. (Lewisburg, PA: Bucknell, 1998). The controversy about the composition of water in the 1780s is discussed in David Philip Miller, *Discovering Water: James Watt, Henry Cavendish and the Nineteenth-Century 'Water Controversy'* (Aldershot, Hants. and Burlington, VT: Ashgate, 2004), Chapter 3. The literature on the chemical revolution of the 1780s is vast. Good starting places are Bernadette Bensaude-Vincent, *Lavoisier: mémoires d'une revolution* (Paris: Flammarion, 1993), and the still valuable Édouard Grimaux, *Lavoisier 1743–1794* (Paris: Félix Alcan, 1888).
30 Perrin, "Joseph Black and the Absolute Levity of Phlogiston", 135.
31 Louis Bernard Guyton de Morveau, Antoine-Laurent Lavoisier, Claude-Louis Berthollet and Antoine François Fourcroy, *Méthode de nomenclature chimique* . . . (Paris: Cuchet, 1787). A.-L. Lavoisier, *Traité élémentaire de chimie* (Paris: Cuchet, 1789). A.-L. Lavoisier, trans. Robert Kerr, *Elements of Chemistry, in a New Systematic Order* (Edinburgh: William Creech, 1790); although this translation is dated 1790, it was actually released for sale in 1789 (see John D. Baird, "Note on the Date of Publication of the English Translation of Lavoisier's *Traité Élémentaire de Chymie*", *Ambix*, **47** (1789), 47–48.
32 C. L., 7.
33 Ibid., vol. I, 50.
34 C. L., vol. I, 37
35 C. L., 37, 39. For all his claims for the chemical philosopher, he felt, as Robert Anderson observes, insecurity about the status of chemistry. John Robison quoted Black: "Chemistry is not yet a science. We should avoid every thing that has pretensions of a full system. The whole of medical science should, as yet, be analytical, like Newton's Optics." CJB, I 44, quoting Black, ed. Robison (1803), I, 547.
36 See Chapter 4.
37 *A Dissertation on Elective Attractions: By Torbern Bergmann . . . Translated from the Latin by the Translator of Spallanzani's Dissertations* (London: John Murray, 1785). Beddoes was also involved in the translation and making additions to *The Chemical Essays of Charles-William Scheele, Translated from the Transactions of the Academy of Sciences at Stockholm: With Additions* (London: John Murray, and Edinburgh: Gordon and Elliot, 1786). See Chapter 4.
38 James Kendall, "The First Chemical Society, the First Chemical Journal, and the Chemical Revolution", *Proceedings of the Royal Society of Edinburgh: Section A, Mathematical and Physical Sciences*, **61** (1952), 234–245; Pt. 2, **63** (1952), 385–393.
39 *Dissertations Read before the Chemical Society, Instituted in the Beginning of the Year 1785*, EUL (see abbreviation list), Special Collections, MS 2748.
40 Cavendish, "Experiments on Air", *Philosophical Transactions*, **74** (1784), 119–169, and **75** (1785), 372–384. Watt was indignant that Cavendish and Lavoisier both obtained credit for this, while he did not; Miller, *Discovering Water*, 27–57 describes the dispute and its interpretation.
41 James Kendall, "The First Chemical Society, the First Chemical Journal, and the Chemical Revolution" (Part II), *Proceedings of the Royal Society of Edinburgh* Sect. A, 63 (1952), 385–392 at 391.
42 Ibid., 392.
43 Miller, *Discovering Water*, 29. See also Hasok Chang, *Is Water H_2O?* (Dordrecht: Springer, 2012), 1–50.

44 Beddoes in EUL MS 2748, 5.
45 See e.g. Black to Watt 26 June 1784 in CJB I, 720 and Watt to Black 4 February 1788 in ibid., II, 939–940.
46 Joseph Black discussed manure in his *Lectures . . .*, ed. Robison, II, 175–176, 215, 261, and 3, 410, and Richard Kirwan subsequently devoted a book to it: *The Manures Most Advantageously Applicable to the Various Sorts of Soils, and the Causes of Their Beneficial Effect in Each Particular Instance* (London: Printed for Vernon and Hood, 1796). William Cullen repeatedly discussed the role of manures in agriculture;, see e.g. John Thomson, *An Account of the Life, Lectures and Writings, of William Cullen*, 2 vols. (Edinburgh and London: William Blackwood, 1832), I, 63, 592, 594, 598. An early account of manure is Thomas Liveings, *Manure for Land: His Majesty Has Been . . . Pleased to Grant His Royal Letters Patents to Thomas Liveings . . . for the Sole Making and Vending an Artificial Compost or Compound Manure* ([ca. 1730] single sheet, in the British Library, shelfmark 1505/177).
47 Joseph Priestley, *Experiments and Observations on Different Kinds of Air* (London: Printed for J. Johnston, 1775). Jean Senebier, *Mémoires physico-chymiques, sur l'influence de la lumière solaire pour modifier les êtres des trois règnes de la nature, et surtout ceux du règne vegetal*, 3 vols. (Geneva: Barthelemi Chirol, 1782). Jan Ingenhousz, *Experiments upon Vegetables, Discovering their Great Power of Purifying the Common Air in the Sunshine, and of Injuring It in the Shade and at Night: To Which Is Joined, a New Method of Examining the Accurate Degree of Salubrity of the Atmosphere* (London: Printed for P. Elmsly and H. Payne, 1779).
48 Beddoes in EUL MS 2748, 8.
49 Beddoes to Joseph Banks, 1808, in Stock, 15. The letter is not in *Scientific Correspondence of Joseph Banks, 1765–1820*, ed. Neil Chambers, 6 vols. (London: Pickering and Chatto, 2007), and I have been unable to trace the original.
50 Victor D. Boantza, *Matter and Method in the Long Chemical Revolution: Laws of Another Order* (Farnham, Surrey: Ashgate, 2013), part ii; Boantza, "Collecting Airs and Ideas: Priestley's Style of Experimental Reasoning", *Studies in the History and Philosophy of Science*, **38** (2007), 506–522.
51 Beddoes to Black, London 6 November 1787, EUL MS Gen 875/III/52, 53875/III/52, /53, published CJB II, 918–921.
52 Stansfield, 27. Stock, 86, 135, refers to their subsequent interactions.
53 The old library of the Georg-August University of Göttingen has kept the original record of its acquisitions, including their date of purchase. Beddoes' major works were usually acquired within a year of their publication. Beddoes refers with strong approval to this library (see Chapter 4), and it is hard to believe that he never visited it, but I have not found evidence that he was ever there; the library's lending records do not show that he borrowed books from Göttingen. His knowledge may have come through Girtanner.
54 Beddoes to Trye 1784 or 1785, see note 14.
55 Trevor H. Levere, *Poetry Realized in Nature: Samuel Taylor Coleridge and Early Nineteenth-Century Science* (Cambridge and New York: Cambridge University Press, 1981), 262–263. Neil Vickers, *Coleridge and the Doctors* (Oxford: Oxford University Press, 2004), Chapter 2 *passim*.
56 John Brown, *Lectures on the Elements of Medicine by Dr. Brown and First of His Aphorisms: The Brunonian System 1785*, Royal College of Physicians of Edinburgh MS Brown J1, bound MS, 4to.
57 Hubert Steinke, *Irritating Experiments: Haller's Concept and the European Controversy on Irritability and Sensibility, 1750–90* (Amsterdam and New York: Rodopi, 2005). Dominique Boury, "Irritability and Sensibility: Key Concepts in Assessing the Medical Doctrines of Haller and Bordeu", *Science in Context*, **21** (2008), 521–535.

58 Vickers, *Coleridge and the Doctors*, Chapter 2, 37–78.
59 F. W. Gibbs and W. A. Smeaton, "Thomas Beddoes at Oxford", *Ambix*, 9 (1961), 47–49 at 48.
60 Memorial (see abbreviation list). This document refers to Beddoes' activities in Oxford over the past three months. See Trevor Levere, "Dr. Thomas Beddoes at Oxford: Radical Politics in 1788–1793 and the Fate of the Regius Chair in Chemistry", *Ambix*, **28** (1961), 61–69, from which parts of the following section are taken. See Chapter 4.
61 [Guyton de Morveau et al.], *Elémens de chymie, théorique et pratique, rédigé dans un nouvel ordre, d'après les découvertes modernes, pour server aux cours publics de l'Académie de Dijon*, 3 vols. (Dijon: L.N.Frantin, 1777, 1777, 1778).
62 W. A. Smeaton, "The Contributions of P.-J. Macquer, T. O. Bergman and L. B. Guyton de Morveau to the Reform of Chemical Nomenclature", *Annals of Science*, **10** (1954), 87–106. Guyton de Morveau, Lavoisier, Berthollet and de Fourcroy, *Méthode de nomenclature chimique* (Paris: chez Cuchet, 1787).
63 W. A. Smeaton, "Guyton de Morveau, Louis Bernard", DSB (see abbreviation list) 5 600–604 at 602.
64 W. A. Smeaton, "Guyton de Morveau's Course of Chemistry in the Dijon Academy", *Ambix*, **2** (1961), 53–69.
65 Stansfield, 33–35. William A. Smeaton, "Platinum and Ground Glass: Some Innovations in Chemical Apparatus by Guyton de Morveau and Others", in *Apparatus and Experimentation in the History of Chemistry*, ed. Frederic L. Holmes and Trevor H. Levere (Cambridge, MA: Massachusetts University Press, 2000), 210–237.
66 Stock, 17. W. A. Smeaton, "Louis Bernard Guyton de Morveau, FRS (1737–1816) and His Relations with British Scientists", *Notes and Records of the Royal Society of London*, **22** (1967), 113–130 at 118.
67 Beddoes to Black, 6 November 1787, EUL MS Gen 875/III/52, 53, published CJB II, 918–921.

> I sent Dr. Hutton's theory of rain to Morveau at Dijon & he was very much pleased with it; & I afterwards gave it to Lavoisier; I am very sorry I had not at that time a copy of his theory of the earth. If he has a copy left & wd sent it to me at Murray's No:32 Fleet Street, I will undertake to forward it to Lavoisier. Mr Lumsden told me he had brought it to Paris; but I suppose that was only the abstract; none of the French Philosophers with whom I conversed know any thing of it. You may procure a copy of the nouvelle nomenclature chymique from Elm[sly?] if you have not one already. The continuation of Morveau's dictionary will be published in the course of the winter; Madame Lavoisier has translated Kirwan's book on Phlogiston, to which several of the best chemists are add [*sic*] their notes.

68 *A Letter to the Right Honourable Sir Joseph Banks . . . on the Causes and Removal of the Prevailing Discontents, Imperfections, and Abuses, in Medicine* (London: Printed for R. Phillips, by T. Gillet, Crown Court, Fleet-Street, 1808), 45.
69 Smeaton, "Louis Bernard Guyton de Morveau, FRS (1737–1816) and His Relations with British Scientists", 118.
70 Beddoes to Black, London, 6 November 1787, EUL MS Gen 875/III/52, 53875/III/52, 53, published CJB II, 918–921.
71 The contents of Beddoes' library after his death, as listed for sale in 2131 (!) lots by Leigh and Sotheby, suggest that this is a slight exaggeration, but at his death he owned more than 35 chemical "modern elementary books" in English, French, German, and Latin. See Chapter 4. The auctioneer's annotated copy of the sale catalogue is listed as being in the BL (see abbreviation list), Sotheby 60 (2),

henceforth CBL. This item can no longer be found in the British Library; fortunately, we have been able to work from a photocopy of the Catalogue.

72 C. L., vol I, 11.
73 Beddoes to Black, 6 November 1787.
74 Beddoes to Black, 21 April [1789], EUL Special Collections, Black Correspondence Gen 873/III/f. 129–130, published in CJB II, 1011–1013.
75 Beddoes to Watt, 25 February 1778, Watt MSS, BCL.
76 Beddoes to Black, 6 November 1787.
77 Lavoisier, *Elements of Chemistry* (1789), xiii–xiv, from Etienne Bonnot and abbé de Condillac, *La logique, ou, Les premiers développemens de l'art de penser. . .* (Paris: chez L'Esprit . . . et Debure ainé, 1780).
78 Beddoes to Black, 6 November 1787.
79 Guyton de Morveau sent Beddoes news of chemistry in France: Guyton to Beddoes 19 September 1788, Bodley (see abbreviation list) MS Dep.c.134/2.
80 Beddoes MSS, Bodley MS Dep. C134.
81 CRO MS DD DG (see abbreviation list) 42/18. See also Bodley MS Dep. C.134.1 & 5.
82 Beddoes to James Watt junior 25 February 1798, BCL JWP MivB. Isaac Milner, *A Plan of a Course of Lectures Introductory to Chemistry and Other Branches of Natural Philosophy* (Cambridge: printed by J. Archdeacon for the University Press, 1780; further editions 1784 and 1788).
83 Black to Beddoes 24 November 1787, EUL MS Gen 873/III/54, published in CJB II, 924–925.
84 Beddoes to Black 23 February 1788, EUL MS Gen 873/III/71, 72, published in CJB II, 949–951. The content of another letter from Beddoes to Black dated 2 February 1788 has been lost, leaving only the cover sheet, EUL MS Gen 873/III/68.
85 He did, however, become an active promoter of pneumatic medicine based on pneumatic chemistry, when neither Black nor Watt were interested in publishing; even Watt did so only in association with Beddoes. See Chapter 3 and Beddoes to Black, 23 February 1788, MS Gen. 873/III/71, 72.
86 Beddoes to Black, 23 February 1788.
87 Ibid.
88 Maxwell Garthshore to Black, 7 October 1788. EUL MS Gen 873/III/103,104, published in CJB II, 983–934.
89 Davies Giddy Diary, entry dated 1826 facing entry for 18 July 1791, CRO MS DD DG. But Black, in a letter to Watt of 7 August 1788, published in CJB II, 972–973, alludes to William Thomson but makes no mention of Beddoes.
90 C. Willoughby, Baldon House, 21 July 1792 to [?Henry Dundas], NA (see abbreviation list) MS HO.42.208 (see abbreviation list).
91 Beddoes to Thomas Wedgwood, 3 August 1793, WMB W/M (see abbreviation list) 35. See also Simon Schaffer, "The Consuming Flame: Electrical Showmen and Tory Mystics in the World of Goods", in *Consumption and the World of Goods*, ed. John Brewer and Roy Porter (London: Routledge, 1993), 489–526.
92 Henry Richard Fox, afterwards Vassall, Baron Holland, *Further Memoirs of the Whig Party, 1807–1821, with Some Miscellaneous Reminiscences* . . . Edited by Lord Stavordale (London: John Murray, 1905), 324.
93 Ian Hacking, "The Self-Vindication of the Laboratory Sciences", in *Science as Practice and Culture*, ed. Andrew Pickering (Chicago: University of Chicago Press, 1992), 29–64.
94 A.-L. Lavoisier, *Traité élémentaire de chimie*, 2 vols. (Paris: Cuchet, 1789), I, xxix–xxx.
95 WMB WC (see abbreviation list) MS 563-I. Beddoes to Black, 21 April 1789, EUL Gen 873/III/129,130, published in CJB II, 1011–1013. This burning glass

is the only piece of Beddoes' laboratory apparatus that has been unequivocally identified as his in the Museum of the History of Science in Oxford University, which once housed the University's chemical laboratory. The Museum also houses ceramic retorts made by Wedgwood, which are hard to date; although these may be from the mid or late nineteenth century, it is just possible that they were acquired and used by Beddoes, who ordered apparatus from Wedgwood. Parker was the leading vendor of scientific glass, but not of optical instruments (telescopes and microscopes).

96 Beddoes to Josiah Wedgwood, quoted in R. E. Schofield, *The Lunar Society of Birmingham* (Oxford: Clarendon Press, 1963), 373

97 T. H. Levere, "Lavoisier's Gasometer and Others: Research, Control, and Dissemination", in *Lavoisier in Perspective*, ed. Marco Beretta (Munich: Deutsches Museum 2005), 53–67.

98 Beddoes to Black 15 April 1791, EUL Gen 873/III/200, 201, published in CJB II, 1122–1124.

99 See Hugh Torrens's article on James Sadler, ODNB. See also James Sadler to Thomas Beddoes, 14 January 1791, Bodley MS Dep. c.134/2 for Sadler's own statement of the case.

100 Stock quoted in David Philip Miller and Trevor H. Levere, " 'Inhale It and See?' The Collaboration between Thomas Beddoes and James Watt in Pneumatic Medicine", *Ambix*, 55 (2008), 5–28 at 8.

101 John T. Stock, *Development of the Chemical Balance* (London: Her Majesty's Stationery Office, 1969). T. H. Levere, "Balance and Gasometer in Lavoisier's Chemical Revolution", in *Lavoisier et la Révolution Chimique: Actes du Colloque tenu à l'occasion du bicentenaire de la publication du 'Traité élémtaire de chimie' 1789*, eds. M. Goupil with the collaboration of P. Bret and F. Masson (Palaiseau: ABIX-Ecole Polytechnique, 1992), 313–332.

102 Beddoes to James Watt, letter 1795 undated, BCL JWP MS 3219/4/27:9. *Considerations* (see abbreviation list), part III (London and Bristol, 1795).

103 Levere, *Lavoisier in Perspective*, ed. Beretta (2005), 53–68 at 65.

104 See Barrie Trinder's entry on William Reynolds in ODNB.

105 Joseph Reynolds, letter to Thomas Beddoes via Davies Giddy, 26 August 1791, CRO DD/DG 41.30

106 Beddoes to Black 15 April 1791. Robison considered that Black had been hoodwinked by Lavoisier (Black, ed. Robison II, 215–222).

107 Beddoes to Banks 3 January 1791, British Museum (Natural History) Dawson Turner Collection, 7 ff. 189–190, in *Scientific Correspondence of Joseph Banks*, IV, 30.

108 Beddoes to Black 15 April 1791. Beddoes to William Withering, nd, typescript of letter in McGill University, Bibl. Osler no. 1988.

109 A. C. Todd, *Beyond the Blaze: A Biography of Davies Giddy* (Truro: D. Bradford Barton Ltd., 1967), 21, referring to Giddy's Diary 18 July 1791 and note of 1826.

110 Beddoes to Giddy [26] May 1791, CRO (see abbreviation list) MS DG 41.13.

111 Beddoes to Giddy 4 July 1792, CRO MS DG41.16.

112 Thomas Beddoes, *A Letter to a Lady on the Subject of Early Instruction, Particularly That of the Poor*, 1792.

113 Beddoes to Giddy [1793], CRO MS DG 41.22.

114 John Cooke to Frederick North, second Earl of Guilford (known as Lord North) 16 July 1792, PRO MS HO 42.208.

115 C. Willoughby to Dundas 21 July 1792, PRO MS HO 42.21.

116 Lavoisier, *Traité élémentaire de chimie*.

117 Joseph Priestley, *Experiments and Observations on Different Kinds of Air*, 2nd. edn., 3 vols. (London: Printed for J. Hohnson, 1775–1777), I, xiv.

118 Home Office, "Disaffected & seditious persons. Goodmans fields 28 July 1792", NA HO 42.21. There were others of significance on the list. See Chapter 3.
119 M. P. Crosland, 'The Image of Science as a Threat: Burke versus Priestley and the "Philosophic Revolution" ', *BJHS*, **20** (1987), 277–307.
120 Edmund Burke, *A Letter from the Right Honourable Edmund Burke to a Noble Lord, on the Attacks Made upon Him and His Pension, in the House of Lords, by the Duke of Bedford and the Earl of Lauderdale Early in the Present Sessions of Parliament* (London: J. Owen and F. and C. Rivington, 1796), 65.
121 Edmund Burke, *Reflections on the Revolution in France, and on the Proceedings of Certain Societies in London Relative to that Event: In a Letter Intended to Have Been Sent to a Gentleman in Paris* (London: Printed for J. Dodsley, 1790), 8.
122 Beddoes to Giddy, CRO MS DG 41.2.
123 See Introduction, and Beddoes to Giddy 11 May 1792 [?1793], CRO MS DG 41/3.
124 Beddoes to Black 23 February 1788, EUL MS Gen 873/III/71, 72 published in CJB II, 949–951.
125 Joseph Priestley, *Directions for Impregnating Water with Fixed Air; in Order to Communicate to It the Peculiar Spirit and Virtues of Pyrmont Water, and Other Mineral Waters of a Similar Nature* (London: J. Johnson, 1772).
126 Darwin to Beddoes October 1787, in *The Collected Letters of Erasmus Darwin*, 296–297.
127 Edmund Goodwyn, *The Connexion of Life with Respiration; or, an Experimental Inquiry into the Effects [sic] of Submersion, Strangulation, and Several Kinds of Noxious Airs, on Living Animals: With an Account of the Nature of the Disease They Produce; Its Distinction from Death itself; and the Most Effectual Means of Cure* (London: Printed by T. Spilsbury for J. Johnson, 1788).
128 [Mayow] trans. and ed. Beddoes, *Chemical Experiments and Opinions . . .* (Oxford: Clarendon Press, 1790), p. v.
129 Ibid., iii, xxi.
130 Beddoes to Erasmus Darwin [?–1790], in Stock (1811) 25.
131 John Wedgwood to Josiah Wedgwood 27 October 1785, Wedgwood Archives, MS 2002–2028. For Wedgwood's pyrometer, see J. A. Chaldecott, "Josiah Wedgwood (1730–1795) Scientist", *British Journal for the History of Science*, 8 (1975), 1–16, and Alan Q. Morton and Jane A. Wess, *Public & Private Science* (Oxford: Oxford University Press, 1993), 479.
132 See chapter 2, this volume.
133 Erasmus Darwin to Beddoes October 1787 in CED (see Abbreviation list), 296–297.
134 Darwin to Beddoes, [October–November 1788?], Stock, Appendix 6, xxxvii–xxxviii, reprinted in CED, 326–327. Darwin was not alone in his interest in science, medicine, and enlightened improvement; see Paul A. Elliott, *The Derby Philosophers: Science and Culture in British Urban Society, 1700–1850* (Manchester and New York: Manchester University Press, 2009).
135 J. L. Moilliet and B. M. D. Smith, *A Mighty Chemist: James Keir of the Lunar Society* (privately printed, 1982); J. L. Moilliet, "Keir's 'Dialogues on Chemistry' – an Unpublished Masterpiece", *Chemistry and Industry*, 19 December 1964, 2081–2083.
136 Pierre Joseph Macquer, *A Dictionary of Chemistry [. . .] Translated from the French by James Keir [. . .] The Second Edition: To Which Is Added, as an Appendix, A Treatise [by James Keir] on the Various Kinds of Permanently Elastic Fluids, or Gases* (London, 1777–1779). Keir's translation is of the first

edition of Macquer's *Dictionnaire de Chymie* (Paris: chez Lacombe, 1766). Macquer published a second edition (Paris: Didot, 1778). The publication history of these and other editions and translations are given in Roy G. Neville and W. A. Smeaton, "Macquer's Dictionnaire de Chymie: A Bibliographical Study", *Annals of Science*, **38** (1981), 613–662, and in William A. Cole, *Chemical Literature 1700–1860: A Bibliography with Annotations, Detailed Descriptions, Comparisons and Locations* (London and New York: Mansell Publishing Ltd., 1988), 349–360. Macquer did not include plates in his *Dictionnaire*, regarding detailed descriptions of apparatus as sufficiently informative; the contrast with Lavoisier and his splendid plates illustrating instruments is striking. Guyton de Morveau wrote to Richard Kirwan on 13 October 1785: "Mr Keir n'a rien écrit sur la chymie depuis 8 ans; ses notes sur le Dictionaire de Macquer sont toutes tirées de Cramer, Gellert, Newman, Margraf et d'autres auteurs anciens, et ne contiennent rien qui puisse vous interesser" ("Mr. Keir has written nothing on chemistry for eight years. His notes on Maquer's Dictionary are all taken from Cramer, Gellert, Newman, Margraf and other ancient authors, and contain nothing which could interest you."): letter published in *A Scientific Correspondence During the Chemical Revolution: Louis-Bernard Guyton de Morveau & Richard Kirwan, 1782–1802*, ed. Emmanuel Grison, Michelle Goupil and Patrice Bret (Berkeley, CA: Office for History of Science and Technology, University of California at Berkeley, 1994), 133–136 at 135.

137 Beddoes to Giddy 21 November 1791, CRO, MS DD DG 41/48.

138 For a lively account of the riots, and their significance for Beddoes, as well as his pneumatic experiments and the Pneumatic Institution, see Jay. See Chapter 3, this volume.

139 Beddoes to Giddy 21 November 1791, CRO MS 48/41/1.

140 Barbara M. D. Smith, "Keir, James (1735–1820), Chemist and Industrialist", ODNB; B. M. D. Smith and J. L. Moilliet, "James Keir of the Lunar Society", *Notes and Records of the Royal Society of London*, **22** (1967), 144–154; B. M. D. Smith, *James Keir, 1735–1820: Entrepreneur and Scientist* (1966). Further accounts of Keir's life are Moilliet and Smith, *A Mighty Chemist* (n.d. [1868]; [A. Moilliet and J. K. Moilliet], *Sketch of the Life of James Keir, Esq., FRS* (privately printed, n.d. [1868]);On Keir and Wedgwood, see Larry Stewart, "Assistants to Enlightenment: William Lewis, Alexander Chisholm and Invisible Technicians in the Industrial Revolution", *Notes and Records of the Royal Society*, **62** (2008), 17–29.

141 J. L. Moilliet, "Keir's Caustic Soda Process – an Attempted Reconstruction", *Chemistry and Industry*, 5 March 1966, 405–408.

142 Beddoes to Giddy 21 November 1791, CRO MS 48/41/1.

143 James Keir, *The First Part of a Dictionary of Chemistry* (Birmingham: Pearson and Rollaston, 1789).

144 This copy is in Trevor Levere's library. [P. J. Macquer], *A Dictionary of Chemistry*, trans. Keir (1777).

145 Keir, *The First Part*, ii.

146 Johann Gottfried Leonhardi, *Macquer's chymisches Wörterbuch*, 7 vols. (Leipzig, 1788–1791); James Keir, *The First Part of a Dictionary of Chemistry* (Birmingham: Pearson and Rollason, 1789), ii.

147 Keir, *The First Part*, 5.

148 Beddoes to Giddy 21 November 1791, CRO MS DG 48/41/1.

149 James Keir to Josiah Wedgwood 17 February 1791, W/M (see abbreviation list) MS 688.1.

150 Beddoes to Giddy 21 November 1791, CRO MS DD/DG 41/48.

151 Keir, *The First Part*, iii.

152 Joseph Priestley, *Experiments and Observations on Different Kinds of Air*, 2nd edn., 3 vols. (London: Printed for J. Johnson, 1775), **1**, xiv. The growth of metaphors concerning Watt and Burke is also discussed in Chapter 3, this volume.
153 Keir, *The First Part*, v.
154 Burke, *Reflections*.
155 Beddoes to Giddy 3 April 1792, CRO MS DG 41/50.
156 James Keir to Josiah Wedgwood 27 October 1791, WMB WC MS 692.1.
157 Bond for performance of Covenant. Dr. Beddoes to Jas. Keir and Wm. Reynolds Esqrs, NA MS C (see abbreviation list) 104/41, and Tripartite Indenture, op. cit., C104/41. The bond was dated 27 August 1793, and was subsequently revised on 16 April 1794. It stated that if Beddoes, his father-in-law Edgeworth, and Anna Maria, his bride-to-be, observed the terms of a separate tripartite indenture, then the bond would be forgiven; and if not, not. The tripartite indenture, like the bond, is in the Chancery rolls in the National Archives. The three parties are: first Beddoes, second Richard Lovell Edgeworth and Anna Maria Edgeworth, and third James Keir and William Reynolds. The terms relate to Anna's marriage portion, the disposition of land and money if Beddoes predeceased Anna, and the inheritance and provisions to be made to Anna and to any issue of their marriage. The indenture bears five seals, one for Beddoes, another for Richard Lovell Edgeworth, a third for Anna Maria Edgeworth, and two others intended for Keir and Reynolds. Signatures accompany the first three seals; Keir and Reynolds did not sign the document. Unsurprisingly, Beddoes' death was followed by a series of lawsuits in the Court of Chancery. The lawsuits went slowly forward for decades; they seem straight out of the Court of Chancery in Dickens's *Bleak House*. See Levere, "Dr. Thomas Beddoes: Chemistry, Medicine, and the Perils of Democracy", *Notes and Records of the Royal Society*, **63** (2009), 215–229, from which this note and part of the text is taken.
158 Watt's letters to Beddoes in BCL JWP begin in 1794, but Watt's correspondence with Erasmus Darwin shows that he was aware of Beddoes' work as a chemist and physician from 1793 at the latest.
159 Watt to Black 17 July 1793, BCA JWP 4/12/29, published in CJB II, 1207–1209. The frog's electricity was Galvani's discovery.
160 Thomas Beddoes, *A Letter to Erasmus Darwin, M.D. on a New Method of Treating Pulmonary Consumption, and Some Other Disease Hitherto Found Incurable* (London: Printed by Bulgin and Rosser, Broad-Street; Sold by J. Murray, No. 32, Fleet-Street; and J. Johnson, No. 72, St. Paul's Church-Yard, London; also by Bulgin and Sheppard, J. Norton, J. Cottle, W. Browne, and T. Mills, Booksellers, Bristol, 1793). Pulmonary consumption was tuberculosis.
161 Ibid., 61–67, reprinted in CED, 413–415.
162 Beddoes, *Letter to Erasmus Darwin*, 29–30.
163 Ibid., 50.
164 Ibid., 56–58. Haemoptysis: Spitting of blood; expectoration of blood, or of bloody mucus, etc., from the lungs or bronchi. Anasarca: A dropsical affection of the subcutaneous cellular tissue of a limb or other large surface of the body, producing a very puffed appearance of the flesh. Hydrothorax: A disease characterized by an effusion of serous fluid into one or both of the pleural cavities; dropsy of the chest. Schirrus (scirrus): A hard, firm, and almost painless swelling or tumour (OED).
165 Beddoes, *Letter to Erasmus Darwin*, 59–60.
166 Georgiana, Duchess of Devonshire, letter to second Earl Spencer, 1 June 1795, quote in Amanda Foreman, *The Duchess: Georgiana: Duchess of Devonshire* (London and New York: Harper Perennial, 2008), 293.

167 Stock (1811), 100–101, quoting Beddoes to Darwin, February 1793.
168 Ibid.
169 Ibid., 42.
170 Miller, *James Watt, Chemist.*
171 David Philip Miller, "Seeing the Chemical Steam for the Historical Fog: Watt's Steam Engine as a Chemical Device", *Annals of Science*, **65** (2008), 47–72; Miller and Levere, " 'Inhale It and See?' ", 5–28.
172 Richard L. Hills, *James Watt*, 3 vols. (Ashbourne: Landmark Publishing Ltd., 2002–2006), IV: *Triumph through Adversity, 1785–1819* (2006), 152 argues that Watt's personal grief was his prime motive for working with Beddoes. We take a much broader view, showing there were many concerned in the search for pneumatic remedies with whom Beddoes was in touch, even if Watt was the most significant.
173 Darwin to Watt 11 June 1794, BCL JWP MS 3219/4/28/35, published in CED, 440. Darwin to Watt, 3 June 1794, BCL JWP 3219/4/28/36. See also D. P. Miller, *Discovering Water*.
174 Jan Golinski, *Science as Public Culture: Chemistry and Enlightenment in Britain, 1760–1820* (Cambridge: Cambridge University Press, 1992), 170, re. Beddoes, *Observations on the Nature of Demonstrative Evidence; with an Explanation of Certain Difficulties Occurring in the Elements of Geometry: And Reflections on Language* (London: Printed for J. Johnson, 1793), iii, viii–ix.
175 See e.g. Watt to Black, 9 October 1796, BCL JWP MS 3219/4/124:497, published in CJB II, 1290–1293.
176 Edmund Goodwyn, *The Connexion of Life with Respiration; or, and Experimental Inquiry into the Effects of Submersion, Strangulation, and Several Kinds of Noxious Airs on Living Animals* (London: J. Johnson, 1788). For Thornton, see Chapter 3, this volume.
177 *Considerations* (see abbreviation list).
178 Beddoes, *Letter to Erasmus Darwin*, 40–48. It is not clear from the evidence (see note 160), but Beddoes may have had some kind of breathing apparatus prior to the development of Watt's instrument.
179 Ibid., 55.
180 Erasmus Darwin to Beddoes, 6 February 1794, Bod. MS Dep.C.134, published in CED, 430–431.
181 Darwin to Watt, 17 August 1794, BCL JWP MS 3219/4/28/37, published in CED, 448–450.
182 Watt in Beddoes to James Watt, letter 1795 undated, BCL JWP MS 3219/4/27:9. Considerations, part III (London and Bristol, 1795).
183 Watt to Beddoes 15 July 1794, BCL JWP MS 3219/4/124:322.
184 Watt to Beddoes, 20 August 1794, BCL JWP MS 3219/4/124:333.
185 *Considerations*, 38. Watt wrote to Beddoes on 4 September 1794:

> My opinion I have hinted before that water in specie forms the greater part of all airs, & I think the heavy inflammable air should be named Hydro carbonate for if it can be produced in so small a mixture as 1/12 of fixt air as Dr Priestley affirms what becomes of the oxygen if the water is supposed to be decomposed[?]
>
> BCL JWP MS 3219/4/124

186 Beddoes and Watt, *Considerations* (1796) Part I, 9.
187 Ibid., 9–10.
188 See notes 127 and 176 for Goodwyn: notes 119, 125, 152, and 185 for Priestley; and notes 29, 31, 40, 62, 67, 77, 94, 97, and 101 for Lavoisier. See also John [Jan] Ingenhousz, *Experiments upon Vegetables, Discovering Their Great*

Power of Purifying the Common Air in the Sun-Shine . . . to Which Is Joined, a New Method of Examining the Accurate Degree of Salubrity of the Atmosphere (London: Printed for P. Elmsly and H. Payne, 1779), and Jean Senebier, *Recherches sur l'influence de la lumière solaire pour métamorphoser l'air fixe en air pur par la vegetation* (Geneva: Barthelemi Chirol, 1793).

189 Joseph Black, *Lectures on the Elements of Chemistry* (Philadelphia: Printed for Mathew Carey, 1806–1807), I, 143–146.

190 Maurice Crosland, "The Image of Science as a Threat: Burke versus Priestley and the 'Philosophic Revolution'", *British Journal for the History of Science*, 20 (1987), 277–307.

191 Watt to Black, 17 July 1793, BPL JWP MS 4/12/29.

192 James Watt to Joseph Black, 17 July 1793, BCL JWP 4/12/29, published in CJB II, 1207–1209.

193 Miller, *James Watt: Chemist*, and Miller and Levere, "'Inhale It and See?'", 21, 24.

194 Beddoes to Watt, 4 March 1795, cited in Miller and Levere, "'Inhale It and See?'",5–28.

195 Thompson Cooper rev. Patrick Wallace, "Lind, James (1736–1812), Physician", ODNB.

196 Lind to Watt, 20 February 1795, BCL JWP 4/65, MS 3219/4/27.

197 "It is Right to be Taught, Even by the Enemy", Ovid, *Metamorphoses*, IV, 428.

198 Watt to Beddoes, 2 March 1795, BCL JWP MS 3219/4/124:377. For Thornton, see Chapter 3, this volume.

199 Watt to Beddoes, 28 November 1795, BCL JWP MS3219/4/124:415.

200 Beddoes to Watt, 26 November 1795, BCL JWP MS 3219/4/28:23:

> I know very well that my politics have been very injurious to the airs. Mr. Keir today furnishing me with a striking proof, if any were wanting. Yet as every stroke aimed at liberty, equally threatens science, morals, & humanity, it requires great self-denial to look on patiently & silently, when such great interests are at stake. On the last occasion the adversaries of Pitt stood on more popular ground than ever before; & I therefore think I did no harm to the other cause. I hope too that in endeavouring to animate the public spirit I have done something to repress vengeance & calm violence. At least I can say that no Bristol publication has been so eagerly & generally read.

201 Watt to Black, 9 October 1796, BCL JWP MS 3219/4/124:497, published in CJB II, 1290–1293.

202 S. T. Coleridge, review of Beddoes, *A Letter to the Right Hon. William Pitt*, in *The Watchman* (1796), *The Collected Works of Samuel Taylor Coleridge*, II, ed. L. Patton (London: Routledge and Kegan Paul, and Princeton: Princeton University Press, 1970), 100. Beddoes was regularly reviewed by and contributed to *The Watchman*.

203 Beddoes, *A Letter to the Right Hon. William Pitt* (London, 1796), 11.

204 Ibid., 14–15.

205 Beddoes, *A Letter to Erasmus Darwin*, 59–60.

206 Beddoes, *A Proposal for the Improvement of Medicine*. There is a copy in WMB W/M 35, dated Clifton 24 July 1794. In this copy, July is crossed out, and September is written in. There is another copy in CRO MS DG 41/35.

207 Anon., *The Golden Age, a Poetical Epistle from Erasmus D – n M.D. to Thomas Beddoes M.D.* (London and Oxford: Printed for F. Rivington & Co., and J. Cooke, 1794).

208 CRO DG two-page broadsheet, 1.

209 *The Golden Age*, 1. The author of the pamphlet by Erasmus D – n, M.D., has written "Liberty" where Beddoes had "Philosophy".

210 Thomas Beddoes, *Observations on the Nature and Cure of Calculus, Sea Scurvy, Consumption, Catarrh, and Fever: Together with Conjectures upon Several Other Subjects of Physiology and Pathology* (London: J. Murray, 1793), 109.
211 George Canning was the prime mover for *The Anti-Jacobin: or, Weekly Examiner* (1797–1798), succeeded by *The Anti-Jacobin Review* in 1798. Darwin's *Loves of the Plants* (Part II of *The Botanic Garden*) was satirized in Canning and John Hookham Frere, "Loves of the Triangles", *The Anti-Jacobin Review* (1799), reprinted in *Poetry of the Anti-Jacobin* (London: For J. Wright, 1799), 113 et seq. See also Patricia Fara, *Erasmus Darwin: Sex, Science, & Serendipity* (Oxford: Oxford University Press, 2012), 259–279, Appendix.
212 *The Golden Age*, 3.
213 Beddoes to Giddy, 31 July 1791, CRO MS DG 41/7.
214 Beddoes, *A Proposal for the Improvement of Medicine*, July <Sepr.> 29 1794, WMB MS W/M 35.
215 Beddoes, *A Proposal towards the Improvement of Medicine*, 29 July 1794, Keele University Library B125.12, no publisher identified.
216 Banks to Georgiana, Duchess of Devonshire, 30 November and2 December 1794, British Museum, Natural History, Dawson-Turner collection MS 9, ff. 125, 128–129: the latter is published in *Scientific Correspondence of Joseph Banks* (2007), IV, 337.
217 Beddoes to T. Wedgwood, 17 March 1795, WM, W/M 35 B1.483. On Birmingham, see above.
218 Beddoes and Watt, *Considerations*, 3rd edn. (1796) Fig. 1. See also Levere, "Dr. Thomas Beddoes: The Interaction of Pneumatic and Preventive Medicine with Chemistry", *Interdisciplinary Science Reviews*, 7 (1982), 137–147, Fig. 1.
219 Beddoes to Giddy, 23 August 1796, CRO MS DG 42/7.
220 Beddoes to Giddy, 25 January 1797, CRO, MS DG 42/32.
221 See e.g. Beddoes to Watt, 24 October 1797, BCL JWP MS 3219/4/29, and Beddoes to James Watt junior, 12 December 1797, BCL BWP (see abbreviation list) MS MivB
222 Beddoes to Watt junior, 24 December 1797, BCL BWP MS M iv B.
223 Advertisement in *Bristol Gazette*, 21 March 1799.
224 W. D. A. Smith, "William Clayfield's Mercurial Airholder", in *Essays on the History of Anaesthesia* eds. A. Marshall Barr, Thomas B. Boulton, and David J. Wilkinson (London: Royal Society of Medicine Press, 1996), 13–15. See Figure 1.12 and June Z. Fullmer, *Young Humphry Davy: The Making of an Experimental Chemist* (Philadelphia: American Philosophical Society, 2000), 197. Mark Davies, *King of All Balloons: The Adventurous Life of James Sadler, the First English Aeronaut* (Stroud, Gloucestershire: Amberley Publishing, 2015). Davy got the credit.
225 See, e.g., David Knight, *Humphry Davy: Science & Power* (Oxford: Blackwell, 1992), 21–23.
226 Beddoes to Watt, 24 October 1797, BCL JWP MS 3219/4/29 no.9.
227 The following paragraphs are taken from Levere, "Dr. Thomas Beddoes: Chemistry, Medicine, and Books", in *New Narratives in Eighteenth-Century Chemistry*, ed. L. Principe (Dordrecht: Springer, 2002), 157–176.
228 Beddoes to James Watt junior, 12 December 1797, BCL BWP MS MivB.
229 J. A. Scherer, *Beweis, dass Johann Mayow vor hundert jahren den grund zur antiphlogistischen chemie und physiologie gelegt hat* (Wien: Wappler, 1793).
230 See J. R. Partington, *A History of Chemistry*, 4 vols. (London: Macmillan, 1962), III, 598.
231 Beddoes to James Watt, 12 December 1797, BCL BWP MS MivB. For Higgins, see Stansfield, 17.

232 Beddoes to James Watt junior, 2 January 1798, BCL BWP MS MivB.
233 A reverbatory furnace, now known as a reverberatory furnace, is one where the material being heated is not in contact with the fuel. A muffle or retort furnace similarly keeps the material being heated from contact with the fuel and with the products of combustion, and often involves operates at high temperatures.
234 Beddoes to James Watt junior, 2 January 1798, BCL BWP MS MivB. This is a copy, in a less sure and literate hand than TB's, of a letter sent to Boyd.
235 Cadwallader Boyd from Dublin appears in the MSS Minutes of the Bath and West of England Chemical Committee (Bath Record Office, Bath & West of England Archives [B &W of E], VII) from 1805–1808. This was set up as a result of Davy's activities at the Royal Institution. At their annual meeting of December 1805, Sir John Coxe Hippesley suggested that a B & W of E chemical laboratory should be equipped, so a subscription was launched and a committee of chemical research was set up. Dr. Clement Archer (1748–1806) was appointed Superintendent and Professor, and he lectured to the B & W of E in Spring 1806 (*Bath Chronicle*, 13 February 1806). Boyd, "a very ingenious and intelligent chemist whose real knowledge and acquaintance with the science is accompanied by that unassuming modesty generally attendant on true merit"" ("Bath & West of England Letters and Papers", **11** (1807), xiv) was Archer's assistant operator, and he continued the work of this chemical committee after Archer died. It seems probable that Boyd was vouched for by Davy.
236 Beddoes to James Watt junior, 25 February 1798, BCL JWP MivB.
237 June Z. Fullmer, *Young Humphry Davy: The Making of an Experimental Chemist* (Philadelphia: American Philosophical Society, 2000), 197. Mark Davies, *King of All Balloons: The Adventurous Life of James Sadler, the First English Aeronaut* (Stroud, Gloucestershire: Amberley Publishing, 2015).
238 Beddoes to James Watt, 15 [January] 1798, BCL MS 3219/4/29:32.
239 David Knight, *Humphry Davy*, 25, and Harold Hartley, *Humphry Davy* (London: Thomas Nelson, 1966), 151.
240 Davy, "An Essay on Heat, Light and the Combinations of Light", in *Contributions to Physical and Medical Knowledge, Principally from the West of England, Collected by Thomas Beddoes, M.D.* (Bristol: Biggs and Cottle, 1799), 4–147.
241 Davy to Henry Penneck, 26 January 1799, APS MS BD315.1/1969.1821. The Pennecks were an old family from West Cornwall.
242 Davy to Henry Penneck, 26 January 1798, American Philosophical Society, MS file B.D315.1.
243 Mattrass: a glass vessel with a round or oval body and a long neck, used by chemists for digesting and distilling. OED.
244 The Argand lamp was invented in 1780 by Aimé Argand and promoted by Boulton & Watt. It has a cylindrical wick: air passes through the inside of the wick and around its outside, and is then drawn into a cylindrical glass chimney, which improves the flow and makes the flame more stable.
245 APS MS BD315.1/1969.1821.
246 Ibid.
247 Beddoes to James Watt, 7 February 1798, BCL JWP MS 3219/4/29:28. R. G. W. Anderson, "Joseph Black and His Chemical Furnace", in *Making Instruments Count*, eds. R. G. W. Anderson, J. A. Bennett and W. F. Ryan (Ashgate: Aldershot, 1993), 118–126.
248 Beddoes to Watt, 9 March 1798, BCL BWP MS M IV B.
249 Beddoes to J. Watt junior, March 1798, BCL BWP MS M IV B.
250 Beddoes to J. Watt, junior 25 February 1798, BCL BWP MS M IV B.
251 Beddoes' work on mineralogy and geology is thoroughly discussed in Chapter 2.
252 Beddoes to J. Watt, junior 25 February 1798, BCL BWP MS M IV B.

253 Thomas Beddoes, *Contributions to Physical and Medical Knowledge, Principally from the West of England, Collected by Thomas Beddoes, M.D.* (Bristol: Biggs and Cottle, 1799), 210 et seq.

254 Beddoes, *Contributions to physical and medical knowledge, principally from the West of England, collected by Thomas Beddoes, M.D.* (printed by Biggs & Cottle for T. N. Longman and O. Rees Paternoster-Row London), 213.

255 Benjamin Thompson, *Experiments on Heat* [London, 1792], reprinted from *Philosophical Transactions* (1792), 48–80; *New Experiments upon Heat* (London: Printed by J. Nichols, 1786), 273, from *Philosophical Transactions* (1786). Beddoes, *Contributions* (1799), 212.

256 The identity of static electricity with the electricity produced in the pile or battery was not convincingly demonstrated until the 1830s (see Michael Faraday, "Experimental Researches in Electricity", 3rd Series, *Philosophical Transactions*, 1833, 23–54), although it was much earlier suspected by William Hyde Wollaston, *Philosophical Magazine*, **3** (1801), 211.

257 Victor D. Boantza, "Light in the Pneumatic Context: Aspects of Interplay between Theory and Practice in Early Photochemical Research", *Historia Scientiarum*, 2nd Series, **16** (2006), 105–127.

258 This was Lavoisier's invention. He argued that since all acids contained oxygen (which they don't), therefore the acid from sea salt (our hydrochloric acid) must contain oxygen, combined with a not yet isolated radical, the muriatic acid radical. Davy was to disprove this hypothesis: see "On the fallacy of experiments in which water is said to have been formed by the decomposition of chlorine", *Philosophical Transactions*, **108** (1818), 169–171, and *Philosophical Magazines*, **53** (1819), 326–328.

259 Beddoes, "Letter from Dr. BEDDOES on Certain Points of History, Relative to the Component Parts of the Alkalis, with Observations Relating to the Composition of the Bodies Termed Simple", *Journal of Natural Philosophy, Chemistry, and the Arts*, ed. William Nicholson, **21** (1808), 139–141:

> As an incentive and a clew to experiment (which is the only use of hypothesis) I beg leave to repeat, that *metals and other combustibles may be formed of hidrogen and azote*. . . . One cannot proceed far in this train of speculation without getting the prospect of all nature as consisting of two elements, oxigen and hidrogen.

260 Beddoes to James Watt junior, September 1798, BCL JWP M IV B. H. Davy, "An Essay on Heat, Light and the Combinations of Light", *Contributions* (1799), 4–147; "An Essay on the Generation of Phosoxygen, or Oxygen Gas; and on the Causes of Colours of Organic Beings", *Contributions*, 151–198. Davy was soon to be embarrassed by these papers, which claimed more than they could prove.

261 Joseph Priestley, *Experiments and Observations on Different Kinds of Air*, 3 vols. (London, printed for J. Johnson, 1794, 1795, 1797).

262 Joseph Priestley to H. Davy, Northumberland, Pennsylvania 31 October 1801, published in *A Scientific Autobiography of Joseph Priestley (1733–1804): Selected Scientific Correspondence Edited with Commentary*, ed. R. E. Schofield (Cambridge, MA: MIT Press, 1966), 313–314: "It gives me peculiar satisfaction, that as I am now far advanced in life, and cannot expect to do much more, I shall leave so able a fellow-labourer of my own country in the great field of experimental philosophy."

263 S. L. Mitchill, *Remarks on the Gaseous Oxyd of Azote and on the Effects it Produces* (New York: printed by T. and J. Swords, printers to the faculty of physic of Columbia College, 1795), a work known to Beddoes, who published Mitchill's views in Appendix I to *Considerations* (1795).

264 *The Collected Works of H. Davy, Bart.*, ed. John Davy, 9 vols. (London: Smith Elder & Co., 1839), 3, 327, 329–30; Beddoes to J. Watt junior 27 June 1799 and Beddoes to Matthew Boulton [1799], BCL BWP MS M IV B, and Box B2, 19 and 24.
265 For the significance of these experiments, see F. F. Cartwright, *The English Pioneers of Anaesthesia (Beddoes, Davy, and Hickman)* (Bristol: Wright, 1952).
266 Beddoes, *Notes of Some Observations Made at the Medical Pneumatic Institution* (Bristol: Printed by Biggs and Cottle for T. N. Longman and O. Rees, London, 1799), 7–9, 13, 15–16.
267 Watt to Black, 6 November 1799, BCL MS JWP 4/12/10 (renumbered 3219/4/), published CJB II, 1376–1378.
268 Beddoes, *Notes* . . . (1799), 16, 26–27, 32–34.
269 *The Notebooks of Samuel Taylor Coleridge*, ed. Kathleen Coburn, 5 vols. (New York: Pantheon Books for the Bollingen Foundation, 1957–2002), I, entry 1034 [December 1801]; original "eurekas" in Greek letters.
270 Watt to Beddoes, 6 May 1803, BCL JWP MS 3219/4/118:538.
271 Their regularity was encouraged by Beddoes' policy of charging them a moderate fee at the beginning of a course of treatment, and refunding it only when the course was completed. Beddoes to [Josiah] Wedgwood [November 1801], WMB W/M 35 LH
272 Bristol Record Office, *Bristol Infirmary Biographical Records*, IX, 72, Beddoes to [unknown, possibly James Watt] 27 November 1804. BCL BWP MivB. Levere (1982), 137–147.
273 Levere, "Dr. Thomas Beddoes (1750–1808): Science and Medicine in Politics and Society", *British Journal for the History of Science, 17* (1984), 187–204 at 203. Letter quoted from A. Treneer, *The Mercurial Chemist: A Life of Sir Humphry Davy* (London: Methuen, 1963), 113.
274 S. T. Coleridge, *The Watchman*, in *The Collected Works of Samuel Taylor Coleridge*, no. 2, ed. Lewis Patton (London: Routledge & Kegan Paul, and Princeton: Princeton University Press), 100.
275 *Collected Letters of Samuel Taylor Coleridge*, ed. Earl Leslie Griggs, 6 vols. (Oxford: Clarendon Press, 1959), III, 171.

2 Thomas Beddoes and natural history, especially geology

Hugh Torrens

Introduction

In a perceptive article in 1997, David Allen wrote how, in Britain, geology had become "the lost limb" of natural history, and "slipped out of touch with zoology and botany", and yet how little had been "written on this massive and far-reaching split".[1] He dated the division as between "the arrival of Huttonian theory [1785–1795]", which first "wrenched geology loose", and "the middle of the last century [1850]", when the divorce became final. Thomas Beddoes worked across this divide, which has helped his activities across natural history to be neglected, in favour of his work in the more un-natural science of chemistry. As a result, geology has remained the only one of the many scientific worlds in which Beddoes was active not to have been previously analysed.

Beddoes was born in 1760 in Shropshire, the English county most famous for advancing the world's first "industrial revolution".[2] Many believe, with this author, that the link between the new science – geology – and industrial activity was of fundamental importance. But too little attention has also been paid to this topic. Beddoes' activities within geology clearly confirm how close links were here between geology and industry.

Another aspect of any geo-analysis of Beddoes' work is the problem of the absence of any significant personal archive. Both Cartwright and Porter have noted this. Cartwright wrote how Beddoes' "literary remains [had proved] even scantier than his financial . . ., nor was there any important unpublished material. It is probable that Beddoes destroyed a vast quantity of papers".[3] Porter instead, claimed it was "Anna [Beddoes' wife, who having] commissioned a highly doctored biography, then, it seems, had most of his papers destroyed (she clearly feared she had much to hide. . .)."[4] Given the state trials for treason and sedition in the 1790s, and the climate of alarm and fear that accompanied them, the latter claim is at least plausible.[5] Whoever caused the destruction, it has made work on Beddoes much harder.

Beddoes' early science

First, we must explore Beddoes' early activities across natural history. Shropshire's beautiful countryside had already encouraged many naturalists and collectors, and Beddoes was no different. During his time as an Oxford student, 1776–1779, we know that he attended natural philosophy lectures there, probably those offered by Thomas Hornsby, who had been elected Oxford's professor in 1763, as an undated letter from Beddoes to his father during his time there, shows. This listed the costs of future attendances at "Natl Philosophy lectures, 3 guineas" and "Mathematical do, 1 guinea".[6]

Edinburgh, 1784–1786

Beddoes' interests in natural history were really aroused, however, in Edinburgh. In summer 1784 Beddoes moved there to prosecute his medical studies. He soon attended the courses of the Professor of Natural History at Edinburgh, John Walker. His first was held in November 1784, for which Beddoes paid £3–5–0, to include the second, which was specifically on mineralogy, given in June 1785.[7]

Beddoes soon joined the flourishing student medical, chemical, and natural history societies then active here, which were so different from any available at scientifically uninterested Oxford. The first was the Natural History Society, founded earliest, in March 1782,[8] by a group of five. These included the first two, Worcester-born, later associates of Beddoes, Jonathan Stokes[9] and William Thomson, and James Edward Smith, the later founder of the Linnean Society. Beddoes was elected a member on 9 December 1784, and soon elected its president, on 1 December 1785, until his resignation on 9 March 1786. He also read it two papers; (1) "On the Sexual System" and (2) "On the Chain of Beings", both reproduced by Stock.[10]

It was the Natural History Society, from which the Chemical Society split off early in 1785, and with which it thereafter, chameleon-like, played a complex role in competition with the earlier Society, until chemistry returned to dominate in January 1800.[11] Beddoes was a founding member of the Chemical Society, from 1785,[12] to which again he read papers.[13] Averley reminds us that this may well be the world's first chemical society.[14] Finally Beddoes also joined the, by then Royal (from 1778), Medical Society in 1785,[15] and became its president that same year.[16]

Beddoes' botanical work

Beddoes was also an early student of botany and had already written a, sadly now lost, manuscript titled *Flora Britannica*. Of this, Stock noted that it "appears to have [been] written at Oxford, and the elegance and beauty of the penmanship of this little volume would excite admiration and surprise".[17] That this interest continued at Edinburgh is clear. This notice appeared in the *Morning Chronicle*:

> We are informed in a letter from Edinburgh, dated 24 August [1785], that Mr. Dickson from London, a skilful and expert botanist is making

> a tour through the northern parts of Scotland in search of rare and curious plants. He is accompanied with Mr Watt, gardener, also from London, Henry Grimston and Thomas Beddoes Esqs, both of this University, and Mr. Henry Porteous from the Botanical Garden of Edinburgh. As the object of this tour is the discovery of new and useful plants, it is not to be doubted, that those gentlemen will receive civilities and attention from a country which has always been distinguished by its hospitality and politeness to strangers.[18]

The Scottish-born nurseryman James Dickson was soon to be one of the founders of the Linnean Society, in 1788. Henry Grimston[19] was then a student at Edinburgh. He had come from a landed family at Etton, near Beverley, Yorkshire.[20] Henry had also attended two of John Walker's courses, in May and, with Beddoes, that on mineralogy in June 1785. He had been earlier been elected a corresponding member of the student Natural History Society on 20 May 1784.[21] Henry Porteous was an employee of the Edinburgh Botanic Garden.[22]

We learn more of their botanical tour in the Dumbarton-postmarked letter that Beddoes wrote to his father on 11 September 1785. This noted they had set out about three weeks earlier from Edinburgh, going northwards, via Perth, Dunkeld, to Blair in Athol, where he had climbed Ben Lawers, and Luss.[23] Later, in an undated letter to Sir John Sinclair from Bristol, Beddoes noted how, "when a student at Edinburgh, I walked through much of the Western Highlands; and although I was then intent on natural history, I paid what attention I could to the personal condition of the Highlanders."[24] The condition of the poor, rural and urban, was always Beddoes' concern.

His interest in botany certainly continued. In August 1787 the infamous second edition of William Withering's *Botanical Arrangement of British Plants* was published.[25] To this had been added a "New set of references to Figures" by Jonathan Stokes, which Withering regarded as "one of the most valuable parts of the present Edition". Vol. 1 duly listed "Dr Beddoes, lecturer in Chemistry, at Oxford, [as another] who had favoured this Edition with their assistance".

Toward the end of a very long letter in January 1792 to Davies Giddy, mainly on politics, Beddoes wrote "will you laugh or will you frown, if I tell you that I have begun to write a set of botanical dialogues? I wish the thought had struck me three months ago" (when he had been with Giddy, in Cornwall).[26]

While in Edinburgh, Beddoes came into close contact with, and must have frequently met, both the University Professor of Chemistry, Joseph Black, whose lectures he attended, and geologist James Hutton, who did not give University lectures, but who was well known in academic circles. Black's course had already included, by 1767–1768, a good deal of what we would today call geology, with discussions of naturally occurring materials, like chalk, flint, oil, coal, and so on.[27] We know much less of Beddoes' direct contacts with Hutton. In a letter from London to Black, dated 6 November 1787, Beddoes wrote how he had earlier that autumn, when in France already

> sent Dr. Hutton's theory of rain [read 1784] to Morveau at Dijon & he was very much pleased with it & I afterwards gave it [also] to Lavoisier. I am very sorry I had not at that time a copy of his theory of the earth [1787]. If he has a copy left and could send it to me at Murray's . . . I will undertake to forward it to Lavoisier. Mr. Lumsden told me he had brought it to Paris; but I suppose that was only the Abstract [1785]; none of the French philosophers with whom I conversed know any thing of it.[28]

Beddoes had travelled on the continent in the autumn of 1787 and he next planned a specifically "mineralogical tour of the northern part of Germany" for a subsequent year, but the soon-to-be disturbed state of European politics meant this never took place.[29]

Back in Oxford, 1786–1793

Beddoes returned to Oxford in December 1786 and was appointed Reader in chemistry there in the spring of 1788.[30] In another letter to Black, dated 23 February 1788, from Oxford he asks Black to thank Hutton and notes that

> I shall take the opportunity of bringing his admirable paper on the theory of the earth [read 1785 and published 1787/1788] before the class; and I am very desirous of being able to illustrate it as much as possible. Now could he without infirming his own collection give me a specimen of the granite which he has figured?[31] If he could spare me a morsel of anything else I should be infinitely obliged to him . . . of you I wish to beg a specimen of the petrified wood from Loch Neagh.[32]

Beddoes duly sent his published geological Royal Society paper (see section on Beddoes on basalts and granites) to Black on 15 April 1791, noting

> I wish you may think the observations contained in this accompanying paper a prosecution of the subject as it is entered upon in Dr. Hutton's theory of the earth, worthy of such a beginning. I have in reserve another paper of which I gave you some information before & of which I shall make a point of sending you a copy as soon as it is printed, as I hope it will be in the course of ye summer. . . . [He ends] If you or Dr. Hutton have duplicates of specimens which afford striking illustrations of his theory or my syllabus, I solicit the communication of them. . . . This summer I project a mineralogical excursion of some length. I think it will be into Wales.[33]

Beddoes certainly undertook this tour, as a letter to Giddy starts "on my return from Wales".[34] He was by then a firm convert to Huttonian theory and thus an early contributor to the split that allowed the "lost limb" of geology to grow as a separate science.

Oxford geology lectures

Beddoes' conversion to Huttonian theory was reflected in his Oxford lectures. In November–December 1789 he gave a course of lectures entitled "On certain Parts of the Natural History of the Earth and Atmosphere", while William Thomson was lecturing on both anatomy and mineralogy. Beddoes' friend Davies Giddy was one of those who attended his 1789 geology lectures.[35] In November–December 1790 Beddoes repeated this course, which was "of about 16 lectures, and the subscription a Guinea".[36] In 1791, Beddoes offered a completely new geology course. This was now to be "On the Natural History of the Strata, Rocks and Mountains of the Earth". This was to continue daily from 17 June to either 2 July or 4 July.[37] The two lecturers, Beddoes and Thomson, were clearly those who attended the Lunar Society meeting at Birmingham with Cyril Jackson, the Dean of Christ Church, Oxford, on 3 August 1789.[38] Their lectures had helped produce at Oxford, as Stock put it, "a taste for scientific researches which bordered on enthusiasm".[39] Some of Beddoes' geological lecture notes survive in the Bodleian Library[40] and would repay greater attention, in view of Beddoes' clear, if forgotten, role in fostering the spread of geological, and especially Huttonian, ideas.

Beddoes certainly transmitted his enthusiasms to some of his students. One was Richard Edwards, who, with Beddoes and the balloonist James Sadler, was involved in sending up a balloon from Pembroke College Garden in June 1790.[41] The *English Chronicle* reported that the three of them had

> sent up a balloon filled with hydrogen gas from marshes, to represent and account for the production of meteors, which it did in a very satisfactory and pleasing manner, although it later suddenly caught fire when it had gained a certain height.[42]

Edwards was later much involved with geology, clearly inspired by Beddoes. Another such student was the Scottish barrister Charles Maitland Bushby from Dumfries, who graduated B.A. from Pembroke College in 1792 and then M.A in 1794, and who later sailed for India, where he died in Madras on 5 February 1799, shortly after his arrival.[43] Bushby had written from London to Beddoes at Shifnal, on 14 December 1791, in a letter that Beddoes later forwarded to Davies Giddy in Cornwall.[44] In this, Bushby wrote of his recent time in Scotland:

> I am sure you will give me the great credit for having spent a morning in examining the Basaltic Columns at Duddingston, when I might have

> been on the race ground at Leith. From the enclosed letter too you will see that I have not forgot your commission to the Leadhills, it is not a bad thought of Taylor's to write to you[45] & I send you specimens as they occur. . . . When I was in Edinburgh, my friend Mr. Young offered me a collection of Icelandic specimens which by some unaccountable means or other had come into his possession. I declined taking them myself lest it should give me a rage for collection – but if they can be of any service to you, I have no doubts but I can still obtain them.[46]

Other such students almost certainly included Abraham Mills of Macclesfield and Thomas Weaver junior, the later geologist and mining consultant of Gloucester, during Beddoes' last lecture course on the Natural History of Fossils (i.e. rocks) in 1792. On 11 April 1792, Beddoes had written to Giddy:

> next term, I intend to read publicly the papers I have drawn up on the Natural History of Fossils. I wish to lay myself under the necessity of carefully reflecting [on] them all over once more. Then perhaps I may print them. Perhaps they may give a popular, and not altogether an imperfect, view of Hutton's subterraneous system with many additional proofs, elucidations and applications.

He continued

> I am under great trepidation for a most valuable box of specimens from Mr Mills, who is a warm advocate for my system of Granite and Basaltes and therefore an excellent mineralogist. Will you enquire if any box for me arrived lately . . . from Macclesfield ultimately.[47]

Mills was, from 1783, a partner in the Macclesfield Copper Company, which, from 1787, had started to work copper mines in Wicklow, Ireland. In 1790 he published two articles on collieries in the *Transactions of the Royal Irish Academy*[48] and others in two letters on strata and volcanic appearances in the north of Ireland and west of Scotland in the *Philosophical Transactions*.[49] Weaver, however, had studied mining under Werner in Freiburg, Germany, between 1790 and 1792.[50] On 16 October 1792, Mills wrote to fellow mineralogist White Watson in Derbyshire how their "friend Weaver has been returned some time to Gloster, and means to attend Dr. Beddoes lectures at Oxford, previous to his going to Ireland",[51] where he was to take up management of copper mines in the Wicklow mountains for Mills' company.[52] But, in the absence of any surviving lecture lists of Beddoes' students, it is impossible to prove that both were his students. In addition, Beddoes' intended Huttonian publication never appeared.

Henry R. V. Fox (later Vassall), third Lord Holland, who never actually attended any of Beddoes' lectures, was still able to report, unfavourably, on his abilities as a lecturer. He also noted that Beddoes had "above once, announced discoveries which he was obliged to retract".[53] One such must be the geological matter later referred to by one of Beddoes' anonymous biographers, who recorded how

> a gentleman had brought to Oxford from the summit of one of the mountains surrounding Coniston lake in Lancashire, some specimens which had evidently undergone the operation of fire, but which happened to abound near a hollow on the top of the mountain, which some Italian gentlemen had not long before pronounced to be the crater of an extinct volcano. Upon shewing them to Dr. Beddoes, he was so persuaded of the fact, that he even summoned a particular assembly of the members of the University by an extraordinary notice, before whom he delivered a long lecture on the specimens supplied, as indicative of the natural operations of fire in those parts of England. A very short time after, he declared that they were evidently nothing better than mere slags from some old furnace, and that he had since discovered a criterion by which he could distinguish between the productions of natural and artificial fire; but this discovery, and the consequent change of his sentiments, he could not be prevailed on to announce as publicly as he had delivered his former opinions.[54]

Sadly, the early history of geology in the English Lake District has yet been hardly examined. Oldroyd's fine study mentions only how James Hutton's friend John Playfair had confirmed Hutton's 1788 observations of true fossils occurring in Cumbrian "schistus" near Windermere, in 1791 to confirm this was not a Primitive rock.[55] The initially anonymous author of a series of "Observations, natural, oeconomical and literary, made in a Tour from London to the Lakes in the summer of 1791" certainly thought he had also seen "the mouth of an extinguished volcano, in the hills near Keswick".[56]

Porter rightly regarded any idea that the British Isles had harboured ancient volcanoes as then a "revolutionary conception".[57] Pembroke College's historian later noted how Beddoes had "incurred much ridicule" over this Lakeland incident, "by agitating the University over the discovery of an extinct crater at Coniston".[58] The date of this event may be connected with the letter Beddoes wrote to Sir Joseph Banks from Oxford on 30 May [1792]. Beddoes wrote that "I hope, in the course of a month, to convince you of the existence of Volcanos in England, at some period posterior to the formation of our rocks and mountains."[59] So this Lakeland debacle may well have provided another reason, besides his well-known political activities,[60] behind Beddoes' resolve to resign from Oxford by June 1792.

William Thomson at Oxford

Beddoes' time at Oxford, until September 1790, was shared with William Thomson (1760–1806), physician and mineralogist, and later, after 1790, a man we should now call England's first vulcanologist. He was born in 1760 in Worcester, eldest son of Dr. William Thomson (c.1722–1802), M.D. Leyden 1751, physician to the Worcester Infirmary between 1757 and 1793, and his wife Ann. Because Thomson's career in so many ways mirrored Beddoes', we need to know more. Thomson first attended King's School, Worcester, and matriculated at Queen's College, Oxford, in 1776.[61] In 1779 he was awarded a studentship at Christ Church and graduated B.A. in 1780. Thomson's first interests were archaeological, but from 1781 to 1782 he studied medicine at Edinburgh University, with chemistry under Joseph Black. He was here elected to the Royal Medical Society (in 1781) and was a founder member of the Natural History Society (in 1782), and the fifth to attend John Walker's new university course in that subject. In 1782 he too went on tour to the Highlands, collecting minerals on Staffa. In all this he presaged Beddoes.

On his return to Oxford, he graduated M.A. in 1783. In London in 1784, Thomson joined the London-based Society for Promoting Natural History and recommended that his friend James Macie (later Smithson) should join the tour that took Barthelemy Faujas de St. Fond to Scotland. Thomson graduated B.Med. in 1785, and was elected Dr. Lee's Lecturer in Anatomy at Oxford, in April 1785. His M.D. followed in 1786, when he was also elected physician to the Radcliffe Infirmary and FRS. In 1787 he gave probably the first Oxford lecture course on mineralogy, and these lectures – with his on anatomy, and Beddoes' on chemistry and geology – were among those later described as having "produced [there] a taste for scientific researches which bordered on enthusiasm",[62] In 1788, Thomson helped found the Linnean Society and tried to find true fossils for the Edinburgh geological friend, he now shared with Beddoes, James Hutton (1726–1797). In 1789 Thomson helped Beddoes with those metallurgical experiments, and those on the agency of heat in geology, which impressed on Beddoes the "very strong conviction of the truth of Hutton's Theory of the Earth", since "Dr Thomson admits the facts . . . and thinks my specimens justify the inferences."[63]

Suddenly, in September 1790, Thomson had to leave Oxford and resign from the Royal Society. In a letter from London, dated 25 September 1790, he explained that he had suffered "a most scandalous imputation from an Experiment he had performed on a man 4 years ago",[64] Oxford University's minutes of their Hebdominal Board confirm this accusation, and record how, on 18 October 1790, the Vice-Chancellor reported that

> several members of Convocation had expressed a desire that the most public academical censures might be inflicted on Dr. Wm Thomson, late Anatomy Reader of Christ Church, on a charge of suspicion that the said Dr. W. T. had been guilty (about 4 years ago) of some unnatural & detestable practices with one Willm Parsons, then a Servant Boy to Mr. Fletcher, Bookseller.[65]

This was an offence of which London gossip noted "happily the public opinion in this country never forgives",[66] although the London surgeon and anatomist John Hunter (1728–1793) tried to rally to Thomson's defence. Thomson was next ordered to appear before the Vice-Chancellor, but failed to appear. So, on 2 November 1790, his decree of degradation and banishment was issued, since Thomson was "in suspicionem foedissimi et sodomitici criminis inciderit".[67]

The imputation must be that Thomson stood, at least, accused of sodomy, an offence then punishable by death.[68] But his failure to appear at Oxford, in these turbulent, revolutionary times (since the start of the French Revolution in July 1789), which equally affected Beddoes, need not be seen as an admission of guilt. Thomson's problems may have lain in his having carried out medical "experiments" on a human. Beddoes, too, was soon to suffer from similar imputations. In part 3 of his *Considerations on the Medicinal Use and Production of Factitious Airs* of 1795, written with James Watt, Beddoes' added a spoof letter, in which one of his medical enemies discussed Beddoes' methods, with the potential parents of a consumptive daughter, who might soon become Beddoes' patient:

> but would you Sir, or you, my lady, have experiments [here] made upon your daughter? My Lady shuddered at the question; she even started, nor could her countenance have expressed more horror, [than] if she had actually held her daughter in the arms of an assassin.[69]

Beddoes and Thomson had both been vilified for having carried out supposed "medical experiments" on humans. And although Henry Fox recorded that "the memory of [Thomson's] ready eloquence and extraordinary perspicuity survived[this] ruin of his moral character,"[70] Thomson now left England for his final destination in Italy, never to return.

Shropshire and William Reynolds

Another of Beddoes' friends had taken an equal interest in geological matters. This was William Reynolds, whom Beddoes had met back in Shropshire in 1787 or 1788, and who became one of his firmest friends.[71] Reynolds, who was soon to be disowned by Quakers for marrying his first cousin in 1789, was an ironmaster at Ketley.[72] Reynolds had long been interested in promoting geology, since he that saw its knowledge would advance his mining, quarrying, and manufacturing concerns. He had already started his major Shropshire geological collection by 1776.[73] His influence on Beddoes' geology is clear. One particular phenomenon much interested them both. This was the occurrence of natural bitumen in Shropshire, which had been discovered there from 1694.[74] This was heightened by the discovery of a natural bitumen well in Reynolds' Tar Tunnel at Ironbridge in October 1786.[75] This became the subject of some fascinating correspondence between Erasmus Darwin and Beddoes from 1787.[76] This debated the origin of both coal and oil, which they both thought gave early confirmation to Huttonian theory.[77]

Reynolds was much involved in such geological matters from his role as a leading Shropshire entrepreneur. We learn this from his single surviving letter to Beddoes, dated Ketley 4 March 1789. It reads:

> It gives me pleasure to hear that Dr. [William] Thompson [Thomson] thinks the specimens obtained from the Bear [the mass of metal found below hearth level in a blast furnace] at Donington [Shropshire] are instructive & I cannot help joining him in the wish that thy thoughts upon the subject were put together for the Royal Society. Speculations respecting the theory of the earth may appear to some of little consequence to those who dwell upon it, but they are often connected with & lead to speculations which are relative to the arts & manufactures upon which depend many of the comforts of the more polished parts of this globe.
>
> The articles mentioned in thine for Dr. Thompson together with several others are packed & will be sent by the waggon on Saturday. I wish they may be acceptable. If he has by him any duplicates of Iron ore or coal, I shall esteem it a favor [if] he will send me some, as I wish to obtain as good a collection of those articles as I can, it being more in my line than any other part of mineralogy. . . .
>
> I am surprized to hear thy Books are not come to hand, they were sent from hence by the waggon near a month since except one was omitted, but will now be sent in one of the boxes for Dr. Thompson together with the 2 specimens of Scotch Granite.[78]

This Scotch granite must refer to specimens of the graphic granite from Portsoy, which Hutton had already illustrated in his 1787–1788 version of his *Theory of the Earth* (plate 2).

These Shropshire researches were soon referred to by Beddoes in his next letter to Black from Oxford, in April 1789.

> Happening to be furnished by Dr. [William] Thomson with a Frank, I write a few lines to renew your recollection of me. Since I had the pleasure of seeing you in Shropshire, I have had an opportunity of making some observations on the agency of heat which have impressed upon my mind a very strong conviction of the truth of Dr. Hutton's Theory of the Earth. Dr. Hutton will have received a hint of it from Dr. Arnemann.[79] & I shall soon myself send a paper upon the subject of it to the Royal Society. I have seen not merely basaltic pillars, but petrosilex dendrites & as I think asbestos formed by artificial heat. Dr. Thomson, who as you know is very intelligent on these subjects & by no means hasty in giving his assent to miraculous transmutations, admits the facts which I shall communicate to the public and thinks

> that my specimens justify the inferences. Unhappily this kind of conviction is not very capable of being diffused universally, both on account of the difference between inspection & description, & of another person's report & one's own perception. As far as you & Dr. Hutton are concerned, I hope to have it in my power to send you specimens of the principal changes produced if I shall understand that you desire it. The public at large must depend on my narration & a plate or two which will tend to illustrate it.[80]

Beddoes wrote twice to Sir Joseph Banks, from Shifnal, warning him that he was soon to send this paper to the Royal Society. The first, on 3 January 1791, noted he had been unable to complete it, "owing to repairs at the Oxford Laboratory". The second, dated 27 January 1791, noted that the paper had now been sent, but that although he had intended to send with it a series of specimens to be exhibited when the paper was read, these were now required for the lectures he was now delivering.[81] His new Oxford lecture course was that announced on 23 October 1790, as "on certain points of the Natural History of the Earth & Atmosphere".[82] The next week Beddoes announced the course would now consist of about 16 lectures and that the subscription would be one guinea.[83] His Royal Society paper was duly read that same day in London.[84]

Another letter from William Reynolds' half-brother Joseph Reynolds shows further connections with geology and with Edinburgh. This, dated 26 August 1791 from Ketley, read:

> Our good friend T[heophilus] Houlbrook[e][85] is at my father's and having taken up the pursuit of mineralogy, with a warmth bordering on Enthusiasm has been a journey in to the Highlands in company with Dr. Hutton and there finds many circumstances tending to confirm Dr. H's Theory, particularly at Blair Atholl, where the Granite is, to use his own expression, "driven up into the Schistus." This may perhaps be nothing new to thee but he mentioned it as being similar to the manner in which Lava is forced thro' the incumbent Strata and I think it had a considerable share of weight as an argument to prove that from the similarity of their situation they both were placed in the same circumstances by the same cause and probably were in a similar state of fluidity at the time of their being driven thro' solid masses.[86]

This is a fine description of this critical exposure where this important phenomenon had been first observed by James Hutton and John Clerk, before Hutton had been able to publish it himself.[87]

Beddoes on basalts and granites

Beddoes' paper on the affinity between basalts and granites was read to the Royal Society on 27 January 1791. It was naturally a thoroughly Huttonian offering. Beddoes argued that all opinions on such subjects must be based on analogy, since there was no direct testimony and, so reasoning, he "assumed the origin of basalts from subterraneous fusion" as having been "thoroughly established by authorities", which he then listed (in French, English, and German). But opinions on granite "were generally loose conjectures", with the exception of John Strange's paper on two Giants Causeways in the Veneto in 1775,[88] and Hutton's 1787–1788 thoughts on the Portsoy granite in his "Theory of the Earth".[89] These both confirmed the once fluid/fused state of granite. Beddoes then noted that such basalts included trapp and whinstone, which passed "by gradations both to the porphyries with which it coincides in appearance, in composition and doubtless also in [fused] origin, and to the hornstein [flint] of Germany".[90] He has found such "siliceous, *semi-granitic*, porphyritic and common whinstone" at Lilleshall Hill, near Shifnal, in Shropshire, where he could trace basalts and granites gradually approaching and changing into each other, as at Fairhead, in northern Ireland. If so, they had to have had a common origin, through fire.

Apart from giving much confirmation of the correctness of Huttonian theory in explaining such, now igneous, rocks, Beddoes' paper was significant in another way: in it, he wrote that "one consequence of these observations is too important to be omitted. They lead us to reject the common division of mountains into primary and secondary."[91] Here we have Beddoes' confirmation of what Francois Ellenberger has rightly called Hutton's most major geological contribution.

Hutton was the first who dared to affirm that the "Primitive" [or Primary] terrains, far from being an association of rocks formed right at the start, in conditions entirely different from those of the modern world, were merely ordinary ancient sediments, some melted, others transformed. The entire nineteenth century exploited this great notion.[92] Beddoes too had realised its great significance.

Sarjeant and Harvey, who wrongly regard Beddoes as "a visitor to Shropshire", claim this paper was "surely one of the earliest to suggest a metamorphic origin for granites, but is not even mentioned in any of the standard histories of geology".[93] This was not in fact the case, as a thought-provoking paper had appeared in 1969, written by Cyril Smith.[94] This discussed ideas on crystallization in the eighteenth century. Here Smith noted that although Hutton does not mention Thomas Beddoes, his [Hutton's] mature ideas were probably stimulated by the latter's [1791 Royal Society] paper . . . which was an astute discussion of crystallization . . . which was badly needed in geological thinking at the time. While Beddoes had warned that his ideas did not test the igneous hypothesis any more than the opposite, aqueous, one, he did believe that both granite and basalt had arisen from the injection of a fused mass raising, rending and shivering the incumbent strata, while its heat hardened them into laminated stone.

Good analyses of the Neptunist (water) vs. Plutonist (fire) debate are now available from Fritscher[95] and Newcomb.[96] A significant means by which this 1791 Beddoes publication was later transmitted throughout Europe, and perhaps beyond, came in the *Bibliothèque Britannique*. This was a Swiss publication in which Horace Benedict de Saussure (1740–1799) strongly criticized Beddoes' 1791 memoir, on 22 March 1798, after an equally strong criticism of Hutton's whole *Theory of the Earth* had been published in that journal in 1797. This last was a virulent refutation of the whole of Huttonian theory by Jean-André Deluc. All this then provoked an outraged response from John Playfair, after Hutton's death, to the editor of the journal, Marc-Auguste Pictet, dated 29 December 1797. Playfair pointed out that Hutton's final published theory (1795) had not even reached Geneva when De Luc's review had been written.[97]

In his January 1791 paper, Beddoes had wondered

> whether the basaltes, like the Rowley ragstone, proceeds southward by such interruptions till it join the Elvin or whinstone [basalt], and granite of Devonshire and Cornwall, where I imagine they may be found incorporated, and I wish for an opportunity to examine.[98]

In a letter dated 5 May 1791, Beddoes had written to Giddy, asking him to find in London mineral specimens of schoerl (tourmaline) and of amber for his forthcoming Oxford lectures on the natural history of the Earth; in another of 16 May 1791[99] he asks "if Mr Hawkins could give me a specimen of true Wernerian Wakke (or Wacke), I should be glad." Wacke was a "rock-like clay, formed by the decomposition of basalts *in situ*". It was a German miners' word introduced to geology by Abraham Werner of the Freiburg Academy.[100] Both letters were addressed to "Giddy, c/o Mr Hawkins, Chandos Street, Cavendish Square, London". This was John Hawkins, Giddy's neighbour in Cornwall. It seems certain that Giddy had by now introduced Beddoes to Hawkins, who was soon to help Beddoes plan his next geological expedition to Cornwall.

A political Cornish interlude, August–September 1791

In the summer and early autumn of 1791, Beddoes was able to made a significant visit to Cornwall, to see if he could garner any Hutton-supporting geological facts. Of the tour's geological aspects, we learn more from the letters of several Cornish mineralogists[101] and Cambridge University's John Hailstone. A major bone of contention in Cornwall then, as elsewhere, was the battle between Neptunists, who believed (with Abraham Werner in Germany), that water had played the crucial role in the formation of most rocks, against Vulcanists, like Beddoes, who believed that fire and heat had played as dominant a role.

The first we hear from this debate on this tour is in a letter of 26 August [1791] from mineralogist Rev. William Gregor[102] to Giddy's friend, John Hawkins, who had just returned from a day trip with Giddy and Beddoes.

Hawkins had been the first English student to attend Werner's Bergakademie in Freiberg, in 1786,[103] and had there become a confirmed Neptunist, or Wernerian. Gregor wrote how Beddoes "is a great stickler for the cause of Fire in the formation of Loads [productive mineral veins], basalt &c. He is certainly a very ingenious man & the world is indebted to him for some very useful publications".[104] Hawkins immediately wrote to the mineralogist Philip Rashleigh to report how

> Professor Hailstone, who sides with me in referring the general strata of the earth to an aquatic formation, met with the Fire-spouting Mineralogist Dr. Beddoes at Mr. Giddy's. Mr. Edwards,[105] who received these gentlemen at the Copper House [at Hayle] wrote me a witticism of Dr. Smith of Pembroke,[106] one of the party, "that they had Hailstones & Coals of Fire" . . . A pupil of Werner's, young Watt of Birmingham, intends to publish in the next volume of the *Manchester Memoirs* an impartial statement of the two doctrines [fire versus water] of the formation of basalt. Although I know that he is become a convert to the Igneous Theory, we may expect something much [more] satisfactory upon this subject than what is contained in Dr. Beddoes paper.[107]

This introduces us to two more protagonists in this, then current and highly political, debate on the origin of basalt. The first was John Hailstone, Woodwardian professor at Cambridge University since 1788.[108] He too was soon to be a student at Abraham Werner's Freiberg Bergakademie, in 1792,[109] and had clearly already become an enthusiastic Wernerian, or "watery", geologist. He had just made a separate visit to Cornish mines, before Beddoes and his party, in July–August 1791. Hailstone soon got involved in this geological controversy and afterwards, in September 1791, wrote to Hawkins, of how

> the story of my conversion at Cooks Kitchen is as you observe a trick of the Fire-spouters. My observations during 5 hours that I was underground at that work were of an opposite tendency. Dr. B[eddoes] is much attached to Hutton's theory & it was the report current of your Countrymen that he was come to prove their country made by Fire & that my Errand was to shew it was made by Water. He fancies he sees a confirmation of his ideas in the Copper Gossans [decayed outcrop veins] many of which seem to have undergone the Fire. . . . Dr. B. says that he has imitations of all sorts of Stones from Granite to Flint from the Iron works in Shropshire so like the originals that they cannot be distinguished. Every kind of blue Rock & especially your Freestone he calls by the comprehensive name of Basaltes. You may suppose we had some differences in opinion upon these subjects but not withstanding I passed some days very pleasantly in his company. There is a Freedom of Thinking about him which inspires one with esteem.[110]

The other protagonist was James Watt junior, whose father's steam engines were now revolutionising both Cornish mine outputs and stimulating coal-less Cornish ingenuity in attempts to improve on such highly taxed Midlands technology, since coal was a highly expensive fuel in Cornwall. On 12 August 1791 mineralogist John Wedgwood, one of Beddoes' more enthusiastic supporters,[111] had written to Hawkins:

> I am much obliged to you for your present of Volcanic fossils [clearly again rocks] which arrived very safe & were very acceptable to me. Young Watt was with us at Etruria [Staffordshire] when they came & was much pleased with them after inspecting them. Tho' a pupil of Werner's, he is become a convert to the igneous formation of Basalt. He tells me he means in the next *Manchester Memoir* to give a Statement of the two doctrines extracted from the works of ye different supporters of each system. He will afterwards leave every one to form their own opinion on the subject without any bias from his reasonings for he will give none. If this is executed with judgement & impartiality it will be a very acceptable present to all lovers of mineralogy for such a thing has long been wanted.[112]

It was clearly James Watt junior[113] who was now to set to work on the origin of Basalt. He had been inspired, in part, by Beddoes' own paper to the Royal Society read in January 1791. But Watt junior was then, like Beddoes,[114] an enthusiastic republican and supporter of the French Revolution. This was to cause both of them great problems.[115] Watt junior had been based in Manchester when his two papers were read to the Manchester Society, and he was still there at the time of the Joseph Priestley riots in Birmingham, in July 1791. In March 1792 Watt escaped on a "business trip" to France during which, his political views being well known, he was named by Edmund Burke in Parliament as a known Jacobin in April 1792. These proved dangerous times in Paris as well, so Watt junior moved secretly on to Naples in September 1792. As Robinson says, "there were only two choices [open to him] – to keep silent or emigrate."[116] Watt chose the first and so was, after fleeing France, ultimately able to return quietly to England in 1794. But this meant that James junior was meanwhile unable to continue publishing his scientific work in Britain. His geological work would be taken up by his younger half-brother Gregory Watt, who was in turn later to recommend Humphry Davy to Beddoes in 1797. Gregory had taken an early interest in geology, but he took up only the problem of the origin of basalt, with some enthusiasm, after his own return from continental travel, in 1804.[117]

But this Beddoes-Giddy Cornish tour had had other aspects, both humanitarian, as well as scientific objectives.[118] An anecdotal description of it was later given by the Cornish physician John Ayrton Paris in 1831.[119] A young Devon lawyer friend of the Cornish historian, and Anti-Jacobin writer, Richard Polwhele had also written of Beddoes' tour, after Beddoes had left.

This friend was the later Exeter-based barrister and solicitor John Jones (1768–1821), who was then studying law at the Inner Temple in London,[120] and who was later elected a Fellow of the Society of Antiquaries of London, in January 1795, when he had moved to study at the Middle Temple. On 19 November 1791, Jones had written to Polwhele:

> In my journey from Exeter to Bath I had the company of a very silent gentleman, whom I discovered, after my arrival in town [London], to have been Dr. Beddoes. I am really surprised at the pains he took to conceal himself; for, observing an Italian book in his hands (Strange's account of some volcano in Italy),[121] I asked him some questions about it, to which he answered, that he knew nothing of the subject, and that he carried the book for some friend of his. In the book were some loose plates belonging in Sheldon's "*Patella*".[122] I spoke of Sheldon, but he made no answer. I find since that Sheldon gave him the book. At Cullumpton, I afterwards read some pages of this book and intimated that I had heard a Doctor Beddoes, a naturalist, had been to explore Cornwall, but the Doctor was still silent, and, though we were *tête à tête* for an hour before we arrived at Bath, and although we passed the evening together at Pickwick's [Bath hotel], and slept in the same room, yet I had not the least idea who he was till I reached town [London].[123]

We can see from this how well known both Beddoes' scientific, and his highly political, reputations were established. Beddoes wrote amusingly of this very same occasion, in November 1791 letters to Davies Giddy, his accompanist during their geological tour in Cornwall:

> At breakfast, I had the gratification of hearing an account of myself, *incognito*. A young man, a templar I think [clearly Jones], said I was gone to town with Sir C. Hawkins,[124] that I had discovered three volcanoes in Cornwall, and was to explore Devonshire next summer. A lady asked if this was Dr. B[eddoes] and if the author of this intelligence knew him. He replied in the negative. She added neither did she; "but I have heard, excepting what he may know about fossils [clearly a word still used here in the old sense of anything dug up] and such out of the way things, that he is perfectly stupid and incurably heterodox. Besides he is so fat and short that he might almost do for a show." At first, I encouraged the conversation, as supposing that my appearance would ill correspond to the grave and dignified idea of a professor. But now, I was afraid of being detected, especially as I had Strange on Basaltes in my hand, the plates of which they had been admiring. However it passed off & I heard a good deal more news of myself.[125]

Clearly Beddoes' "heterodox activities" here were both well known, already seen as highly political, and so were to be held against him. In confirmation, another friend wrote to Polwhele in 1791 of Beddoes. This was Rev. John Collins, Shakespearean scholar and latterly rector of Ledbury in Herefordshire. From David's Hill, Exeter, he wrote:

> to caution you against entering into any formal engagement with the Doctor and Professor, who addressed you from the London or Oxford Inn (I forget which) when I was with you [in Exeter], at least till you know something more of him than either of us did at that time. . . . I learn he does not pass for any great things there [Oxford], where he is best known. And his journey into Cornwall was visibly as much in search of *gold* as any other baser metals, by his making a sort of trading voyage of it, doling out prescriptions for guineas, professionally, and professedly.[126]

The medical problem here had been Beddoes' highly political publication, *Considerations on Infirmaries*, dated Tredea, 1 September 1791, which he had published whilst in Cornwall.[127] This, clearly equally controversially, had urged the formation of a series of small house infirmaries there. Beddoes wrote of this to Giddy in January 1792:

> As to the unfavourable opinion entertained at Oxford with regard to my sentiments on hospitals, you will readily believe that anticipated disprobation falls light. You know as well as I do, the orbit in which Oxford heads move. I suppose one might trace a chain of ideas from the French Revolution to doubts concerning the extensive usefullness of hospitals, and one might venture to foretell that neither the one nor the other would be well received in the house adjacent to the divinity school or the tower of St Angelo. Establishments would be equally respected in both.[128]

As a finale to this interlude, Richard Polwhele, later a reviewer for the *Anti Jacobin*, and a political opponent of Beddoes, added a note to his own 1797 *History of Devonshire*:

> The basis of the county . . . is limestone, schist or shillet [shale], lava in the different forms of granite, basaltes and whinstone, see Dr. Beddoes, in the 81st vol. of the *Phil. Transactions*. His Hypothesis will account for all the irregularities that are to be found in any part of the earth." For this and many other notes on my little sketch, I am indebted to the ingenious Mr. Jackson of Exeter.[129] What this gentleman modestly calls Dr. Beddoes Hypothesis was, many years ago, his own. All his friends at Exeter can attest the fact: and for the honor of Devonshire, I must vindicate Mr. Jackson's fame, without attempting to detract from Dr. Beddoes reputation. Mr. J. indeed, is one of those chosen few who have the firmness to think for themselves. And we are obliged to him for many striking originalities in philology and science.[130]

Clearly Polwhele had earlier asked Beddoes to help him with this Devonshire project. In an undated letter Beddoes later wrote to Giddy that he had been asked by Polwhele "to correct a manuscript, but finds this difficult, because he disagrees".[131] Clearly this concerned their differing views, both political and scientific, on the geology of Devonshire.[132] The Beddoes-Giddy tour to Cornwall thus had both personal, as well as scientific, objectives.[133]

Davies Giddy, in a letter to Beddoes in Oxford, written in June 1792, had been able to send him details of some geological investigations he had since made for John Hawkins, regarding how an elvan (the Cornish name for a mass of intrusive rocks, filling up a fissure and sometimes rising from the original strata as a dike, now known to have had an igneous origin) had intersected a copper lode (or vein). He had also made limited investigations for Hawkins into how, and when, the blasting of rocks had been introduced into Cornwall. His letter ends, "Your boxes [of geological specimens collected in autumn 1791 for Beddoes' next geological lecture course at Oxford] have been sent off at last this very day."[134] Beddoes' last word on Hawkins himself came in a letter to Giddy: "Hawkins is here. I think more highly of him than I was led to do by the first impression of his manners. He has still the same intolerance in mineralogy. It pervades the whole Wernerian School."[135]

Beddoes had been unable to convince Hawkins and, as Paris rightly says, had been equally unable to find as much evidence as he had hoped to prove the igneous origins of many relevant Cornish rocks. All this may be why he also abandoned the only geological example of his many translation projects. Eddy has noted how, in the 1790s, "Beddoes had begun to read Werner's work in the *Bergmännisches Journal*."[136] In fact Beddoes had hoped to do more. A letter from Hailstone, now at Hamburg, to Hawkins, dated 14 September 1791, where Hailstone was on his slow route to Freiberg to study with Werner, had noted "I am sorry Dr. Beddoes has relinquished the translation of Werner's work. Was it owing to his favourite Prejudices [in favour of Hutton] that he became disgusted?"[137] In further confirmation, Beddoes replied to Hawkins, from Shifnal on 16 September [1792]. He apologised for some previous misunderstanding and agreed that it had been Hawkins who had asked him to translate more than only parts of Werner's book. Beddoes now stated that it was only "particular circumstances which had suggested a wish to lay the design [of translating Werner] aside". He ended "When and where am I to return your Werner [the copy of the book which Beddoes was to have translated]?"[138]

These "particular circumstances" must surely be connected to the sudden appearance of Beddoes' name on a Government list of "disaffected and seditious persons" on 28 July 1792, which also meant Beddoes had lost all hope that he might ever occupy a chair at Oxford.[139] This loss in turn inspired Beddoes' highly political *Reasons for believing the friends of liberty in France not to be the authors or abettors of the crimes committed in that country*, dated Shifnal, 9 October 1792.

Geology at Bristol

Beddoes now left Oxford and moved to Bristol. His last geological contributions during this period were two papers read to the Manchester Literary and Philosophical Society. The first was slight, dated Bristol Hot Wells, 17 October 1793, and read 29 November. It related to a "blue [chemical] precipitate" that he had earlier observed in "a green, glassy mass" at the bottom of one of Reynolds' smelting furnaces at Ketley, Shropshire. These furnaces had earlier inspired Beddoes' thoughts on the significance of Hutton's theory on the power of fire in past geological activity.

The second paper was much more geological. It was entitled "Some observations on the Flints of Chalk-Beds", and read that same day.[140] It concerned observations Beddoes had earlier made in Chalk quarries whilst at Oxford, and especially those near Henley, which he had "examined with great attention". He had wanted to understand how these had formed. The perplexing origin of flints had long been one of the big puzzles in Huttonian theory,[141] and Hutton's idea that these must have been somehow injected into Chalk by subterraneous heat was soon questioned. Beddoes had earlier corresponded with Erasmus Darwin on this very problem. Darwin, writing late in 1788, had discussed the origin of coal, which he thought was non-marine.[142] Beddoes' reply discussed his own view that coal had at least originated in water and added,

> I design a paper for the Royal Society on the formation of flint, in which I hope to render it probable that it has been produced from subterraneous heat: but why flint should be found in its present position I cannot comprehend. Dr. Hutton's supposition of its having been injected into chalk and wood does not assist my comprehension; and yet such as it is, each piece has derived its form from a state of fusion.[143]

Beddoes' 1793 paper was clearly the remains of this planned Royal Society paper on flints, which he must have decided to send elsewhere. Beddoes too remained quite unable to frame any adequate hypothesis for the formation of flint, and so he sent only his earlier observations on the subject to Manchester. He still concluded that from their "glassy texture and fracture", flints must once have been fused, but he could not "agree with Dr. Hutton that flints were [then] spouted into the body of the chalk from subterraneous fires." Beddoes later wanted to send a copy of this Manchester paper to Josiah Wedgwood junior in October 1794, noting how "it will not, I know, be well received by you philosophers of the Wernerian [i.e. watery] school."[144] The complex history of the origin of flint, which long remained a geological mystery, has since at least been discussed, in a paper revealing the contribution of a completely forgotten Wiltshire collector, Henry Shorto III.[145]

Beddoes needed to earn a new living, so it is no wonder that the final page of *The Bristol and Hotwell Guide*, in 1793, contained among recent "occurrences" a notice that "amongst the medical advantages at the Hotwells must be mentioned the extensive apparatus constructed by Dr. Beddoes for infusing into the lungs of sick persons factitious air of any degree of purity that may be required. This attempt to reach the source of consumptive and asthmatic diseases, will, it is hoped, for the benefit of mankind, be attended with success."[146] Beddoes' attentions now moved almost entirely into medicine, and little more is heard of any of his geological interests for another five years.

Geology is renewed

Interest by Beddoes in Bristol seems to have been aroused again in 1798. In April he was giving chemical lectures there and wrote to Giddy that now "I think to add the geological" ones.[147] In November, shortly after Humphry Davy had joined Beddoes, Davy wrote to Giddy:

> I suppose you have not heard of the discovery of the native *sulphate of strontian* in England. I shall perhaps surprise you by stating that we have it in large quantities here. . . . We opened a fine vein of it about a fortnight ago at the Old Passage near the mouth of the Severn.[148]

In 1791 Thomas Charles Hope in Edinburgh had started investigation of a rare mineral that had been found at Strontian in Scotland.[149] This soon stimulated an enthusiastic search for similar minerals, and, at Bristol, a dispute as to by whom, and how, such had been discovered there. This was to start a small local mining industry.[150]

Dr. George Smith Gibbes of Bath wrote to *Nicholson's Journal* early in 1799[151] to announce that the sulphate of strontian had now been found near Sodbury and that Gibbes had been able to confirm this by the analysis of specimens given him by Rev. Benjamin Richardson.[152]

This immediately encouraged Beddoes' assistant, William Clayfield,[153] to give a full account of the veins of sulphate of strontian that he had earlier found in the neighbourhood of Bristol, which was then published in March 1799. He stated that the first such specimen had been found both there by James Webbe Tobin of Bristol, the abolitionist, botanist, zoologist, and radical politician, "about three years since" [thus in 1796], but had then been mistaken for sulphate of barytes, "and at Redland, a short time before that". Then, in June 1797, Clayfield found the same material at Aust Passage, which Gibbes had then analysed. Clayfield described this outcrop, how the mineral occurred here, and how it had been assessed. Clayfield noted that at Aust, "a variety was observed there about three months since by Mr. Deriabin (inspector of the Russian mines) who was immediately struck with its resemblance to a substance of the same nature found in Pennsylvania." Thus is Andrei Fedorovich Deriabin,[154] who had become another student at Werner's Bergakademie in Freiberg in 1793, and later became Director of the St. Petersburg Mining

Institute from 1811 to 1817),[155] and who was then on a visit to England. Clayfield added that "Mr. Bright has, since the first discovery, furnished me with specimens of another variety from the neighbourhood of Ham Green, near Bristol."[156] This was the Unitarian, Richard Bright, a Bristol banker, merchant, and another major geological collector.[157]

Beddoes added an editorial note to this, in which he showed how complicated a story these discoveries had been, and noted other localities at which this mineral had been found. These included another Deriabin record of its occurrence, in a coal pit near Dumfries. Beddoes next added how, earlier,

> While on a professional journey [he had been] struck at Keswick, with another specimen in Mr. Hutton's collection,[158] labelled *striated gypsum*. It is an exceedingly beautiful white sulphate of strontian, from Alston. From Mr. Hutton, I have also received a blueish specimen . . ., which I took to be sulphate of strontian, but Mr. Clayfield finds to be barites. . . . Many specimens, supposed to be baritic, will doubtless . . . prove to be strontianitic. But the distinction will require nice inspection, even for those most versed in the external characters of fossils [i.e. still of rocks – on which Abraham Werner's system of mineral recognition then depended].[159]

We can confirm that Beddoes' visit to Hutton's museum was before late July 1800. Edward Adolphus Seymour and his brother Webb had then visited it, when they were told that "Dr. Biddors [sic] had said that the Sattin Spar from Aldstone Moor contained Strontianite."[160]

Beddoes next wrote Nicholson a further letter, dated 24 March 1799 explaining how Gibbes' material, which he had got from Benjamin Richardson, had been obtained.[161] Beddoes' note read:

> In the collection of the Rev. Mr. Richardson, at Bath, a friend of mine (Mr. Notcutt), instructed by Mr. W. Clayfield's specimens, pointed out some sulphate of strontian. Mr. Richardson [then] gave a piece to Dr. Gibbes in January last [1799]. I had exhibited this substance, as found in other places near Bristol, to a large audience, nine months before, viz in Spring, 1798, and had sent specimens to Mr. W. Henry,[162] who communicated the fact to the Philosophical Society, at Manchester. Dr. Gibbes does not, undoubtedly, mean to claim the discovery. He knew last year that Mr. Clayfield was analysing the fossil. Hundreds of persons might have anticipated Mr. C. in announcing the fact. It is distinctly noticed in the *Appendix* to the *Monthly Review*, vol. xxv. p. 580, June 1780.[163]

This introduces another forgotten Bristol mineralogist, the dissenter William Russell Notcutt, FLS elected 1796.[164] He had taught chemistry at the dissenting academy, New College, Hackney, from that same year. By 1799 he was in Bristol, but he soon set off on an exploring expedition to Surinam, in South America, where, after only six days in the country, he succumbed to yellow fever on 25 April 1800.[165]

Beddoes and metallurgy

At the same time (January 1791) as he had sent his granite paper to Banks for reading to, and publication by, the Royal Society in London, Beddoes had also sent two notes on the manufacture of iron. These were "an Account of some Appearances attending the Conversion of cast into malleable Iron", and "further Observations on the Process for converting cast into malleable Iron".[166] Beddoes' metallurgical work was done with Reynolds and has been discussed by the late John R. Brown.[167] He noted Reynolds' "willingness to experiment with the latest scientific ideas", inspired by his friendship with Beddoes. But Brown considered that Beddoes' attempts to explain the chemistry behind the puddling process in these Royal Society papers were unsuccessful, as the chemistry of gas-metal reactions was not fully known at this time.

Then, on 6 December 1799, Reynolds took out his only patent, "for preparing iron for its conversion into steel".[168] This initiative is significant as Reynolds was normally against the whole concept of patenting, which was still then causing so many problems with the improvement of the steam engine.[169] Reynolds' attitudes to this industrial secrecy were recorded by the American mineralogist Thomas P. Smith while travelling in Europe in 1801–1802.[170] Smith had asked Reynolds when he had visited him in Shropshire, at what point he wished to keep secret the processes that he there employed. He recorded Reynolds' reply (here translated from the original in French): "I have no secrets, and I wish no-one to have any where the success of humanity is concerned."[171] Smith then commented how such frank generosity in giving any similar response was so rare, but he also asked what advantages would there then be for the progress of the arts and the sciences, and for the satisfaction of those who cultivate them.[172] The reason why Reynolds patented this only process was presumably because of the war-torn state of Europe in 1799.

Brown claims that the idea that Reynolds was then trying to improve steel is unlikely. "It is more likely that Reynolds wanted to provide a quality of wrought iron suitable to be converted into steel, and so break into the market occupied by Swedish ironmakers."[173] Brown quoted figures published by John Randall in 1880,[174] and correctly suggested that the idea of adding manganese had come from Beddoes, via the Swedish chemist Torbern Bergman. Eric Svedenstierna, a Swedish industrial spy, who was not allowed to see Reynolds iron-making experiments when he visited them in 1803, had already noted Smith's earlier description of Ketley, adding that "Mr. Raynolds had set up several large-scale experiments to make steel with an addition of manganese."[175] Trinder has confirmed that steel was made at Ketley "by a process . . . that involved adding 40lb of manganese to a charge of 270lb of iron in a reverberatory furnace" and how in 1799 "John Wilkinson had expressed delight that Reynolds had succeeded in making steel from iron smelted in Britain, rather than Swedish iron."[176]

There is evidence from the *Select Committee on Artizans and Machinery* of 1824 that these trials had had only limited success. The engineer Peter Ewart was asked about the comparative merits of English and Swedish iron for conversion into steel. He replied that he had "been a good deal connected with the making of steel . . ., and had seen many attempts, for upwards of 30 years [up to 1824] to make good steel from English iron, but that all had failed and never made good steel during that time". The only exception he recorded was that "there was as good steel made 30 years ago [as in 1824] by Mr. Reynolds of Ketley, with the assistance of Dr. Beddoes, as there has been since." But when asked about these experiments to use manganese in converting English iron into steel, he reported that "they were repeatedly tried at Ketley, but not successfully."[177]

Whatever their outcome, there is no doubt that those who later improved English steel were convinced of the importance of these Reynolds-Beddoes experiments from 1799. In 1839, Josiah Marshall Heath's patent to add manganese in the casting of steel ran into trouble because of "prior use", citing this Reynolds patent.[178] Henry Bessemer's autobiography later included a whole chapter on "Manganese in Steel Making", in which he pointed out that the addition of manganese, whether as the metal or the oxide, had become public property ever since Reynolds' patent of 1799.[179]

Beddoes' last geology with stratigrapher William Smith (1769–1839)

Beddoes' interests in geology were again aroused in 1802. He was then consulted medically by his old friend William Reynolds, who stayed at the York Hotel, Clifton. Here Reynolds wrote an undated note to William Smith (1769–1839), the English pioneer of mineral prospecting and stratigraphy:

> William Reynolds of Coalbrookdale, with his respectful compliments to Mr. Smith, informs Mr. S. that he had the pleasure of spending one day last week with Mr. Davies of Longleat,[180] who strongly recommended W.R. to call on Mr. Smith to see his projected mineralogical Map: and, as from circumstances, W.R. had it not in his power to call on his return, he hopes it will not be inconvenient to Mr. Smith to shew it to him this evening & proposes to wait on Mr. S., with a friend [Beddoes], between 5 & 6 o'clock this present Sunday.[181]

On the verso is a copy of Smith hand-written reply:

> Mr. Smith's Compliments to Mr. R. and will be very happy to show him and his Friend anything that lies in his power at any time this evening as may be convenient.
>
> Trim Bridge, Bath, August 8 1802.[182]

Later, in June 1818, when Smith was in dire financial straights, he issued a lithographed *Statement of his Claims*, regarding his work on the geology of England.[183] In this he noted how his

> Discovery [of the stratification] and the confirmation thereof by numerous Organised Fossils . . . was fully explained at Bath to the late Mr. Wm. Reynolds . . . who [then] showed a Copy of my original account [of the strata round Bath] drawn up in 1799, which, he said "was not secret, but to his knowledge had been widely circulated." Dr Beddoes, and several other Scientific Gentlemen were of the Party.

John Phillips, Smith's nephew, later confirmed the date of this visit as 8 August 1802, and added, of Reynolds' copy of Smith's *Table of strata* drawn up in manuscript in 1799, that copies of it, had, to Reynolds' knowledge, "been sent to the East and West Indies",[184] Reynolds also duly subscribed to Smith's intended *Geological Map* but was long dead (as so many other subscribers) by the time it appeared in 1815. Smith's surviving notes on stratigraphy, still contain this manuscript note: "Bath, August 8 1802, Dr Beddoes says there are Nodules of Granite to be found in all parts of England."[185]

Another connection of Beddoes with the new stratigraphic geology had come from his work as a medical practitioner. In January 1804, the draper and mercer, William Cunnington, later the archaeologist and geologist of Heytesbury, Wiltshire, wrote to a friend explaining how his new scientific interests had arisen. He noted

> it was only two years ago that I first saw Old Sarum, when Doctors Fothergill[186] and Beddoes told me I must ride out or die, I preferred the former, and, thank God, that though poorly, I am yet alive. Should (but God forbid) your health ever be the same like mine, may you find a similar stimulus.[187]

This medical instruction became the driving force behind Cunnington's scientific work,[188] and it would be instructive to learn of other field scientists so advised by their doctors.

When the Geological Society of London was founded in November 1807, those then elected as original Honorary Members included William Clayfield, with Beddoes later elected on 1 April 1808, when proposed by Humphry Davy,[189] in the year of Beddoes' death. Such delay may well have again been due to the continuing politics of the time, which certainly caused Beddoes problems with his mysterious Bristol Philosophical Institute. The first we hear of this is between 1798 and 1804, in letters from Beddoes to Giddy, when Beddoes himself was planning such an Institute, but which he then had to abandon.[190]

> The Bristol historian John Evans (1774–1828) wrote in 1816, how among its institutions, Bristol *had* a Philosophical Society, of which one

> of the objects seems to have been the dissemination of science by means of public lectures. Of the history of this Society, the author knows nothing, and among its transactions the only remarkable circumstance of which he has heard is, that the late Dr. Beddoes was refused the honour of ranking among its members on account, as it is said, of his opinions in regard to religion.[191]

A subsequent, anonymous, writer to the *Bristol Mercury* in 1824 discussed the Bristol Institution for the Advancement of Science, Literature and the Arts, founded in 1823, which made major contributions to science, and especially to geology. At this former Society a toast had at last been held "to the memory of the late Mr. Beddoes".[192] Another writer in 1833 confirmed how

> About thirty years ago [1803] a literary and scientific institution was established in Bristol, and it bid fair to prosper, until Dr. Beddoes was objected to as a member by the Ultra Tories, principally consisting of clergymen of the Established Church, on account of his political principles. This wild conduct disgusted the men of science – they resigned, and the institution was broken up.[193]

The Bristol Society referred to here by Evans was probably that founded in 1807 or 1808 (it had to have been, for Beddoes to have been refused membership, since he died on Christmas Eve, 1808). Michael Neve has noted that this had existed from "about 1809, and met in St Augustine's Place and then 1 Trinity Street".[194] Michael Fryer,[195] the mathematician from Newcastle-on-Tyne, was its first secretary, and a lecturer "for many years", and his 1808 printed *Syllabus of a Course of Lectures on Natural Philosophy to be delivered in the Philosophical Society's rooms in Bristol*[196] seems to have been its first printed production. In October 1810, Fryer was still its secretary, and proposed to publish *A General History of Mathematics* in three volumes,[197] but this failed to find enough subscribers and the Society had become defunct by 1816.

Beddoes' final geological project

Stock noted,

> as mineralogical researches were among Beddoes' earliest, so were they among the latest objects of his pursuit. In the last year of his life he was engaged in an extensive investigation of the different materials made use of in the construction or repair of the public roads. Large parcels of unopened specimens, which had been collected from various parts of England, were found in his laboratory. To what purpose he had intended to apply the information then collected does not appear.[198]

Stock may have been puzzled, but it seems certain this was another philanthropic gesture by Beddoes towards the road-building pioneer John Loudon McAdam (1755–1836),[199] who had moved permanently to Bristol in 1801.[200] Beddoes had clearly remained a philanthropist to his end.

Conclusion

In 1792 Beddoes issued *Alexander's Expedition*,[201] a critique of the East India Company's policy in India, which was published anonymously. Beddoes may well have been reacting directly to the trial of Warren Hastings, which ran from 1788 to 1795.[202] Hastings had been the first Governor-General of Bengal. This was after Tom Paine's *Rights of Man* had been published in 1791, at that point becoming the best-selling book in the history of publishing. By May 1791, 50,000 copies had been sold and there was an estimated sale in Britain alone of between 400,000 and 500,000, within 10 years.[203]

It is a pity Beddoes' book had instead to be published under such circumstances.[204] In his fourth Appendix, on the "Antiquity of the Hindoos", Beddoes had written how

> In the first place, the system of subterranean Nature [i.e. Geology], which is beginning to be understood, and which exhibits, as well as the system of the heavens, an arrangement worthy of admiration, proves the earth to have existed for millions of years, perhaps of ages.

This extraordinarily prescient comment certainly exceeded anything that his aged mentor James Hutton managed to write, who noted only there was "no prospect of a beginning, or of any end" to the earth's antiquity. But Beddoes then continued:

> For I cannot scruple to apply a rule, similar to one of Newton's rules for philosophising, to this subject and to *take it for granted* [emphasis added] that the same causes operate in the same manner and in the same time now, as they ever did. Secondly, nothing in art opposes this result from nature.[205]

This espousal of apparent Lyellian thinking predates the completion of Charles Lyell's classic *Principles of Geology* (1830–1833) by more than forty years. Lyell's subtitle had explained his was an "attempt to explain the former changes of the Earth's surface by references to causes now in operation", in words which echoed, but knew nothing of, Beddoes'. We can only wish, in view of these words, that Beddoes had been able to make a greater commitment to geological matters, during his multifaceted life.

The contrast with John Kidd who became, what Beddoes never could, namely first Aldrichian Professor of Chemistry at Oxford University in 1803, is extraordinary. In 1815, Kidd wrote that

> nothing like probability of any high order has been yet attained in geological reasoning; from all these considerations we may at least be convinced, that the science of geology is at present so completely in its infancy as to render hopeless any attempt at successful generalization, and may therefore be induced to persevere with patience in the accumulation of useful facts.[206]

But Beddoes', and also Thomson's, geological contributions had been recognized in their own lifetimes. In April 1804, the Scot Thomas Charles Hope, who had followed in both their footsteps as a medical student at Edinburgh University, where he was by then Professor of chemistry, wrote to Gregory Watt

> For though the hint of Beddoes, and the more correct views of Dr. Thomson preceded Sir James Hall's decisive paper,[207] they are the Emanations of Huttonism. Both B[eddoes] and T[homson] were in Edinb[urgh] when Dr. Hutton's *Theory* was brought into view and both knew that Dr. H. ascribed the stony manure of the fossils [i.e. rocks] which he imagines to have been fluid, to their Crystallization.[208]

Despite this, the geological contributions of both have been too long forgotten. The strife-torn circumstances of both their lives and publications have too long obscured Beddoes' and Thomson's geological contributions.[209] Beddoes deserves recognition as the first English Huttonian.

Notes

1 D. E. Allen, *Naturalists and Society: The Culture of Natural History in Britain, 1700–1900* (Aldershot: Ashgate, Variorum Series, 2001), Chapter XIV, 203.
2 H. S. Torrens, *The Practice of British Geology, 1750–1850* (Aldershot: Ashgate, Variorum Series, 2002), Chapter IX, 112, and Chapter X.
3 F. F. Cartwright, *The English Pioneers of Anaesthesia (Beddoes, Davy, and Hickman)* (Bristol: Wright, 1952), 161–162.
4 Roy Porter, "Taking Histories, Medical Lives: Thomas Beddoes and Biography", in *Telling Lives in Science: Essays on Scientific Biography*, eds. M. Shortland and Richard Yeo (Cambridge and New York: Cambridge University Press, 1996), 219.
5 See Appendix 1, this volume.
6 Bodleian Library, MSS Dep. C 135/2.
7 Edinburgh University Library class lists, pressmark DC.1.18; M. D. Eddy, "The University of Edinburgh Natural History Class Lists 1782–1800", *Archives of Natural History*, **30** (2003), 97–117.
8 Anon., *Laws of the Society Instituted at Edinburgh [1782] for the Investigation of Natural History* (Edinburgh, The Society: 1788), and Douglas Mckie, "Some

Notes on a Students' Scientific Society in Eighteenth Century Edinburgh", *Science Progress*, **49** (1961), 228–241.

9 H. S. Torrens, "Another Quaker 'Lunatick': The Worcester Origins of Jonathan Stokes Junior (1754–1831)", *Journal of the Friends Historical Society*, **61** (2010), 196–220.

10 On Stock, see Appendix 1.

11 D. E. Allen, "James Edward Smith and the Natural History Society of Edinburgh", *Journal of the Society for the Bibliography of Natural History*, **8** (1978), 483–493, & for the 1800 takeover, see Henry Peter Brougham, *The Life and Times of Henry Lord Brougham* (Edinburgh: Blackwood, 1871), I, 550–551.

12 J. Kendall, "The First Chemical Society, the First Chemical Journal and the Chemical Revolution", *Proceedings of the Royal Society of Edinburgh*, **63A** (January, 1952), 346–358, 385–400 and Kendall, "Some Eighteenth-century Chemical Societies", *Endeavour*, **1** (1942), 106–109.

13 See Chapter 1.

14 Gwen Averley, "The 'Social Chemists': English Chemical Societies in the Eighteenth Century and Early Nineteenth Century", *Ambix*, **33** (1986), 99–128.

15 James Gray, *History of the Royal Medical Society 1737–1937* (Edinburgh: Edinburgh University Press, 1952), 63 and 316.

16 *St James Chronicle or the British Evening Post*, 1 December 1785.

17 Stock (1811), 9.

18 *Morning Chronicle and London Advertiser*, 31 August 1785.

19 *Lancaster Gazette*, 4 November 1820.

20 Burke, *Landed Gentry* (1972), 405.

21 EUL, Class lists DC.1.18, and 8.

22 See https://sites.google.com/site/historicaltimelines/home/botanical-cottage-leith-walk under 7 February 1782.

23 Bodleian Library, MS Dep. C 135.

24 J. Sinclair, *The Correspondence of . . . Sir John Sinclair, Bart* (London: Colburn & Bentley, 1831), I, 427.

25 Withering, *An Arrangement of British Plants; According to the Latest Improvements of the Linnaean System, to Which Is Prefixed, an Easy Introduction to the Study of Botany* (Birmingham: for three publishers, 1787), I, v and xi.

26 Beddoes to Giddy, 9 January 1792. CRO, DD/DG41.

27 D. Mckie, ed., *Notes from Doctor Black's Lectures on Chemistry, 1767/8* (Wilmslow: ICI Industries, 1966).

28 Beddoes to Black, 6 November 1787 in R. G. W. Anderson and Jean Jones, eds., *The Correspondence of Joseph Black* (Farnham: Ashgate, 2012), II, 920.

29 Stock (1811), 17–18.

30 See Chapter 1.

31 In his 1787/1788 *Theory*, plate 2.

32 Beddoes to Black, 23 February 1788, EUL, MS Gen. 873/III/72; reprinted in CJB, II, 949.

33 Beddoes to Black, 15 April 1791, reprinted in CJB, II, 1122.

34 Beddoes to Giddy, 21 October 1792. CRO DG 41/20.

35 *Jackson's Oxford Journal*, 7 November 1789, 3 (introductory lecture) and 12 December 1789, 3, and see Todd, *Beyond the Blaze*, 21.

36 *Jackson's Oxford Journal*, 23 and 30 October 1790, 3.

37 *Jackson's Oxford Journal*, 28 May and 11 June 1791, 3. Some of the notes for this course still survive in Oxford; see Stansfield, 43

38 A. E. Musson and Eric Robinson, *Science and Technology in the Industrial Revolution* (Manchester: University Press, 1969), 177.

39 Stock (1811), 24.

40 These include notes on coal and its origin, on bitumens, and naphtha, with discussions of a naphtha lake in Sicily and the famous Bitumen Well at Broseley, and other tar springs in Shropshire. Another note dealt with strata and the production of coal, with a further fragment on theories of the Earth. Bodley, MS Dep. C 134–135.
41 *English Chronicle* 15–17 June 1790.
42 Ibid.
43 *True Briton* 16 July, and *Courier* 19 July 1799.
44 CRO, DD/DG 41/46–47. Bushby was then was at Lincoln's Inn, London. Bushby's letter was later forwarded to Giddy, who had also been a Pembroke student with Bushby, in January 1792, DD/DG 41. There were two Bushbys of this name. This one went out to India as a barrister, on the ship Walpole, early in 1798 (*Bell's Weekly Messenger*, 18 March 1798) and a cousin of the same name (1798–1835 – for whom see Burke, *Landed Gentry* [1952], 330–331), who was also in India, by 1815, and who must have been named after Beddoes' student (who had died there early in 1799).
45 Taylor's letter does not survive.
46 CRO, DD/DG 41/46-47.
47 Beddoes to Giddy, 11 April 1792, CRO DG 41/12.
48 *TRIA* (1790), 3, 50–54.
49 *Philosophical Transactions*, **80** (1790), 73–100.
50 For Freiberg students, see the list published by C. G. Gottschalk, "Verzeichniss Derer, welche seit Eröffnung der Bergakademie und bis Schluss des ersten Säculum's auf ihr studirt haben", in *Festschrift zum hundertjährigen Jubiläum der Königl. Sächs. Bergakademie zu Freiberg am 30 Juli 1866* (Dresden: Meinhold & Söhne, 1867), s. 234, student no. 349.
51 White Watson letters, bound in one MS vol., Bateman archive, Sheffield City Museum.
52 For this company, and Mills and Weaver's roles in it, see D. B. Smith, *A Georgian Gent: The Life and Times of Charles Roe & Co. of Macclesfield* (Ashbourne: Landmark, 2005).
53 H. R. Vassall, *Further Memoirs of the Whig Party 1807–1821 with Some Miscellaneous Reminiscences* (London: Murray, 1905), 324–325.
54 This story first emerged in the notice of Beddoes given in Alexander Chalmers, *General Biographical Dictionary* (London: Nichols & Son, 1812), vols. 3–4, and reprinted in *Gentleman's Magazine*, **82** (July, 1812), 42–44.
55 D. R. Oldroyd, "Earth, Water, Ice and Fire: Two Hundred Years of Geological Research in the English Lake District", *Geological Society of London, Memoir*, **25** (2003), 4–6.
56 *Whitehall Evening Post* (September, 1791), 20–22. The actual author of these observations was not revealed until the second edition of these letters was published in 1792, as *Remarks made in a tour from London to the Lakes . . . originally published in the Whitehall Evening Post* as being by the Lakeland-born itinerant science lecturer Adam Walker (1730 or 31–1821). See *ODNB*.
57 Roy Porter, *The Making of Geology: Earth Science in Britain 1660–1815* (Cambridge: University Press, 1977), 162–163.
58 D. Macleane, *A History of Pembroke College, Oxford* (Oxford: Historical Society Publications, 1897), XXX, 391–393.
59 Beddoes to Banks, 30 May [1792], in Neil Chambers, ed., *The Scientific Correspondence of Joseph Banks, 1765–1820* (London: Pickering and Chatto, 2007), IV, 126–127.
60 See Chapters 1 and 3, this volume.
61 For more on Thomson, see also H. S. Torrens, "The Geological Work of Gregory Watt, His Travels with William Maclure in Italy (1801–1802), and Watt's

'Proto-geological' Map of Italy (1804)", in *The Origins of Geology in Italy*, eds. G. B. Vai and W. G. E. Caldwell, *Geological Society of America Special Paper*, **411** (2006), 179–197.
62 Stock (1811), 24.
63 Beddoes to Black, 2 April 1789, EUL, reprinted in CJB, II, 1011.
64 Thomson to George Paton, National Library of Scotland Adv. MS 29.5.8(ii) f. 80, quoted in Heather Ewing, *The Lost World of James Smithson: Science, Revolution, and the Birth of the Smithsonian* (New York: Bloomsbury, 2007), 146.
65 Oxford University Archives, WPy/24/2, 45–51.
66 Blagden to Banks, 26/9/1790, BL Add MS 33272.
67 Oxford University Archives, NEP/Subtus/Reg Bk, 437–438.
68 See also H. Ewing, *The Lost World of James Smithson. Science, Revolution, and the Birth of the Smithsonian* (London: Bloomsbury, 2007), 145–146. Ewing also notes that Smithson, then Macie (who was another Pembroke College student), "must have been acquainted, in the tiny world of science at Oxford, with . . . Beddoes" (68). This can be confirmed from Davies Giddy's diary (CRO). The entry for 15 June 1791 shows that Giddy then gave a dinner to a party including both Beddoes and Macie.
69 *Considerations*, part 3, vii (June 1795).
70 Vassall, *Further Memoirs*, 340.
71 Reynolds, along with James Keir, was to be one of the two main contributors to Beddoes' career and his Pneumatic Institution. See Chapter 2.
72 His industrial activities are well reviewed by B. Trinder, "William Reynolds: Polymath – a Biographical Strand through the Industrial Revolution", *Industrial Archaeology Review*, **30** (May, 2008), 17–32.
73 H. S. Torrens, "The Reynolds-Anstice Shropshire Geological Collection 1776–1981", *Archives of Natural History*, **10** (1982), 429–441.
74 H. S. Torrens, "300 years of Oil: Mirrored by Developments in the West Midlands", *The British Association Lectures* (London: Geological Society, 1993), 4–8.
75 B. Trinder, *The Industrial Revolution in Shropshire* (Chichester: Phillimore, 1981) and Trinder, *'The Most Extraordinary District in the World': Ironbridge and Coalbrookdale* (Chichester: Phillimore, 1988).
76 CED, 296–297, 326–327. See letters 87–20 and 88–20 with Beddoes' replies to Darwin printed in Stock (1811), Appendix 6.
77 H. S. Torrens, "Erasmus Darwin's Contributions to the Geological Sciences", in *The Genius of Erasmus Darwin*, eds. C. U. M. Smith and R. Arnott (Aldershot: Ashgate, 2005), 259–272.
78 Bodleian Library, MS Dep. C 134–135, see also Torrens, "Reynolds-Anstice Shropshire Geological Collection", 430.
79 Dr. Justus Arnemann (c.1763–1806) was a medical professor at Göttingen, Germany, who had married an English woman and was, in 1789, on tour in Britain. Thomson's letter to George Paton, dated 7 August 1789 (National Library of Scotland Advocates, MSS. 29.5.8) records he had carried a Thomson box, containing fossils, to Black in Edinburgh from Oxford. Arnemann committed suicide in 1806. *Gentleman's Magazine*, **76** (December, 1806), 1177.
80 Beddoes to Black, 21 April 1789, EUL, reprinted in CJB, II, 1011–1012.
81 "Letter copies are in Dawson Turner Collection", 7, 180–190 and 182, *Natural History Museum*, London.
82 *Jackson's Oxford Journal*, 23 October 1790, 2.
83 *Jackson's Oxford Journal*, 30 October 1790, 3.
84 Stock (1811), 28.
85 Houlbrooke (1745–1824) was born at Lichfield, went to Shrewsbury School, and graduated LL.B. at Cambridge in 1769. He then became an Anglican priest,

until 1784 when he left the church and became a Unitarian. He married Mary, daughter of the Midlands ironmaster John Wilkinson, in 1785 at Market Drayton, where he now worked as a creator of water meadows. He next studied chemistry at Edinburgh, where he matriculated in 1790 and soon published a vitriolic anti-slavery tract, *A Short Address to the People of Scotland*, in 1792. By 1795 he was based in Liverpool, where he became highly active, for example as chairman of its Athenaeum and Literary and Philosophical Society. See also Desmond King-Hele, *The Collected Letters of Erasmus Darwin*, 2nd edn. (Cambridge: University Press, 2007), 280.

86 Joseph Reynolds to [Beddoes via Davies Giddy], 26 August 1791, CRO DD/DG 41.

87 G. Y. Craig, D. B McIntyre and C. D. Waterston, *James Hutton's Theory of the Earth: The Lost Drawings* (Edinburgh: Scottish Academic Press, 1978), 30–35.

88 John Strange, "An Account of Two Giants Causeways, or Groups of Prismatic Basaltine Columns, and Other Curious Vulcanic Concretions, in the Venetian State in Italy; with Some Remarks on the Characters of These and Other Similar Bodies, and on the Physical Geography of the Countries in Which They are Found", *Philosophical Transactions*, **65** (1775), 5–47.

89 James Hutton, "Theory of the Earth", *Transactions of the Royal Society of Edinburgh*, **1** (1788), 255–257 and plate 2.

90 Beddoes, "Observations on the Affinity between Basaltes and Granites", *Philosophical Transactions*, **81** (1791), 48–70 at 50.

91 Ibid. at 68.

92 F. Ellenberger, *History of Geology* (Rotterdam: Balkema, 1999), vol. 2, 327–328.

93 W. A. S. Sarjeant and A. P. Harvey, "Uriconian and Longmyndian: A History of the Study of the Pre-Cambrian Rocks of the Welsh Borderland", in *History of Concepts in Pre-Cambrian Geology: Geological Association of Canada Special Paper*, eds. W. O. Kupsch and W. A. S. Sarjeant, **19** (1979), 184–175.

94 Cyril S. Smith, "Porcelain and Plutonism", in *Toward a History of Geology*, ed. C. J. Schneer (Cambridge, MA: MIT Press, 1969), 317–338.

95 B. Fritscher, *Vulkanismusstreit und Geochemie* (Stuttgart: Franz Steiner Verlag, 1991).

96 S. Newcomb, "Contributions of British Experimentalists to the Discipline of Geology 1780–1820", *Proceedings of the American Philosophical Society*, **134** (1990), 161–225.

97 The whole episode is discussed by A. V. Carozzi, "Histoire des sciences de la terre entre 1790 et 1815 vue à travers les documents inédits de la Société de Physique et d'Histoire Naturelle de Genève", *Mémoires de la Société de Physique et d'Histoire Naturelle de Genève*, **45** (1990), 113–119.

98 *Philosophical Transactions* (1791), 85.

99 Beddoes to Giddy, 5 May and 16 May 1791. CRO, DG 41/6 and 15.

100 W. J. Arkell and S. I. Tomkeieff, *English Rock Terms* (Oxford: University Press, 1953), 123.

101 R. J. Cleevely, "The Contributions of a Trio of Cornish Geologists in the Development of 18th Century Mineralogy", *Transactions of the Royal Geological Society of Cornwall*, **22** (2000), 89–120 and Cleevely, "Collecting the New, Rare and Curious: Letters Selected from the Correspondence of the Cornish Mineralogists, Philip Rashleigh, John Hawkins and William Gregor", *Exeter, Devon and Cornwall Record Society, n.s*, **52** (2011), lxxxv.

102 C. M. Bristow and R. J. Cleevely, "Scientific Enquiry in Late 18th Century Cornwall and the Discovery of Titanium", in *The Osseointegration Book: From Calvarium to Calcaneus*, ed. P.-I. Branemark (Berlin: Quintessenz Verlag, 2005), 1–12.

103 Gottschalk, *Verzeichniss*, 231, student 283.
104 Cleevely, *Collecting*, 76.
105 This was John Edwards (1731–1807), the Cornish industrialist and copper smelter at Hayle. See T. R. Harris "John Edwards (1731–1807), Cornish Industrialist", *Transactions of the Newcomen Society*, **23** (1) (1942), 13–22. His third son, Richard Edwards (1770–1827), had entered Beddoes' Pembroke College, where they must have been in close contact. Edwards first graduated as Bachelor of Civil Law in 1795. But then, clearly inspired by Beddoes, he moved to medicine, graduating M.D. (Oxford) in 1802. He was also an accomplished chemist, on which he lectured at St. Bartholomew's Hospital, before returning to Falmouth where his died in 1827; see W. Munk, *Roll of the Royal College of Physicians*, **3** (1878), 11–12. In 1799 Richard had published a forgotten item of Wernerian geology, probably again inspired by Beddoes. This was his *Werner's Mineral System* (seven pages, but without date or publisher). The North of England Institute of Mining and Mechanical Engineers Library copy, John Bell, *Collection* I, 1 carries this note: "translated by Richard Edwards Esq, of Cornwall, at Edinburgh, January 1799" (probably while he was studying medicine there, as Beddoes before him, having abandoned law). The John Rylands Library copy carries an MSS note in French claiming this translation had not followed the original manuscript given him to be translated. It had many changes, several of which were necessary, with many others that were useless. This again suggests involvement with Beddoes, who had little sympathy with Wernerian methods.
106 This seems certain to have been Dr. John Smith or Smyth (c.1744–1809), who had matriculated at Pembroke in November 1761 and gained his M.A. in 1769, then as a Doctor of Divinity (1796) he became Master of Pembroke College from 1796–1809. He died at Exeter on 19 October 1809. See *European Magazine*, **56** (November 1809), 398 and *Gentleman's Magazine*, **79** (2) (November, 1809), 1079.
107 Cleevely (2011), 78–79.
108 J. W. Clark and T. M. Hughes, *The Life and Letters of the Reverend Adam Sedgwick* (Cambridge: University Press, 1890), **1**, 195–197.
109 Gottschalk, *Verzeichniss*, 235, student 385.
110 Cleevely (2011), 73–74.
111 See Chapter 5, this volume.
112 Cleevely (2011), 75.
113 Francis Steer, followed by Cleevely (2011, 49–50), had wrongly claimed this was Gregory Watt (1777–1804), but who was then a pre-undergraduate teenager, who never published in the *Manchester Memoirs*. But his elder brother James Watt junior had furthermore become another English student of Werner's, at the Freiberg Bergakademie, in 1787. Watt's own copy of Werner's book of 1774, *Von der aeusserlichen Kennzeichen der Fossilien* [On the external characters of minerals] annotated, in a German hand, and signed "James Watt, 1787" survives. See British Geological Survey library, Keyworth, Nottingham, 1/658.
114 Levere, "Dr. Thomas Beddoes at Oxford: Radical Politics in 1788–1793 and the Fate of the Regius Chair in Chemistry", *Ambix*, **28** (1981), 61–69.
115 Cf. Eric Robinson, "An English Jacobin: James Watt, Junior 1769–1848", *Cambridge Historical Journal*, **11** (1955), 349–355.
116 Robinson, "An English Jacobin: James Watt, Junior 1769–1848."
117 Torrens (2006), see note 61.
118 Beddoes' humanitarian purpose is a theme that runs right through his career and throughout this book. See Stock (1811), 32–33 and Todd, *Beyond the Blaze*, 25–26.
119 John Ayrton Paris, *The Life of Sir Humphry Davy* (London: Colburn and Bentley, 1831), **1**, 49–52.

120 See obituaries in *Jackson's Oxford Journal* (17 November 1821) and *Bristol Mercury* (24 November, 1821).
121 A natural philosopher, and Venetian diplomat, John Strange had published in Italian on the volcanic phenomena of the Venetian region in 1778, and of Vesuvius in 1789. The book Jones here refers is that which Strange published in 1778, *De' Monti Colonnari e d'altri Fenomeni Vulcanici dello stato Veneto* (Milano: Morelli) with plates, see L. Ciancio, ed., *A Calendar of the Correspondence of John Strange, FRS (1732–1799)* (London: Wellcome Institute, 1995), but in which no letters from Beddoes are recorded. This actual book was lot 2089 at Beddoes' library sale in November 1809. See CBL and Chapter 4, this volume.
122 For Beddoes' study of anatomy in London under Sheldon, see Chapter 2. Sheldon's *Essay on the Fracture of the Patella* was published in 1789, after he had moved to live in Exeter, where Beddoes must have been able to renew their acquaintance, and be given this book.
123 R. Polwhele, *Traditions and Recollections* (London: Nichols, 1826), 1, 262–263.
124 This is Sir Christopher Hawkins (1758–1829), the political elder brother of John Hawkins. He was then M.P. for Mitchell, in Cornwall. See R. G. Thorne, *The House of Commons, 1790–1820* (London: Secker and Warburg, 1986), 4, 166–170.
125 Beddoes to Giddy, 4 November, 1791, CRO DG 41/41; see also Stock (1811), 35–36.
126 R. Polwhele, *Traditions and Recollections; Domestic, Clerical, and Literary* (London: J. Nichols, 1826), I, 263–264.
127 Stock (1811), Appendix 3, xxv-xxviii, and see Todd, *Beyond the Blaze*, 25–26.
128 Beddoes to Giddy, 9 January 1791, CRO, DD/DG 41/1&2. The tower of St. Angelo in Rome that had been the refuge of popes when Rome was under siege was later to host popes in its prisons.
129 This is the Exeter musician, painter, and author William Jackson (1730–1803; see *ODNB*), whose contributions to, and views on, geology were in complete opposition to Beddoes. From his own notes on his "Theory of the Earth" (see *Gainsborough's House Review* (1996–1997), 102), we know Jackson was merely an opinionated old-style theorist, who saw much reason in the Plutonic [i.e. Beddoes'] System, and more in the Neptunian. This reasoning produced a System of my own, which I gave some slight hint of in *The Thirty Letters*, and which in fact was the original of the Neptuniam Theory that has since been adopted by so many Naturalists.It is clear that such impossible claims by Jackson had misled Polwhele into thinking that Jackson had somehow preceded Beddoes. Jackson had published, in the third edition of his *Thirty Letters on various subjects* (London: Cadell and Evans, 1795), his Letter 24, '*Omnia ab ovo, perhaps not so*'. This was his only published contribution to this subject. Jackson was soon a founding member of the Society of Gentlemen at Exeter, which published a volume of *Essays* in 1796. Certainly the local physician Bartholomew Parr (1750–1810; see *ODNB*) had there discussed James Hutton's "Theory of Rain," which was later published in *Essays of a Society of Gentlemen at Exeter* (London: Cadell and Davies, 1796) as *Essay 17*, but there seems no other suggestion that this Society had been exercised by any such geological matters.
130 Richard Polwhele, *History of Devonshire* (London: Cadell, Johnson and Dilly, 1797), I, 48.
131 CRO, DD DG 40/2.
132 Giddy to Beddoes, 7 January, 1795, CRO DD DG 42/4.
133 Stock (1811), 32–33, and see Todd, *Beyond the Blaze*, 25–26.
134 Giddy to Beddoes, 12 June [1792], Bodleian Library, MS Dep. C 134/2.

135 Beddoes to Giddy, 18 July 1792, CRO, DG 41/14.
136 M. D. Eddy, *The Language of Mineralogy. John Walker, Chemistry and the Edinburgh Medical School* (Farnham: Ashgate, 2008), 130.
137 Cleevely, "The Contributions of a Trio", 100.
138 CRO, Hawkins papers J3/290. It is not clear which of Werner's books Beddoes had planned to translate, but it would certainly have been the first to have appeared in English. See A. O. Ospovat, "Wernerian Influences in the Geological Literature of Western Europe", in *Abraham Gottlob Werner* (Leipzig: VEB Deutscher Verlag, 1967), 221–222.
139 Levere, "Dr. Thomas Beddoes at Oxford" (1981), 61–69, 65.
140 *Memoirs of the Literary and Philosophical Society of Manchester*, 4 (1796), 302–303 and 303–310.
141 See Patsy A. Gerstner, "The Reaction to James Hutton's Use of Heat as a Geological Agent", *British Journal for the History of Science*, 5 (1971), 353–362 and Rachel Laudan, "The Problem of Consolidation in the Huttonian Tradition", *Lychnos. Annual of the Swedish History of Science Society 1977–1978* (1979), 195–206.
142 CED, letter 88–20, 326–327.
143 Stock (1811), Appendix 6, xxxviii–xxxix.
144 Beddoes to Josiah Wedgwood junior, via Tom Wedgwood, 31 October, 1794. WMB, W/582.
145 The flint problem is discussed in H. S. Torrens, "A Wiltshire Pioneer in Geology and His Legacy – Henry Shorto III (1776–1864), Cutler and Fossil Collector of Salisbury", *Wiltshire Archaeological and Natural History Magazine*, 83 (1990), 170–189 at 176–178.
146 Edward Shiercliff, *The Bristol and Hotwell Guide: Containing an Historical Account of the Ancient and Present State of That Opulent City; Also of the Hotwell; The Nature, Properties and Effects of Its Medicinal Water. . .* (Bristol: Bulgin and Rosser, 1793), 146.
147 Beddoes to Giddy, 14 April 1798. CRO, DG 42/2.
148 Davy to Giddy, 12 November 1798 in S. Hutton, *Bristol and Its Famous Associations* (Bristol: Arrowsmith, 1907), 272; and the useful survey of Beddoes' scientific work in Bristol in V. A. Eyles "Scientific Activity in the Bristol Region in the Past", in *Bristol and its Adjoining Counties*, eds. C. M. MacInnes and W. F. Whittard (Bristol: British Association, 1955), 123–143 at 138–142.
149 M. E. Weeks, *Discovery of the Elements* (Easton, PA: Journal of Chemical Education, 1948), 302–305.
150 Eyles, "Scientific Activity", 137.
151 *Nicholson's Journal*, 2 (1799), 535–536.
152 This was the early friend and encourager of William Smith (1739–1839; see *ODNB*), who was then based at nearby Bath. See Torrens, *Practice of British Geology*, Chapter III, 242–244.
153 See Chapter 5.
154 See also Torrens, *Practice of British Geology*, Chapter 4 for more on Deriabin.
155 Gottschalk, "Verzeichniss Derer", 235, student 395; and I. I. Safranovskiy and D. P. Grigor'ev, "Abraham Gottlob Werner in der Geschichte der russischen Mineralogie und Geologie" (1967), 191–203, in *Abraham Gottlob Werner: Gedenkschrift aus Anlass der Wiederkehr seines Todestags nach 150 Jahren am 30 Juni 1967, Freiberger Forschungshefte C 223* (Leipzig: 1967) and A. J. Rieber, "The Rise of Engineers in Russia", *Cahiers du monde russe et sovietique*, 31 (1990), 539–569, at 545–546.
156 Clayfield's piece was first published in T. Beddoes, *Contributions to Physical and Medical Knowledge* (Bristol: Longman and Rees, 1799), 429–444, and

then, somewhat improved, in *Nicholson's Journal*, **3** (April, 1799), 36–39 and 39–41 (the last being another note by Beddoes).

157 See R. M. Kark and D. T. Moore, "The Life, Work, and Geological Collections of Richard Bright M.D. (1788–1858); with a Note on the Collections of Other Members of the Family", *Archives of Natural History*, **10** (1981), 119–151, at 138–142 and plate 2. William Buckland published an informative obituary in *Proceedings of the Geological Society of London*, **3** (1841), 520–522.

158 This is Thomas Hutton (c.1745–1831) who was a mineral dealer there; see M. P. Cooper, *Robbing the Sparry Garniture: A 200 Year History of British Mineral Dealers 1750–1950* (Tucson: Mineralogical Record, 2006), 195–197. Hutton also acted as a tourist guide and kept a competitive museum there; see P. C. D. Brears, "Commercial Museums of Eighteenth-Century Cumbria: The Crosthwaite, Hutton and Todhunter Collections", *Journal of the History of Collections*, **4** (1992), 107–126, at 118–122. The competitive nature of museums here is best revealed by the sadly anonymous author of *Observations, Chiefly Lithological* (London: T. Ostell, 1804), who devoted separate chapters to the two museums there (see 38–47, Mr. Hutton's Museum "with better arranged minerals"). Samuel Parkes later described a single, stupendous, crystal of 29 lbs weight, still supposedly of sulphate of barytes, from Cross Fell, which he had studied in Hutton's museum in 1815. See Parkes, *Chemical Essays: Principally Relating to the Arts and Manufactures of the British Dominions* (London: Baldwin, Cradock & Joy), V, 101–102.

159 *Nicholson's Journal*, **3** (1799), 39.

160 G. Ramsden, *Correspondence of Two Brothers* (London: Longmans Green, 1906), 29–33.

161 *Nicholson's Journal*, **3** (1799), 41–42.

162 This was William Henry (1774–1836); see W. V. Farrar, *Chemistry and the Chemical Industry in the 19th Century: The Henrys of Manchester and Other Studies*, eds. Richard L. Hills and W. H. Brock (Aldershot: Variorum, 1997).

163 Beddoes had written of sulphate of strontium, "we have been informed that this curious fossil is believed to exist far more abundantly near Bristol than in Pennsylvania" in his review of Martin Klaproth's *Beytraege zur Chemischen Kenntniss der Mineralkoerper*, II, for April 1798.

164 See J. Birtwhistle, "An Ardent Naturalist – William Russell Notcutt (1774–1800) Chemist, Botanist and One of Davy's First Nitrous Oxide Subjects", *The History of Anaesthesia Society Proceedings*, **28** (2000), 60–65.

165 *Ipswich Journal*, **2** (August, 1800).

166 *Philosophical Transactions of the Royal Society*, **81** (1791), read 24 March, 1791), 173–181; and **82** (1792), read 3 May 1792, 257–269.

167 On Reynolds the ironmaster, see N. Clarke, ed., *William Reynolds 1758–1803, Proceedings of the Events Held in June 2003 to Commemorate His Life and Achievements* (Telford: Wrekin Local Studies Forum, 2004).

168 William Reynolds (6 December 1799), *Preparing iron for the conversion thereof into steel*, British patent number 2363 (London: Patent Office).

169 H. S. Torrens, "New Light on the Hornblower and Winwood Compound Steam Engine", *Journal of the Trevithick Society*, **9** (1982), 21–41 at 23.

170 Torrens (2006), 179–197 at 185–186, see note 61.

171 T. P. Smith, "An 11 [1802], Sur la fabrication du fer et de l'acier avec la houille, d'apres les procédés de M. William Reynolds, pratiqués a Coal-brook-dale en Angleterre," *Journal des Mines*, **13** (73) (1802), 52–60.

172 Ibid.

173 On Reynolds the Ironmaster see N. Clarke, ed., *William Reynolds 1758-1803, Proceedings of the Events held in June 2003 to Commemorate his Life and Achievements* ([Telford]: Wrekin Local Studies Forum, 2004).

174 J. Randall, *The History of Madeley* (Madeley, Shropshire: Wrekin Echo Office, 1880), 84.
175 E. T. Svedenstierna, *Svedenstierna's Tour: Great Britain 1802–3, The Travel Diary of an Industrial Spy* (Newton Abbot: David and Charles, 1973).
176 Trinder, *The Industrial Revolution in Shropshire*, 20.
177 *Fourth Report from Select Committee on Artisans and Machinery* (1824), reprinted London: Frank Cass, 1968, 253–254.
178 J. Percy, *Metallurgy: Iron and Steel* (London: Murray, 1864), 843.
179 H. Bessemer, *Sir Henry Bessemer FRS: An Autobiography* (London: Offices of 'Engineering', 1905), Chapter 18; and see also Philip W. Bishop, *The Beginnings of Cheap Steel* (Contributions from the Museum of History and Technology, United States National Museum), Bulletin 218, Paper 3 (1959), 29–47.
180 Thomas Davis (1747–1809), here referenced as "Mr. Davies," the steward of the Marquis of Bath's estates at Longleat was one of Smith's earliest, and truest, friends and supporters; see *ODNB* and op. cit. 179.
181 H. S. Torrens, 2003, "Timeless Order: William Smith and the Search for Raw Materials, 1800–1820", in *The Age of the Earth: From 4004 B.C. to A.D. 2002, Geological Society of London Special Publications 190*, ed. Cherry L. E. Lewis and Simon J. Knell (London: The Geological Society, 2001), 61–83, reprinted in John Phillips, *Memoirs of William Smith, LL.D.* (London: Murray, 1844), republished by the Bath Royal Literary and Scientific Institution (Bath: BRLSI, 2003), 153–192.
182 William Smith archive, L. R. Cox Catalogue, Folder "Letters Received by Smith 1799–1803", Oxford University Museum of Natural History.
183 Reproduced by T. Sheppard, "William Smith: His Maps and Memoirs", *Proceedings of the Yorkshire Geological Society*, **19** (1920), 75–253, at 181–182 and 214–220, esp. 216.
184 Torrens, reprint of *Memoirs of William Smith* (2003), 43.
185 William Smith Archive, L. R. Cox Catalogue, folder "Granite etc.", Oxford University Museum of Natural History.
186 This is Anthony Fothergill, then in practice in Bath. Cunnington must have been one of his last patients there before he emigrated to Philadelphia in 1803.
187 R. H. Cunnington, *From Antiquary to Archaeologist: a biography of William Cunnington (1754-1810)* (Aylesbury: Shire Publications, 1975), 4-5.
188 Ibid.
189 H. B. Woodward, *The History of the Geological Society of London* (London: The Society, 1907), 268–270.
190 Stansfield, 189–190.
191 Rev. John Evans, *The History of Bristol, Civil and Ecclesiatical* (Bristol: Sheppard, 1816), **2**, 366–367.
192 *Bristol Mercury*, 1 March 1824.
193 *Bristol Mercury*, 23 February, 1833.
194 Michael Neve, "Science in a Commercial City: Bristol 1820–60", in *Metropolis and Province, Science in British Culture 1790–1850*, eds. I. Inkster and J. Morrell (London: Hutchinson, 1983), 179–204 at 184.
195 *Gentleman's Magazine*, n.s. **21** (April, 1844), 434.
196 Published Bristol: S. Dennis, 1808. (copy in H. S. Torrens collection).
197 *Monthly Magazine*, **30** (1810), 268.
198 Stock (1811), 409.
199 W. J. Reader, *Macadam: The McAdam Family and the Turnpike Roads 1786–1861* (London: Heinemann, 1980).
200 H. S. Torrens, *Men of Iron: The History of the McArthur Group* (Bristol: The Company, 1984), 14.

201 Beddoes, *Alexander's Expedition down the Hydaspes & the Indus to the Indian Ocean* (London: John Murray, 1792).
202 The subsequent impeachment in the House of Commons was orchestrated by MPs, including the remarkable trio of Edmund Burke, Charles James Fox, and Richard Brinsley Sheridan.
203 J. Keane, *Tom Paine: A Political Life* (London: Bloomsbury, 1995), 307. On Paine, see Chapter 3 below.
204 No copy reached the Bodleian Library, which had so much which concerned Beddoes, until digital copies became available. Perhaps it was too controversial for orthodox Oxford.
205 *Alexander's Expedition*, 72.
206 John Kidd, *Geological Essay on the Imperfect Evidence in support of a Theory of the Earth, Deducible Either from its General Structure or from the Changes Produced on Its Surface By the Operation of Existing Causes* (Oxford: Oxford University Press, 1815), 269. Kidd was writing more than a decade after Beddoes had asserted the unimaginably great age of the earth.
207 This refers to James Hall, "Experiments on Whinstone and Lava", read to the Royal Society of Edinburgh in March and June 1798, but only finally published in *Transactions of the Royal Society of Edinburgh*, 5 (1805), 43–75.
208 Hope to Gregory Watt, 24 April, 1804. BCL, C 2/12/44.
209 Thomson had arrived in Naples by April 1792. Here he settled as Guglielmo Thomson. In one of the most volcanic areas of Europe, Thomson now turned his attention to this, and built up fine collections of volcanic specimens, published classifications of volcanic productions, and named several new minerals. Thomson wrote many articles for the *Giornale Litterario di Napoli* and for other German, French, Swiss, and Italian journals, too few of which were translated into English (in the *Monthly Magazine*). In 1797 Thomson also contemplated publishing his own *Theory of the Earth*, clearly inspired by the appearance of Hutton's in 1795, but this never appeared, in a country now terrified by war and by Napoleon. In November 1806, Thomson died in Palermo, at the age of forty-six, two years younger than Beddoes. Thomson had become an important, too long forgotten, point of contact between Continental and British scientists. Thomson's main collections – with 10,000 specimens and 823 books, including a rare copy of Hutton's 1785 abstract of his *Theory of the Earth* – arrived at Edinburgh University in 1808. Here, many books (but few of the specimens) survive, and support an endowment from his estate for a Thomsonian lecturer there. Thomson best deserves the title of Britain's first true vulcanologist.

3 A Jacobin cloven foot

Larry Stewart

In 1789, the Continent showed the early signs of upheaval. An alarm had been sounded. But, in England, Rev. Samuel Parr wrote of the mayhem at the hands of the "bunting, beggarly, brass-making, brazen-faced, brazen-hearted, blackguard, booby Birmingham mob" that could easily arise. Such was then an expectation of workers quick to riot. His comment was only published much later, but written with a common sense of the mob, after much opportunity to reflect on what could too easily be induced by a pamphlet or sermon. Privately, in Birmingham, the industrialists Boulton and Watt also remarked that "Any appeal to the labouring part of the people is always to be dreaded."[1] The flames of the summer of 1791 consumed the library, the laboratory, and the notes of the many experimental trials of the celebrated natural philosopher, chemist and Unitarian minister Joseph Priestley. Over several days, the rampage engulfed the homes, the businesses and chapels of those presumed hostile to Church and King.[2] In an age where riot could break an engine or shake a throne, plenty still had cause for worry. Yet, the Priestley riot was not so unusual, except for its illustrious target. As in many towns throughout the entire century, mobs emerged when the atmosphere was already combustible from a mix of economic, political, and religious tensions. In addition to Edmund Burke a year before, others had seen it coming. In the 1790s, it took little to light British tinder. For those who viewed Birmingham from a distance, who once inclined to hope in the fall of the Bastille exactly two years earlier, there was a frightening parallel to be drawn with French mobs let loose across the Channel. Since 1789 not a few had hailed a rising republicanism, or even before in an earlier adventure in America, ever since acclaimed by some and reviled by others.[3]

Enlightened enthusiasm of many kinds was widely refracted throughout political debate. An alchemy of republicanism, new science, and rebellion erupted in a repeated string of metaphors of explosion and destruction – from Edmund Burke to his nemesis Priestley. Even so, in the elixir of British politics, some still saw more hope than threat even in the ashes left by mobs. Among these were the oft-feared Tom Paine, the naive democrat James Watt, junior, son of the celebrated engineer, and the remarkable Dr. Thomas Beddoes, who saw in chemistry the promise of a wider public good. In an age of riot, this triumvirate was far from alone. Inevitably, cries for reform from some implied rebellion to others.[4]

The attitudes of Thomas Beddoes to chemistry and medicine were formed in this crucible.[5] So too did Birmingham's political laboratory affect many responses to the medical innovations he was to promote among his vast array of correspondents. It is our purpose to suggest that Beddoes' remarkably un-cowed commentary on matters social, political, scientific, and medical were deeply connected in, and complicated by, a promise some believed lay at the heart of the new chemical combinations of the late eighteenth century. As Jan Golinksi has suggested, trials in chemistry attracted many kinds of innovators.[6] Priestley had long argued the diffusion of both experimental and natural philosophy meant the erosion of social barriers and the collapse of what Tom Paine would decry as a lamentable "catalogue of impossibilities" to which America was one answer. A seductive message to some, this was a "tocsin" to others. As we also shall see, not all chemical practitioners shared such proto-democratic visions.

Thomas Beddoes proved to be more than the "little fat Democrat" as his father-in-law Richard Lovell Edgeworth once described him.[7] For his first biographer, John Stock, the image was more than relevant. Beddoes' political notions would surely complicate his own life to no end. Even the apparent utility of his chemistry would be compromised by his public enthusiasm for democracy or by his deep, and sometimes bitter, aversion to the oppressive assertions of the unenlightened and the well-connected. As Paine would put it, these were times that would surely try any man's soul. In an age of incessant alarm, whether about growing aversion to monarchy or strained by war with France, Beddoes to his own detriment simply refused to be circumspect. Much of what he observed was seen through a political lens. Even before the Birmingham riots, he had already developed unrelenting loathing of the ambitious and scabrous scheming of William Pitt, the younger. In 1791, shortly before the Birmingham debacle Beddoes wrote to Davies Giddy, whom he had known at Oxford, that "Pitt & his Parliament bid fair to turn this whole country into one great court of political inquisition." This was as accurate a prediction as any made by Burke. In the emerging chaos in France both Pitt and Beddoes saw a warning, albeit of a very different kind. Beddoes drew Westminster parallels with the failure of the National Assembly in Paris, which he believed would have to answer "not to France only but to mankind" for its inability to establish decent government.[8] So Pitt could readily be blamed.

Beddoes' notion of governing demanded addressing the disadvantages of the poor and the wilful impositions of privilege. Medicine might stand in contrast to socio-political failure rooted in a corrupt constitution. In 1791 Beddoes' views were then in evolution, reflecting the tensions he saw arising from the bare fortunes of the many compared to the advantages of the few.[9] Over the course of the next decade, this was the theme to which he warmed. As the many lived lives of desperation, it was unremarkable that as a physician he would dismiss social elevation as any prophylactic against disease or contamination. Indeed, there was no real medical advantage to the rich except the mirage of access to physicians, many of whom Beddoes regarded with utter cynicism. His own practise revealed no such medical preservative, just different origins and tracks for myriad afflictions. Perhaps illness was ultimately the best metaphor for democratic exchange, moving inexorably across social barriers. Even so, wealth appeared a deep obstruction to improvement of the labouring poor. It became Beddoes' mission to confront these political and economic obstacles. This did not make him an enthusiast of rebellion.

Increasingly frequent riots were, more often than not, the reactions of the desperate and undisciplined, aggravated, some asserted, by the evidently rapid circulation of Tom Paine's *Rights of Man*.[10] The alarm of Beddoes' friend, the ironmaster Joseph Reynolds of Ketley, was probably typical. Reynolds once remarked, within weeks of the Birmingham riots, that Paine uncensored would "destroy that subordination which is necessary in all Societies".[11] Such a view was common among early manufacturers who, of course, relied on the very labourers whom they also occasionally feared. The summer of 1791 was such a time. How far would such upheaval spread, throughout the industrial Midlands, beyond to the collieries of Northumberland, the mines of Cornwall, and the restive slums of East London? Who would be threatened? Beddoes heard enough rumours of a rising tide. Indeed, James Watt of Birmingham, who saw the results firsthand, and had seen Priestley driven out, immediately declared himself an enemy to all tumults and specifically of the very republicanism that Priestley appeared to promote.[12] The aversion to the mayhem, especially as it had directly affected such a renowned chemist as Priestley, was widespread – even more worrisome when it became clear that the magistrates of Birmingham did little to dissuade the mob from continuing in their destruction. Even from the Parisian chemists, who had themselves seen recent rampage, came news there was a subscription afoot to assist the Birmingham dissenters.[13]

There was a great deal of soul-searching among those who could easily have become targets. The partners James Watt and Matthew Boulton, who employed hundreds of workmen, were very much shocked by Birmingham rage. Indeed, Watt wrote to Joseph Black in Edinburgh about the "Hellish miscreants" who committed the havoc and had who virtually broken up the Lunar Society. Watt was deeply worried but highly conflicted. Because he was a friend of many who were otherwise sympathetic to the grievances of labourers and mechanics, his analysis betrayed his occasionally ambiguous position in a world that filled him with so much fear. He wrote of Birmingham,

> this town is divided into 2 parties who hate one another mortaly [sic], that the professed Aristocrats are democrats in principle that is encouragers [sic] of the mob; and that the democrats, are those who have

> always contended for a police & good government of the town, therefore are in fact aristocrats at least would have no objections to an aristocracy of which they themselves are members.[14]

The explosion was so severe that many struggled to understand it. Watt thought that one needed to go much deeper than the alleged reforming "folly of the dissenters". Anticipating Malthus, he even saw in the poor laws a seed bed of profligacy, which surely must seem a bit of a stretch even from someone on the verge of panic. Equally worrisome, "Church & King" had clearly proved no bastion of law and order. The churchmen became promoters of the mob if only because the dissenters were on the other side.[15]

Beddoes, for all his democratic sentiments, writing from nearby Shifnall to the northwest, had the same view of "fanaticism" in industrial Birmingham – especially when it was increasingly the case that many of those such as the nailers were in the double risk of unemployment from mechanization and of the seductions of churchmen ready to raise a mob.[16]

Beddoes understood well enough that Burke's *Reflections on the Revolution in France* could be accused of lighting a fuse. Burke as the "Marat of Pitt" was the view of some.[17] Amid British riots, and the furious unravelling in France, Beddoes nonetheless acknowledged the dangers in the collapse of order. He was not immune to the consequences of upheaval, nor was he prepared to ignore the cause. In a lengthy correspondence with Davies Giddy, he analysed the news from France and especially the "evil spirits" he believed had emerged to "haunt France in bodies". Ever hopeful, however, Beddoes still felt assured that the history of Europe had exhibited "the struggle of Truth & Freedom against falsehood and oppression" that would "resist the attempts of their adversaries in a liberal & enlightened age". But there were many dangers on a rising tide. By 1792, his mood was darkened by further outrage in France: "I flattered myself that the tree of despotism was decaying at its roots – But this infernal club of Jacobins with its mad mob will water it with innocent blood."[18] Throughout 1792 and 1793, at every turn, Beddoes was reminded that he could not be optimistic about the conflict between the republican ideals and the sanguinary practise in the streets of Paris. The French were, he wrote to Watt, "wild beasts broke loose" and in Paris bloody were the boulevards.[19] By then Watt had already reached his own conclusions. The match of metaphors was only just beginning, undoubtedly deriving many from his friend Joseph Priestley although Watt actually had far more in common with Burke than commonly appreciated. Watt told Joseph Black that "The rabble of this country are the mine of Gunpowder that will one day blow it up & violent will be the explosion."[20]

The grains of nitre had been laid. What then might emerge from England's struggling trade, what explosions with the widespread scattering of people from modest employment into uncertainty, poor house, or starvation? With every wind from France, there was a new packet of bad news. In England, republic and regeneration were an increasingly hard sell. Beddoes, throughout 1792, was wrenched from hope to despair – although his real concern was not France but Britain and its powder keg. In that summer, when the government took fright and spies helpfully confirmed suspicions, Beddoes took aim not at mobs but at the looming failure of trade in Birmingham, Sheffield, from Lancashire to Yorkshire and Liverpool, viewed from the leafy Bristol suburb of Clifton.[21] Unemployment raised new dangers. Watt, Boulton, and Wedgwood watched in trepidation. Four hundred men alone had been discharged from the building of the Worcester canal, pointing to a larger crisis. James Keir, chemist and manufacturer, told Beddoes that the toy trades of Birmingham "cannot be said to be languishing; they are dead; & perhaps will never revive". Rebellion and fear poisoned trade as much as the prospects of democrats. Even chemical hopes seemingly evaporated. Beddoes resigned as lecturer at Oxford as, "from an hundred curious facts", he had "become eminently & much beyond my importance, odious to Pitt & his gang".[22] In chemistry or medicine, as in manufactures, politics were inescapable.

Democratic miasmas

Manufacturing towns produced swirling fogs as often as they did mobs. Everyday experience of the murky Midlands' airs deepened resentment. Poisoning the atmosphere was then too common, however, little would be done about it. But there were many motives behind the chemical assessment of urban pollution. In 1777, the democratizing chemist Joseph Priestley had requested the inventive Matthew Boulton send to him, from Birmingham, vials of air "as it is actually breathed by the different manufacturers in this kingdom". His instructions were specific: Boulton should

> do this in those places when you expect the air to be the worst, on account of bad fumes, or a number of people working together &c, and not at your place in the country, but in the middle of the town. I should be glad also to have the air of some of your closest streets, and likewise the best in your neighbourhood, noting the state of the weather at the time.

Priestley actually received thirteen green vials, sealed with mercury, but without any explanation. As he evidently sent many such requests, he complained he could not differentiate the source of one vial from the other. This was significant because it was the particular locations that might allow him to link the sinking salubrity of airs to specific conditions. This was the essence of eudiometry.[23]

Even so, as manufacturing towns were rapidly expanding, noxious fumes were simply regarded a cost to the many of the new wealth of the few. Hence, we find the paeans to progress that described the Arcadia of Boulton's manufactory set up at Soho on the edge of the town. To read Rev. Stebbing Shaw, in 1801, Boulton's bounty arose from the joining of "taste and philosophy with manufacture and commerce; and, from the various branches of chemistry".[24] Thus at Soho, 2 miles from the city centre,

> A barren heath has been covered with plenty and population; and, these works, which in their infancy were little known and attended to, now cover several acres, give employment to more than 600 persons, and are said to be the first of their kind in Europe, [this deemed] a recent monument of the effects of trade on population.

The image was not exactly what we might have expected: "A beautiful garden, with wood, lawn, and water . . . covers one side of this hill" where the "worthy proprietor gradually exercised in the adjoining garden, groves, and pleasure grounds . . . [that] render Soho a much-admired scene of picturesque beauty".[25] From Shaw's pulpit, the view was positively bucolic. There was no satanic mill in sight. He had apparently little experience of Birmingham alleys.

Connections between airs and disease were long a matter of dispute – well before the wave of industrialism spread across Birmingham, Manchester, Sheffield, Bradford, Leeds, and a hundred other towns. The new chemical sense of the compound nature of common air at least suggested that ill effects on health could possibly follow, especially in manufacturing sites.[26] Thus, in a 1786 conversation with Priestley, the physician Thomas Percival had discussed a method, once proposed by Watt, for eliminating the vast amount of smoke from fire engines, furnaces, and manufactures then enveloping Manchester. Arising from the rapid expansion of the cotton industry, especially from burning of fabric, Percival described the cloak of haze as "extremely acrimonious and offensive to the lungs . . . so copious, even from a single chimney, as to scatter an shower of soot over a very considerable space". He dared assert the duty of the magistrates of Manchester to protect the health of the town, to force those who controlled the works "to conduct them in a manner as little injurious as possible to the public".[27] Even when employment was plentiful, there was a price. The public consequences were increasingly hard to ignore, especially among the very labourers on whom manufacturers invariably depended. Air-borne illness obstructed production while shortening the lives of labourers and skilled mechanics. Unhealthy conditions, within the mills and around the alleys of endless towns, could at least raise questions about consequences.

The difficulties of labourers were not limited to urban airs. There were other fears attached to the tendencies of the labouring poor. Indeed, within days of Priestley's close escape, James Watt reflected bitterly on the rot of Birmingham's masses. They were, he lamented, "principally composed of blackguards of no property, as illiterate as horses, & as debauched drunken & insolent as any people . . . be . . . who's [sic] intellectual power extends little further than to know who sells or gives the best ale". It was from this anger, as much as any other, that Boulton and Watt opposed any doctrine which "put *any* power into the hands of the lower class of the people."[28] Not everyone would have agreed. To apportion blame was inevitable after Birmingham rage. Even Thomas Walker of Manchester, a brave proponent of reform whose own house would be attacked in the next year, thought mobs "wretched tools of a most unprincipled faction".[29] Yet, this was also the context in which Beddoes promoted his notions of social reform and medical regeneration. He was not blind to the lack of moral suasion over the poor. While not then closely allied with Watt, Beddoes was really not far from his sentiments. But he also saw the

> inability of sermons to convey any just ideas of morality into the minds of the poor of the danger lest they shd. abuse the modern political doctrines as soon as an opportunity offers, in gratifying their selfish passions, merely because they have no just strong conceptions of the simplest truths of morality . . . unless the knowledge of their daily duties was well impressed upon the minds of the common people.[30]

For Beddoes at least, preventive medicine might yet induce morality in atmospheres so politically charged.[31]

As Beddoes well knew, contempt for the calloused hand was palpable among some manufacturers and righteous churchmen. After 1790, these sentiments did not take long to surface amid the decline of trade or on the tide of political panic. Even early enthusiasts of the Revolution in France, like James Watt, junior, found much to fear. Indeed, the younger Watt ultimately fled from Paris in danger of being arrested, where he busied himself as agent on the Continent of Thomas and Richard Walker, the Manchester fustian and calico manufacturers. But that too meant trouble. Thomas Walker was suspect in England for his democratic notions as much as James junior's Parisian and early Jacobin associations were publicly reported in London. From Genoa in 1793, the young Watt wrote a long letter to his father in which he neatly captured the essence of the British news:

> When I saw a man like Mr. Walker indicted for conspiracy to subvert the state upon the sole consideration of a drunken Irishman and every possible means used to ruin his credit & reputation, a Mr. Frost confined to Newgate for daring to expose the corruption of a minister and to assert that all men are equal, a Dr. Priestley driven from his home and narrowly escaping death, because he could not comprehend that three makes one or that a king is necessary in a state, to say nothing of numberless other prosecutions legal & illegal against men of inferior importance, accused of high-treason for offences of a most trivial manner, I am obliged to consider England as a Country from whence liberty of opinion is banished.[32]

The difficulty for democrats was that once the lid of restraint was raised, then brutal events might well eclipse any promise of reform. This was certainly what Burke had predicted, while Pittite repression helped further deepen hostility. Indeed, if a social morality among the poor was as uncertain as Beddoes believed, then an answer to Burke's *Reflections* was made more imperative. As the nineties wore on, Pitt's reputation was eroding under the weight of every crisis, indictment, and trial. Many had celebrated the failures of the French, but others also saw through Pitt's strategy of pursuing the supposed enemies within.[33] There was much to fear, he believed, following "from a general fermentation among the labouring class". But as far as Beddoes was concerned, a "government by corruption" made Britain a more dangerous place rather than a more secure one.[34]

Democrats were thus trapped between riot and repression. Beddoes' analysis may not have been far off the mark, especially as disturbances became more frequent in Pitt's mendacious reign. Mobs ranged everywhere, among the miners of Cornwall, agricultural workers, and industrial labourers in London and the Midlands. From Birmingham to the outskirts of Bristol there seemed a general ferment.[35] From this Beddoes believed a brutal politics would necessarily ensue. It became a common theme. But here too was a laboratory for democratic idealists. James Watt, junior had adopted the same metaphor in France: "As in a political as a vegetable fermentation the scum always come to the surface and in order to get at the clear liquor you must first skim it well."[36] The froth was only the first indicator of a deeper agitation. The most violent disturbances necessarily attracted attention. But Beddoes had a much more sophisticated sense of the failures of political structure. By the turn of the century, his comprehension had tied politics, social forces, and medicine in a way that defined the contented rich as "diseased supporters of a diseased population".[37] Pitt's suppression of reforming voices simply made the lives of desperate labourers that much worse. Beddoes came to believe oligarchy, whether monarchical or even medical, was no answer. Just has he ventured into pneumatic medicine and the cure of endemic diseases like consumption, an end to "medical imposture" might at least be one answer to social desperation.[38]

The politics of reform defined Beddoes' trajectory in chemistry and his contact with medical men. In the wake of Birmingham's raging mob, unleashed at the wish of local worthies, Beddoes wrote one of his earliest and now almost completely forgotten polemics. It was undoubtedly the Midlands' recent excesses that induced him to write on early education, "Particularly that of the Poor". When mobs ran rampant, he concluded that "the fanaticism and brutality of the common people . . . originate in the wretched condition of governments." Inherent evil of the labouring classes was no explanation. His target was also Burke who, as "the Exterminating Angel . . . would purify the earth at a few strokes from that most mischievous vermin, the French and the Philosophers."[39] Beddoes was consistent in asserting that only knowledge could overcome those beliefs that drove bigotry and fuelled riot. Hence, years later, when he was embarking on a course of lectures in Bristol, he told James Watt that chemistry did not recommend itself purely out of practical utility, but "the effect of a number of people receiving agreeable ideas together may be to soften animosity & that there will be thus a chance of preventing some acts of barbarity in the times that I fear are coming."[40]

Beddoes' social medicine

By the time trade was deeply eroded in the European crisis of the 1790s, Beddoes reflected upon the suffering to be addressed. This, in an important measure, was the foundation of an extensive network of connections that he energetically cultivated amongst physicians, chemists, and manufacturers. The notion of a wisp of hope in pneumatic chemistry was hardly unique to Beddoes. It was clear from the time when Priestley had attempted to assess urban miasmas in the 1770s that it might prove possible to identify aerial poisons and even discover ways of reviving polluted airs. Throughout the next two decades the laboratory methods adopted by French chemistry were attracting increasing attention both in manufacturing and in medicine. As though following James Keir's 1789 dictum of the diffusion of philosophical knowledge, the chemists of the Midlands frequently consulted about apparatus.[41] One such was the young William Henry of Manchester, son of the chemist Thomas Henry. In 1790 William was secretary in Manchester to the physician Thomas Percival, where he also encountered John Dalton and the radical John Ferriar. These were important allies, enlisting James Watt in securing pneumatic devices and furnaces in order to answer the requests of local gentlemen for a course of chemical lectures.[42] Watt's own chemical concerns were then widely known, his correspondence extensive, although he also eschewed any ambitions to publish.[43] This is significant. Although he was acknowledged for his chemical trials, it was primarily in Watt's wide network that his experimental expertise and advice circulated. What little became public emerged very largely through Beddoes, who proved one of Watt's most controversial correspondents. Meanwhile, a network had been born amid the drift toward crises in trade and politics, but also in the relief new airs might bring.

By 1790, James Watt was already active in the range of chemical explorations, of acids and airs – likewise with Josiah Wedgwood, Joseph Priestley, Joseph Black, and Thomas Beddoes, all of whom sometimes collaborated.[44] This was partly the consequence of the interest in chemistry among members of the Lunar Society. Thus we also find James Watt, junior, conducting trials on dogs to determine the medical value of terra ponderosa, literally a heavy earth he obtained from the north of England. Its chemical composition had already been the subject of analysis by the Birmingham physician William Withering in 1784. The assessment of the metallic properties also came to involve Boulton. Watt then believed Boulton might possibly open a "new field of chemistry & will oblige theories to be changed" in ways not unrelated to debates over new airs then manufactured from earths in the laboratory.[45] By the 1780s, the senior Watt's own connections to the experimental chemists were already extensive. He was, moreover, engaged in the phlogiston controversy along with Joseph Priestley, for whom, by 1790, Watt was making "muriate oxygene", or a gas extracted from muriate of potash.[46]

It has, of course, been often noted that the James Watt's chemical curiosity intensified with the desperate efforts to save his young daughter Jessy from the ravages of consumption. The story was deeply personal. But, as we might expect, there were rather more complex undercurrents beyond his immediate household.[47] It might be proposed that Jessy's long illness linked Watt's already extensive chemical experience to much promoted notions of pneumatic medicine. Events surely turned for the worse in the spring of 1794, when Jessy died of fever and convulsions in June despite the interventions of the physicians Withering, Erasmus Darwin, and Beddoes. Watt's description made for painful reading.[48] There was little consolation to be had. As he wrote Darwin, he could but turn his mind to

> the subject of medicinal airs, not from any idea that I understand the subject, but because nobody else does & therefore that any hints might by chance be as good as another means. Where the regular Physician professes his ignorance the quack may safely be called in and have fortune suffered to throw the dice." Darwin acknowledged the attempt, having already heard of the pneumatic project from Beddoes.[49]

The notion that a mechanic, even a startlingly successful one like Watt, might advance the course of medicine was not as unusual as might be thought. Indeed, Dr. John Ewart at Bath, who already had experience with pneumatic remedies, told Watt that "any great revolution to perfect the medical art, was never to be expected from physicians."[50]

Beddoes, of course, had already turned his attentions to airs that might be made in a laboratory and applied to infections in the infirmary. These he had long since contemplated, at least since visiting Guyton de Morveau in France in 1787 and the sick wards in Dijon where oxymuriatic acid gas (chlorine) was used as a disinfectant.[51] Within a decade, encouraged by Watt, Beddoes enthused over benefits derived from chemical airs, from what he had then taken to calling "this grand expt. in behalf of humanity".[52] The foundation of Beddoes' optimism lay not entirely in his own medical trials, but in the many enthusiastic reports he received from others.

One difficulty of pneumatic medicine was the lack of any generally acceptable theory that might explain the physiological actions of acids, alkalis, and new airs, especially those thought to reverse the course of disease. Even medics disposed to the use of factitious airs wished for caution. For example, the physician James Carmichael Smyth, very much attentive to contagious diseases such as typhoid fever, sought out chemical cures. Securing an introduction to Black from Matthew Boulton in 1793, and intrigued by acids and alkalis, Smyth expressed hope that Beddoes would not however assume "too splendid a theory".[53] Chemical effects were none too certain, the benefits dependent on testimony often difficult to sustain. But that also was one reason why the involvement of Watt in the pneumatic enterprise was of real consequence. When Watt began to develop a breathing apparatus, in the summer of 1794, Beddoes acknowledged the empirical would be of everyday significance for chemists as well as to the increasing numbers of pneumatic practitioners.[54] Much thus depended upon experiments in the physician's surgery, such as that of William Saunders who lectured at Guy's Hospital on chemistry and pharmacy.[55] As with the involvement of Watt in the broad pneumatic project, just as Ewart had claimed, chemical training would make all the difference. James Carmichael Smyth decided to launch himself into a medical world intensely engaged with Beddoes' expectations and with the "Ingenuity & Acuteness of Mr. Watts [sic]". Just days earlier, Beddoes was telling Giddy what an advantage it was to publish pamphlets with Watt's name attached.[56] From the credibility in such networks, pneumatic medicine must necessarily thrive.

Triangulations – Beddoes, Watt, and Black

With a nation in perpetual crisis, it was difficult to disentangle political, economic, and medical remedies – and Beddoes did not try. The strength of these threads was perhaps recognizable to those most inclined to reform. Nevertheless, this clearly did not mean that radicals had a monopoly on innovation, in either chemistry or in medicine. Watt, for example, heard from the physician John Carmichael of Edinburgh about the rising taste for chemistry in their generation. Improvement might result, in medicine as in manufactures. It was empirical, everyday experience that mattered most – not, Watt insisted, the influence of those "easily led away by theories or novelties".[57] Watt had certainly played his own role. Yet the tours of the public lecturer Thomas Garnett of Glasgow or those who published like Priestley, and ultimately Beddoes, certainly proved essential. By the late 1790s, when Watt and Beddoes collaborated most fully, Watt and his family were deeply engaged with Beddoes' experimental airs. The desperation of daily life had intervened. Watt instructed his younger son Gregory, apparently in the early stages of consumption, to follow Beddoes' care, while Watt the hypochondriac tried self-medication by oxygen for his apparent asthma.[58] Likewise, the surgeon and electrician Miles Partington of London told Watt he believed that extending pneumatic chemistry to medicine demanded diffusing scientific knowledge as widely as possible, especially among medics like himself.[59] By the end of the century, the desire to engage with new airs had become widespread, even to the point of frequent self-experimentation amongst the curious. Matthew Boulton's son, Robinson, hoped to participate in the new adventures, after inhaling airs under the direction of Dr. Carmichael. He was disappointed. His enthusiasm produced no noticeable effect, "either from mismanagement in the preparation of the gas or the insensibility of our nerves".[60] Personal reactions were unreliable.

In the end, what hope there lay in pneumatic medicine was seemingly left to chemists to decide.

A consequence of patients

Throughout the last few months of 1794, hope flourished among the medical men of Birmingham.[61] Even with Watt's name there were, however, obstacles. Perhaps because of Beddoes' reputation, or just from a resistance to innovation, there was a general lament over the obstinacy of established physicians. It is true that surgeons, apothecaries, and those on the fringes of medical orthodoxy, like the surgeon-electrician Miles Partington, seemed more willing to attempt pneumatic ventures. But there may have been more to new airs than the orthodox resisting invention. Beddoes' immediate reaction was that resistance came from those "almost as inveterate against me as their predecessors were against the discovery of the circulation of the blood".[62] Beddoes' martyr complex may have been overblown, a self-serving notion of a solitary voice, howling against the studied unreason of the powerful and well-connected. The fact was that, partly following his own published efforts, numbers of fellow travellers immediately answered his call, their many letters printed by Beddoes for maximum effect. To take but one example, Walter William Capper provided evidence to Beddoes of the efficacy of airs, while his own brother conducted experiments on animals to determine the consequences of inspirations of hydrocarbonate (carbon monoxide) produced in the laboratory.[63] Such trials were essential and reporting them raised hope.

Beyond the laboratory preparations, pure air was best sought by travel, to the sea shore perhaps, or to the Continent where there were rumours of sun. This was the reason for medical expeditions to Portugal in the late eighteenth century. Such were the pulmonary trails of the long-suffering William Withering, M.D., of Birmingham and of the radical Lord Daer whom we will meet later.[64] What was really needed was to bring new airs to the ill and to the disadvantaged many, rather than rely on the testimony of relief in the voyages of the few. As was typical of Beddoes, he did not mince his words in correspondence with Giddy. New airs meant new relief, to escape the clutches of the quacks and the sullen, virtually prescriptive, obstinacy of the physicians. He was intent that news of his pneumatic grail would result in a revolution in medical practice, such that "Apothecaries will not be able to persuade patients to swallow their slops at their present rate."[65] He intended to serve a general good. The difficulty was in proving results that could encourage a broad adoption of pneumatic therapies. In the laboratory or workshop, it was hardly evident that the qualities of many airs could even be assured any more than their problematic administration or effects. Many trials were required. The reasons were ample. With subjects, or patients, risk needed to be immediately assessed as had Beddoes in the seemingly "desperate" case of the wife of James Keir, chemical manufacturer.

But there were issues of application as well. In the case of Watt's daughter Jessy, Beddoes had attributed part of the failure in 1794 to the lack of a proper apparatus. Within weeks, Watt veiled his grief by sending Beddoes not only a new design but also a list of his own trials to prompt others to do more. The first result was uncertain risk. By August, Watt had been trying one inflammable air from charcoal that produced severe vertigo, "when barely smelling it caused a person to fall down in the sleep of forgetfulness, from which he awakened without pain or uneasiness". At one point Watt was so alarmed that he sent his victim off to Beddoes. The lack of effective devices compromised apparent cures even before they could be fairly tried. This, as Beddoes told Josiah Wedgwood, was the immediate intent behind Watt's design of a domestic device that could be used without requiring great chemical knowledge or skill.[66]

There were enormously varied afflictions to be addressed. – and medics did not readily abandon their usual remedies. Complexity also followed when airs were conjoined with more commonly prescribed medicines. For example, Mrs. Keir did not get much relief from airs, and combinations of hydrogen and oxygen in the case of a Mr. Sheppard near Ludlow had little obvious benefit in his asthma. As much as there had seemed new promise in new methods, there was obviously ample uncertainty. Beddoes had seen a case he deemed desperate, of a "typhoid small-poxd child" in which he resolved to try oxygen. But by the time he had collected the air bags "5 or 6 children had assembled. They grew quite frantic on seeing the air holder as if it had been a coffin."[67] Two days later, Watt echoed some of the same fears from his own experience. Effects varied considerably among subjects, but he was not about to give up. Thus, he told Black, he was intent on trying "charcoal well burnt" as a source for hydrocarbonate "*when I can procure proper patients*".[68] For Watt, chemical subjects were no longer simply that – they had often become his patients, even when not of the Birmingham Infirmary.

Watt's unfortunate failure with Jessy did not end the agony in his household. He was soon very worried about his youngest son, Gregory. Gregory had exhibited various florid symptoms for which both Withering and Beddoes were consulted and airs tried, revealing a particular sensitivity to trials with the otherwise promising hydrocarbonate. But Watt was then also treating his own catarrh with hydrocarbonate, and his "asthma" with oxygen. Too often the results were vertigo and nausea. Watt was willing to countenance many an experiment of "untried airs". He made several recommendations to Dr. Carmichael in Birmingham and sent Beddoes a list of possibilities for use in pulmonary complaints including, variously, "Tallow candle air wax . . . air from a lamp supplied with tar – wool air slightly washed, volatile alkaline air – azote".[69] Invariably there were many disappointments. In the lack of consistent reactions, it was unclear why.

By the end of the century, Watt often claimed to have given up on matters medical and chemical. Despite his frustrations, he nevertheless continued to speculate with Beddoes on the apparent failures of airs among Birmingham practitioners. The uses of nitrous oxide had occasioned general enthusiasm but that too readily evaporated. There were many problems even in describing reactions. In the new chemistry, Watt found much of his pleasure of experimenting further erased by the new terminology, which "we old fellows can never retain in mind." He so objected to the flood of "new words in Chemistry, Mineralogy & Metaphysics [that] I am become and enemy to Neology".[70] Despite every disappointment in projects promoted by Beddoes, Watt simply could not disengage. He continued with trials of airs, prepared from nitrate bought frequently from the local manufacturing chemists, Cope and Biddle. Attending with Dr. Carmichael, Watt reported in great detail on the effects on two volunteers who, upon inhaling, felt as they had been into too much wine although neither, he was most careful to note, had "a disposition to dance".[71] The repeated considerations of the effects of nitrous oxide were also a large part of Watt's final correspondence with Black. In 1799, Watt had written about Beddoes' experience with the air holders. His patients showed no indication of injury to the lungs except when the gas turned out to be especially acidic and even then, it was claimed, without permanent damage. Watt was particularly candid about Mrs. Beddoes' own use of the air in what he believed was an "incipient" consumption. It was not, however, intended that nitrous oxide should be applied as a remedy, especially as Beddoes had warned her off. Beddoes considered it dangerous in some pulmonary complaints. Watt sniffed, "She however chose to have a will of her own & took it at the institution unknown to him, and afterwards continued it for the same reason other Ladies take brandy vizt. the pleasure it gave her."[72]

The Beddoes-Black-Watt nexus had laid one foundation for the expansion of medical experimentation, once Watt's portable apparatus was made available. To a large degree, the spread of pneumatic devices allowed many more medical practitioners such as Darwin to engage with airs. After Black's death, Beddoes was soon following experiments with the young Humphry Davy, on attacking palsy with nitrous oxide.[73] In applying pneumatic remedies, even with enthusiastic correspondents, they remained troubled by inconclusive inspirations on diseased lungs and the repeated instances of headaches, sleepiness, and vertigo noticed also by the French chemists. Close to home, the ailing Matthew Boulton had been put on a course of nitric acid. And Watt was left worrying about Gregory, who had left the miasmas of Italy for the clear air of Innsbruck for his fever, persistent cough, and extreme exhaustion. Laments continued. By January, 1803 Boulton had seemed on the mend. But James Watt, on a trip south through London, had himself turned for the worse with "cough fever expectoration Asthma &c & was quite good for nothing".[74] He caught up with Gregory who, upon his return from the Continent, was nonetheless still very weak and his disease was taking its toll. Despite assurance from many of the continental physicians, his consumption was clearly getting worse. He is an excellent example of how new remedies could attract the desperate when established medicine was helpless even in making an accurate diagnosis.

The Beddoes network

The Beddoes network was one effective result of the continuing communication between James Watt in Birmingham and Joseph Black in Edinburgh. Neither Watt nor Black published much at all. It was left to Beddoes to create a broader public circle from the original connection. Of course, not only did Beddoes learn much from Black while a student at Edinburgh, but Beddoes also easily became part of the existing link between the Scots. Of this, Watt's laboratory work has recently become better known.[75] His chemical interests were part of an experimental enlightenment that can be traced immediately to the work of Joseph Priestley and throughout the phlogiston dispute.

Beddoes' projects emerged from Edinburgh connections. In other words, his links to Watt, if initially incidental, produced a fortuitous convergence of interests. While Beddoes' politics were surely one obstacle to the deepening campaign for pneumatic remedies, they also gave his medical views much wider urgency and even appeal. It should not be assumed that such a trajectory was a result of links to Watt alone. Indeed, at least as early as 1789, a short time after Lavoisier's *Traité élémentaire de chimie* had appeared, Watt reflected on many of the recent discoveries in chemistry, on William Withering's own critical inflammation of the lungs, and reached the conclusion that Adair Crawford, in London, had shown that terra ponderosa was a cure for cancers and "scrophulous diseases".[76] Watt was early involved in the search for chemical remedies in a variety of common ailments, particularly inflammations, which often evolved in apparent consumptives. In 1789 he had been paying special attention to Priestley's experiments on phlogisticated and dephlogisticated airs that employed Volta's eudiometer. Throughout the next year, Watt lamented that the French were so busy with liberty as to be idle in chemistry. Watt implied that reform could exact a price. But, in Britain in the next few years, experimental chemistry was explicitly promoted as medical remedy. Thus, by 1793, Watt was discussing Beddoes' extensive attempts to apply "the antiphlogistic Chemistry to Medicine Azote & other poisonous airs to cure Consumptions & oxygene for spasmodic asthmas".[77]

In all the trials and tests no obvious remedy presented itself, but a sheaf of promises was continuously gathered. Beddoes harvested hopeful tales of inhalations that made for even more promises and then – fell short. His aim was to encourage, then to print, and, in publishing, encourage more. The difficulty was that diseases were too prevalent, too uncertain in their advance or end, and trials too individual and undisciplined to confirm any conclusive therapy. Even among the most knowledgeable and able chemists, no benefit was clear. Such was Beddoes' contact with young Tom Wedgwood who, like others in the provinces, provided further moments of the administration of airs with Watt's. Many patients were needed.[78]

What made the promise of new airs possible? On the one hand there was the obvious: a bird asphyxiated in a bell jar once evacuated but later restored with common air and, if you were Joseph Wright of Derby, painted for posterity. On the other, in illness, there was the commonly regarded remedy of travel to warmer and drier climates on the Continent part, the consequence of a "doctrine of exemption".[79] To withdraw to Lisbon or Lausanne, to avoid the dangers of English airs or of the many new manufactures making breathing miserable, there lay an urgency of escape as the only likely respite. Ironically, it was often industrialists who recognized that the very processes that fouled urban airs also demonstrated the power to purify. The solution appeared scientific and instrumental. Even in the case of the wife of the chemical manufacturer James Keir, who suffered from chronic pulmonary issues, Beddoes suggested hydrogen could be administered safely.[80] Much remained hit or miss.

For this very reason, Watt and Beddoes expanded their net of correspondents. From 1794 it was as though anything was worth a try, at least if airs could be reliably created with a modicum of chemical knowledge. Watt entered into an exchange with the Unitarian physician Thomas Percival of Manchester, to whom he laid out a clear strategy for the expansion of pneumatic medicine. This was a remarkably optimistic exchange in which Watt set out his speculations on various substances, beginning from Priestley's experiments with oxygen, to fixed air, hydrocarbonate, azotic, and inflammable air, the benefits of which he felt unqualified to judge. But, in the workshop, Watt had much experience of all of these, and far more than most. Even so, in the way of experimental physiology, he wrote, "when one thing does not do, let us try another, until we come to the most apposite, & let the doses & manner of exhibition be carried as much as they permit." Watt still professed little knowledge in medicine, even when patients were close to hand. Despite the cynics who preferred Watt's laboratory approach, none of these adventures, he suggested, would prove much without the "direction of experienced Physicians, who can nicely discriminate the ease & avail themselves of the symptoms produced by the medicine". It is a remarkable circumstance that within months, Percival was among those in Manchester who established a Board of Health necessary to track epidemic diseases, especially among the working poor.[81]

Public health and experimental practice urged, in any urban space and in any season, the convergence of medical and social reform. Percival provided a precise reflection of Beddoes' own priorities, notably at a time when Beddoes was canvassing for support of a pneumatic establishment, a narrowly focussed infirmary, in Bristol. At the moment of contact with Percival, Beddoes was also writing to Thomas Wedgwood about new reports of cures from the London surgeon, Jonathan Wathen, and from Rev. Joseph Townsend, who was preparing to undertake factitious airs in administering to his parish charges.[82] The pneumatic net thus widened rapidly, from town to provinces, gathering patients as well as practitioners.

Between themselves, Beddoes and Watt were quite candid about how experiments had produced as much doubt as enthusiasm. It could hardly be claimed that factitious airs were uniformly effective simply on the personal accounts of a small number of practitioners. More evidence was necessary – and promising discoveries were often dashed. Beddoes and Watt were aware that the link between any theory and experimental practice was unable to produce firm conclusions, such as that which had put hyper-oxygenation at the root cause of pulmonary consumption. In 1795 Beddoes declared that early theories were simply "first speculations", deductions or conclusions solely "as might be put to the test of exp[erimen]t".[83] By then Watt was already providing accounts of pneumatic chemistry and useful apparatus with Thomas Henry, one of those advancing medical reform and the health of factory workers in Manchester. Henry hoped to quickly obtain Watt's new breathing device so that he could try oxygen in an epidemic of highly infectious scarlet fever.[84] The cases failed to prove that any of the new airs were effective. Between enthusiasm and inevitable suspicion, every trial was a long shot. Yet, there was ultimately more than enough interest to engage large numbers of physicians, surgeons, apothecaries and even early innovators of electrical therapies.

The project and the patients

The tales of medics were not the only source of hope. Both Watt and Beddoes had firsthand experience of aerial remedies. Both treated patients independently or when consulted by others. By late 1794, the demand for new apparatus left Watt utterly besieged. Almost immediately devices were being sent to Ireland for a patient with hydrothorax, to a Mr. Gladwell in Bristol Hotwells who needed to inspire oxygen, and notably also for the Whig Lord Selkirk, whose son Basil William Douglas, Lord Daer, was consumptive. Daer was a former student of Black and was noted for his reforming sentiments, ultimately coming to be known in France in the early years of the Revolution as Citizen Douglas.[85] But Daer's health was increasingly desperate and Beddoes speculated freely about possible remedies, undoubtedly following from the many reports he received about experiments in consumption. Beddoes heard from the well-connected Dr. John Turton of Oxford, who reported that even among those apparently susceptible in large towns had nonetheless escaped the worst infections. Beddoes received a tale about such a family, of the very poor that concerned Percival and Henry, who lived in a close street "where in some apartments the inhabitants burn candles all day, there being no window". What effect candles had on airs was of much interest. According to the tale, the local parson could scarcely recall "burying a consumptive patient out of it".[86] But, from the working poor to the well-connected, disease made no distinction.

Among the dying crowd, Daer's case was particularly compelling. Daer exemplified the connections Beddoes cultivated between wishes of the constitutional reformers and the rarefied hopes of the air men. In the fall of 1794 Beddoes learned that Daer was in contact with democratic James Watt junior on the Continent, but Daer refused to disclose much about their correspondence. Perhaps it was simply about Daer's plans to escape to a warmer climate in Portugal. Certainly, Beddoes became increasingly exasperated with Daer, whom he was trying to treat with "unrespirable airs."[87] While Beddoes believed Daer had reached a point where he could tolerate a high dose of fixed air, Daer insisted on going to Lisbon, where, in Beddoes' view, "he goes to a certain death from a possibility of help." To seek common relief in the sun and dry air, Daer would still "lose his only chance".[88] Patients could often be difficult to convince. Likewise, Watt sent an unhappy Mr. (possibly Walter William) Capper to Beddoes for similar treatment. Capper was apparently obstinate about aerial remedies. Beddoes thought that by the time Capper left for Sidmouth spa he had recovered "considerably since his impudence".[89] Given the obstacles new therapies inevitably presented, it was hardly surprising that even failing patients would be hard to convince. The way around this, at least in part, was to try to broaden the range of subjects as much as possible – and this would, invariably, require the sanction of the well-connected.

Watt attempted to interest Sir Joseph Banks in supporting Beddoes' pneumatic subscription for an Institute at Clifton. Despite the obvious despair Watt had recently expressed, the plea fell on studiously deaf ears. It proved little good to claim the merits of airs, of those

> new and powerful medicines, which promise to produce uncommon effects, & they should not be rejected untried, or left to Empyricks. . .. [L]et us contribute our money & assistance and from publick trials know the good and bad qualities of these airs. Mankind have been long enough occasionally poisoned by the channel of the lungs, let us try if we cannot receive medicines by the same way
>
> Bright hopes are held out & the chances of doing evil is small, for we cannot much anticipate the fate of consumptive patients nor add to the suffering of the paralytic nor the cancerous.[90]

When Watt attempted to interest Joseph Banks, the experimenters were trading opinions on the viability of Portuguese glass in chemical trials then going on in London. Chemical innovations were certainly not new to Banks. He claimed to have been present at hundreds of such experiments although he professed no particular expertise as an operator.[91] As widespread as the interest in chemistry evidently was, it did not inevitably translate into sympathy with the factitious adventurers. Politics trumped all where democrats and Beddoes lurked. Even the approval of the Duchess of Devonshire made no difference.[92] Banks would go nowhere near the project.

Beddoes was not to be discouraged. Other prominent patronage might yet overcome such politically charged suspicions – and among the Beddoes' patrons there were also many patients. The practice was promoted among acquaintances, chemical and industrial, like Tom Wedgwood and his mother, who inspired oxygen, as well as various notables like the architect Sir William Chambers and William Reynolds who would have their experiences published.[93] Beddoes' medical sources were many and carefully cultivated. Among them were the obscure, like a Dr. Lawrence of Swasham, Norfolk, writing in 1796 to report on the benefits he found in using factitious airs. Otherwise Beddoes heard from the much more well-known German physician and Professor Christoph Girtanner in Göttingen, who had published trials on the cure of consumption by fixed air.[94] Much of the private exchange between Watt and Beddoes involved specific patients, but out of public reports a greater good might emerge. Communication was critical. Beddoes was well aware of the rising tide of debate on the Continent as well, remarking on the German journals, which were full of accounts, and a communication through a Bristol merchant trading to France who carried a report on the interest of Fourcroy and Guyton de Morveau.[95] Beddoes had cast his net wider yet and, as we shall see, even wider than the continental philosophers.

The most repeated and detailed accounts came from many of their most immediate collaborators. These included physicians, surgeons, and chemists, some of whom were clearly more than willing to undergo the risks of self-experimentation. When Tom Wedgwood went to Clifton under the medical direction of Beddoes, Watt sent along further instructions on how to prepare oxygen from manganese and apply it with hydrocarbonate, although Watt claimed whenever he had tried it the result was stomach cramps and "rheumatisms".[96] Apart from their own experience, it was surely the case that their immediate correspondence came from those most frustrated with the limits of treatments. While the greatest attention has often been paid to experiments, as with Humphry Davy on the effects of nitrous oxide, there was a much more mundane programme continued in Bristol and in Birmingham.[97] For example, by the spring of 1799, Beddoes had concluded that the use of ubiquitous digitalis both prevented, and ultimately cured, consumption except in the highest stages of the disease. He told Watt, "Mrs. Beddoes's life has certainly been saved by it," although it was quite evident that she was more than delighted with the adventure. Davy's use of the "properties of gazeous oxyd of azote" was clearly intriguing in both chemistry and physiology. But Davy, Beddoes acknowledged, was a risk taker. While Beddoes then recently had little to do with hydrocarbonate, "Davy made himself (as you may suppose) very ill by having 3 full inspirations of it undiluted, having previously emptied his lungs as well as he cd."[98] It was obvious Davy's misadventures were also an opportunity to promote the drama of the beautiful, misleading, mistaken, and highly amusing effects of airs.

Nitrous oxide seemed overwhelmingly promising across a range of ailments. Beddoes was shortly reporting to Watt about the inspirations of patients in Bristol as the gas seemed especially useful in calming epilepsy.[99]

Special attention was soon being paid where "Nitrous oxyd wd appear as good a medicine for palsy as any in the Mat. Med. for any other diseases."[100] Much, Beddoes frequently remarked, seemed to be thrown into confusion by Davy's experiments on the decomposition of water, from which he was evidently able to produce azote, perhaps something extracted from the presence of metals or chemically "analogous to electy".[101] Investigations were neither limited to Birmingham or Bristol, nor to Paris or Dijon. The network spread even wider. Joseph Priestley, then in exile in America, continued his own experiments on airs, the expense of substances and instruments defrayed by his friends in England. Priestley operated still within a phlogistic chemistry and employed the new Voltaic pile, concerned more about the chemical theory than the therapeutic effects that obsessed Watt and Beddoes.[102]

Testimony was always suspect in some degree – even to Beddoes. A major hurdle was the various responses of patients, without whom any development of therapies would prove impossible. They offered confusion as much as credibility. To carry on the business of factitious experiment and the practise of a Pneumatic Institution, along with lecturing and publishing, proved an enormous burden for Beddoes. But it all had to be done. Opportunity had knocked and Beddoes had no intention of ignoring it. Too many suffered, notably among the households of industrial friends like Watt and Keir. Hence, Beddoes famously treated Gregory Watt with hydrocarbonate and sought a design for a vapour bath to combat his growing consumption. But the brilliant young Gregory, often in Bath under Beddoes' care, was not entirely cooperative. However ill, he still continued with his own chemical trials, soliciting William Henry for pure dry phosphoric acid, the fumes of which could surely not have helped.[103] More significant, however, was the possibility of many more experiments, so long as patients could be induced to come to the Institute. Beddoes explained the demand of patients was already so large he needed more assistants to employ the air more than four days a week. By 1804, not only did he feel able to once again carry on research on airs but he again connected with what he clearly felt his primary purpose. To Watt, Beddoes then declared, "I think I have the full confidence of the great body of our poor – & I am trying to get a young man whose sole business shall be preparing & administering airs."[104] The poor offered bodies for the necessary trials. This was not nearly as cynical as it may now seem. Beddoes saw chemistry as both a medical and a social response to the woeful conditions of the labouring poor.

Pneumatic spaces

Despite the proclamations of Tom Paine, which so frightened the establishment, some might have been forgiven for reflecting upon the blights of men amidst the choking smoke and cinders of rapidly industrializing towns. For some, it was impossible to see pneumatic medicine without the lens of public health. Of course, ideological sensibilities were hardly uniform across factitious enthusiasts or even among the merely curious. But it is remarkable just how many seem to have believed medical innovation implied social renovation. Indeed, the spread of pneumatic trials, especially following the demand for the Beddoes-Watt breathing apparatus, made much of these very implications. As Beddoes told Erasmus Darwin in 1793, despite the claim that "none but beneficial consequences can result to the public," he was certain he would come under "ridicule and obloquy . . . decried at home as a silly projector, and by others as a rapacious empiric".[105] That was not the half of it. Not only was Beddoes' Jacobin cloven foot an issue, but the entire practise of pneumatic medicine attracted the blatant hostility otherwise directed at reformers of many kinds. Those, like Watt, who did not share Beddoes' political notions, had to be extremely circumspect given the highly charged atmosphere of the 1790s. This became increasingly significant by 1794 as Beddoes was invariably linked to Watt and his Lunar friends. By the summer of 1794, Beddoes had so many enquiries from numerous physicians that he applied to Watt to refer mechanics willing to fit up the devices. According to Watt, it was Beddoes who asked that Boulton-Watt manufacture the apparatus for sale.[106]

The experimental aerial practice certainly came with much peril as promise. But public hopes also induced latent suspicions. In a telling correspondence between Watt and the Windsor physician, James Lind, many fears immediately surfaced. In the early spring of 1795, despite his own conviction that many diseases might be addressed by factitious airs, Lind raised the alarm about "such a set of miscreants" as may be associated with French chemical innovations. Referring directly to Watt's friend, the Genevan immigrant Jean-André De Luc, Lind noted that De Luc,

> being no friend to Theories, either Philosophical or *Political* that are founded on falshood [sic]; and are propagated by force, and such He thinks French Chemistry and Politics to be, Indeed there seems to have been such a wonderful coincidence in sentiments of some of the Modern Chemists in this country that I have declined taking a part in the Pneumatical practice of Medicine.[107]

Just as Burke had eviscerated French projectors and calculators, so too did Lind. But Watt was not quite so dismissive. He claimed to give himself "little trouble what are [Beddoes'] theories in Politicks, religion or chemistry". Lind did not change his views of French chemistry and politics, but he did decide to make a trial of the pneumatic apparatus and pneumatic medicine, asking Watt for the device for his son's use on a naval voyage, and thanking Watt for sending it. Watt believed that regardless of whether Beddoes' use of airs like carbonic acid gas (fixed air), or oxygen or hydrocarbonate ultimately proved effective, Beddoes' undoubtedly "engaged in the pneumatick medicine from very disinterested, & Philanthrophic motives". For

Watt, paternal care had overcome political hostility, rooted in his anguish over James junior.[108]

It was always possible to assert a philanthropic principle in matters of public health. Such, for example, one might discover in discussions of contagion such as smallpox or gaol fever. Likewise, the enlightenment doctrine of diffusion, propagated by Priestley and Keir, applied as much to medical remedies or even to preventatives as much as it did to the public debates over experiment. This, surely, was one of the justifications for the promotion of inoculation by John Haygarth of the Chester Infirmary, friend to Thomas Percival, and who must have been known to both Beddoes and Davy. Even before Beddoes was strenuously promoting pneumatic remedies, Haygarth was certainly in contact with Beddoes' allies the Wedgwoods, notably the young chemical adept Tom Wedgwood. Haygarth agreed, for example, with Wedgwood that "the cause of inoculation would be best served by a more general diffusion of knowledge." Thus, the primary aim of any anti-contagionist promotion of inoculation was "to convince persons of superior understanding that contagion may be avoided by practicable regulations". More importantly, such initiatives could not be left to the medical profession alone or to be assumed "out of the province and beyond the comprehension of extraprofessional men".[109] Here, in essence, was a mantra that appealed to Beddoes and his apostles.

It must also be recognized, even for someone who owned Rousseau's *Social Contract* and who could hardly have been unfamiliar with Paine's *Rights of Man*, that Beddoes' interests in public health were neither entirely unique nor were they an enlightenment discovery. Indeed, it is clear that his acknowledgement of the diseases prevalent amongst urban workers had been long foreshadowed at least by the Italian physician Bernard Ramazzini in the seventeenth century. While Ramazzini's work had related to local experience in Padua, there were numerous translations in the following century. We know, from the sale catalogue of Beddoes' library in 1809, that Beddoes had copies of Ramazzini's *de Morbis Artifcuum Diatriba* (1743) and of the German 1783 edition of the *Abhandlung von den Krankheiten der Künstler.*[110] It is obvious that Beddoes' medical sympathies had a long heritage increasingly urgent in the darkening urban landscape.

Infirmaries and hospitals were often receptive to any innovation that might address a multitude of otherwise incurable ailments. Such sites, from Plymouth to Edinburgh, were sources of information that further encouraged medics – at least, once Beddoes published the accounts of the trials that came to him. For example, Beddoes frequently discussed with young Tom Wedgwood the advantages chemistry might bring to medicine. Much evidence lay in the provinces. Thus, he informed Wedgwood of reports he received from Plymouth of the use of nitric acid in venereal disease that seemed to show greater promise than the common application of mercury. Of this there was much discussion, but it was only one of the many accounts by which Beddoes was encouraged.[111] Tom Wedgwood, however, was typical of the private and personal connections Beddoes was to exploit. Equally widespread were sources like the apothecaries of the Manchester Infirmary who were experiencing difficulties with the oiled bags attached to the Watt breathing device. The *Manchester Chronicle* printed reports from William Henry in December, 1794, partly in an effort by Henry to drum up subscriptions for Beddoes' Bristol Pneumatic "Hospital" (Beddoes and friends changed the name several times). Likewise, Watt wrote not only about Henry's apparent success in Manchester with hydrocarbonate in asthma and pertussis, but he also declared the surgeon, John Barr, was successful in using oxygen in a case of a scrofulous ulcer at the Birmingham Dispensary.[112] Public discussion meant public trials.

Beddoes promoted an extensive institutional network among hospitals, infirmaries, and dispensaries immediately in the wake of the Watt apparatus. This diffusion was not entirely about the cure of desperate patients amongst Midlands' smoke. Beddoes' growing web was by 1795 a fundamental objective of the collaboration of Beddoes and Watt. This indeed forced upon Beddoes continued additions and revisions to their *Considerations on the Medical Use and Production of Factitious Airs*.[113] By the spring of 1795 the news of a case in the Birmingham Dispensary seemed promising in pulmonary consumption. Beddoes was so optimistic that he told Tom Wedgwood it was likely that a ward in the Birmingham Hospital would soon be appropriated entirely to pneumatic medicine.[114] It was for this reason that Wedgwood was asked whether "rendering Pneumatic Medicine popular by a device or two & by a pamphlet for readers of fashion worth carrying into practice".[115] All of this, likewise, was to gain the support of the Wedgwood clan for a pneumatic establishment. Beddoes clearly understood that it was not subscriptions that would build the success of factitious airs. Enthusiasm among the fashionable might be as useful as the practice of the medics.

Very quickly, it would appear, there were acolytes of the new airs in Edinburgh, York, Berwick on Tweed, and Hull, as well as among the practitioners of Birmingham and Manchester. This growing interest went along with plans to appeal to MPs once Parliament opened in London.[116] But it was hardly to stop there. Each report of an apparent cure, across a spectrum of ailments, would induce a widening of the practise. For example, in a case reported by Beddoes, a Captain Hemsely, paralysed after yellow fever in the West Indies, was evidently restored by the application of oxygen in Newcastle Upon Tyne when every other remedy had failed. The result was such that the medics at the Newcastle hospital ordered a Watt apparatus.[117] Even such an experienced physician as Robert Cleghorn, of the Glasgow Infirmary, had no doubt the trials of airs needed to continue while his own experiments gave little confidence. Cleghorn was another who told Watt that it was necessary to go on with the effort to "diminish the sum of human misery". He complained that "Those who oppose these trials, & still more those who ridicule them, deserve the reprobation of every lover of mankind." From amongst the workshops and factories, from the likes of the Watts and the young Tom Wedgwood, came much hope. Cleghorn told Watt, "Improvements must come from such men as yourself, or some of your Coadjutors whose circumstances are somewhat different from those of our profession in general."[118] Beddoes would happily agree. A large part of Beddoes' campaign was to present himself as an outsider, to play upon the very reforming sentiments that would otherwise undermine the ability to harness the well-connected to the cause. Of course, the desperate were not confined to the poor. The families of industrialists could often see the wisdom in factitious airs, especially when Watt's name was attached to the scheme. So too with the landed such as Lady Dashwood, whose family seemingly consulted Beddoes in a case of hydrothorax.

The primary vehicle of pneumatic medicine was the testimony of proponents of reform, medical and otherwise. Of these there were many, some obscure and some much more well-known. One of the most illustrious was Erasmus Darwin, part of the Beddoes network like Watt and Josiah Wedgwood of the Lunar circle. From the summer of 1794, Darwin had been encouraging the Derby philosophers to examine Watt's breathing apparatus. Indeed, it was Darwin who proposed a detailed pamphlet be published on how to make the necessary airs.[119] By the following spring, Darwin was on the campaign to get the infirmaries in Nottingham and Shrewsbury to adopt the devices.[120]

There could be no uniform opinion regarding new airs. Even those physicians willing to consider new therapies disputed results, particularly as much knowledge was narrowly derived from self-experiment as well as from patient trials impossible to replicate. Beddoes in 1795 became convinced that even Ferriar was behind doubts published in the *Medical Review*.[121] Likewise, John Haygarth, an active projector of inoculation at Chester, did not believe that Beddoes' proposal for a Pneumatic Institution was a credible medical solution to the spread of consumption or even to the "variolus miasmas" that concerned him.[122] Among the surgeons was Thomas Creaser, of the Bath Infirmary, who knew Beddoes well. While Creaser was intent on the promotion of inoculation, there were many diseases without a response. He accepted that oxygen was effective in a case of herpes when other remedies had continually failed.[123] Whether with Davy in Bristol or Dr. Henry Sully of the Dispensary in Wiveliscombe, Somersetshire, among the adventurous there was evident wish to attempt as many trials of factitious airs or, ultimately, of the new galvanic apparatus as possible.[124]

If we place Beddoes' schemes for his Institution in this same context, it would appear as yet another new infirmary reflecting those in many provincial towns. Birmingham experience, with the surgeon John Barr and with the physicians John Carmichael and Richard Pearson, had already revealed that factitious airs could, in the right circumstances, provide many opportunities for trials. Beddoes' contacts within infirmaries were extensive. Thus, John Seward of the Worcester Infirmary had also decided to pursue the effects of hydro-carbonate on himself, and begged Watt for an air bag and some manganese to prepare the airs. This was typical of how the pneumatic airs followed devices into infirmaries.[125] After the turn of the century, Beddoes pleaded that his own Bristol initiative not "be regarded as a mere local scheme". While he was able to attract the support of the Duchess of Devonshire and Lord Lansdowne, both of whom had offered their help, Beddoes also felt that "in proportion to its notoriety, I think it must prosper."[126] Even then in Beddoes' world, in the decade since 1794, there was no such thing as entirely negative publicity.

The London connection

Beddoes' network reached beyond those whose letters he published. However, it was not to be expected that all his correspondents would carry the same credibility. This was the downside of the demand for the Watt apparatus and the subsequent successes claimed by medics. But the device was surely neither as easily available as Beddoes hoped nor was it inexpensive. Beyond Britain, it is likewise clear that there was considerable continental interest in aerial applications. Besides Girtanner, the Genevan physician Louis Odier was keen to advance the cause, and may well have been behind an attempt to improve upon the Watt instrument.[127] Closer to home, the activities of Dr. Robert John Thornton in London demonstrated that the use of factitious airs had quickly gained traction.

Now very little known, except perhaps as a botanist, Robert Thornton was an early practitioner of pneumatic medicine and one of Beddoes' greatest promoters.[128] Thornton, of Guy's Hospital and the Marylebone Dispensary, represented those physicians prepared to advocate the efficacy of airs. He applied them across a very wide variety of ailments in conjunction with the noted London surgeon Jonathan Wathen. It was at Guy's that Thornton would also have encountered the physician and chemical lecturer William Saunders. In the summer of 1794, in the aftermath of the loss of Jessy, Watt was seeking solace in chemical explorations. According to Beddoes, Saunders then made an unparalleled offer. Saunders was

> so strenuous in the idea that the new remedies will prove of the highest benefit to mankind & open many new lights respecting disorders that he has offered me a share in his practice in Guy's Hospital, i.e. to select for my patients that [?] I may choose for administering the new "remedy". These are the words of a correspondent in town who says he has done much by oxygen air.[129]

The evidence suggests the correspondent was Thornton. Late in 1793, Thornton had already contacted Beddoes about his own medical trials, notably on hyper-oxygenation, with apparent success in some desperate cases. In passing, Thornton not only speculated on the effects of oxygen but also suggested the concurrent application of electricity, a notion that caught on among many medical innovators. A month later, Thornton was reflecting on the role of oxygen in spasmodic diseases and consumption, proposing moreover that vital air would be

> found of great service if it were let loose in mines, in churches, and in crowded rooms, but more especially in the bathing-rooms of Bath, where great faintness is often brought on the patient by breathing a reduced atmosphere from the excitation of azot [sic] out of the waters.[130]

Beddoes' discussions with Thornton were beginning to bear fruit. He reported that Thornton had worked with Wathen in London in what Beddoes described as "paralytic affections of the auditory & optic nerves which surprize me". One to whom Thornton applied airs showed immediate results in a case of *gutta serena*, or a blindness of uncertain cause. Beddoes knew much of the practise at Guy's, to the point that Wathen sent him a young patient. Beddoes reported the boy was "so deaf for 15 months that he could not hear the tower guns, though close to them [but] has since recovered his hearing by oxygenation".[131] This news was quickly passed on to Black in Edinburgh. Watt, moreover, recounted a Birmingham trial of oxygen in *gutta serena*, as well as the case of a consumptive where suffering had been greatly relieved by hydrocarbonate, although without much prospect of full recovery as he was "too far gone".[132] It was quickly the case that both Watt and Beddoes were encouraging consultations with willing practitioners. Even before their alliance, they were hardly alone in pneumatic practice.[133]

The matter-of-fact-ness of failure, mixed as it often was with the fanciful, raised many doubts. Thus new airs were a gift little understood. Watt's garret workshop was a site of constant curiosity. By then, chemistry had become part of a desperate arsenal among aerial practitioners. The reports from London, from Guy's, Saunders, Wathen, and especially Thornton spoke to a growing enthusiasm for further trials of many new gasses vicariously produced. Daily airs revealed the urgency of hopes daily dashed. Failure was too readily apparent. Ironically, even the increasing reports of success actually began to alarm Beddoes. In an expansive correspondence with Watt, he revealed the depth of his concern. Passing on his news from numerous sources, Beddoes took note that Thornton's letters "might bear the interpretation of a wish to monopolize the pneumatic practice in London". There was little doubt Thornton and his associates were rushing ahead with applications of airs in cases that had otherwise proved hopeless. Quoting a letter from Thornton, Beddoes repeated the evidence of cures:

> The first intimation I had of the importance of the oxygen air for the removal of gutta serena was in a Lady, who was patient of mine for nervous head-ach[sic]. By accident, she took up a church prayer-book, & was surprized to find her eye-sight so strengthened that she cd read it, when before, so small a print was perfect confusion. Having

> mentioned this, I was anxious for the trial in the worst case of defective energy of ye optic nerve. The 1st patient Mr. Hill (a surgeon) tried the air with, was the sister of Ld. Walpole. She early lost an eye, though the defect cd not be discovered by a bye-stander & the sight of the other had become very feeble. Electy. was conjoined with the vital air. She cd. not discern, as she told Mr Wathen, the dial-plate belonging to the Horse guards clock, but before she left she cd readily perceive both the hour & the minute hands. I gave you the acct of benefit received in gutta serena by a poor patient, who contended with a moderate sight, left off attending . . . a lad was 7 weeks in St. Thomas's Hospital & turned out incurable. Mr. Wathen, having no success in such a desperate case, sent him to inhale the vital air. In 4 days he [said he] had a glimmering of sight & in 15 discerned objects.[134]

Even before Watt and Beddoes decided to publish, pneumatic medicine had already been adopted by some highly reputable practitioners. While politics could undermine the allure of airs, aerial enthusiasm was not simply left to quacks or to the medical fringe. Indeed, apart from rooms filled with uncertain airs, the evidence of interest clearly followed public pleas for a practical breathing apparatus. To trace the devices is thus to follow the diffusion of factitious hopes. Watt's own design, by the early spring of 1795, of an instrument to manufacture and to apply medical airs turned his engineering reputation to promoting new trials. Watt was initially reluctant to reveal his design, and it soon became clear that he was not the only inventor at work.[135]

During the winter of 1794–1795, Thornton's was using his own contrivance at his surgery in Great Russell Street. He also took a room in Bennet Street, St. James's, to attempt applications of airs. By then fifteen to twenty patients a day were trying inhalations. The correspondence between Beddoes, Thornton, and Watt debated medical protocols as much as continual claims of cures. This was the crucial moment. There were significant problems in transferring chemically created airs as well as dealing with the inevitable confusion of patients. Thornton described his own methods of application, and the consequent reactions:

> I readily conceive that the silken bags are very useful, but being a new process, it requires some trick to prevent a smile, & more so, when not one beside myself in London practices it. A patient is amused at the sinking of the bell, and believes this testimony that some process is going on. Now that friends & facts appear for the airs, I shall try your method, wch I conceive is the best on some accounts, but a patient must be told that the bag is clean each time for fear of INFECTION. Each must have a separate tube, &c. so fearful are the fine folks in London.
>
> A patient of mine having taken the air must soon after have broken out in Erysipelas, the Apothecary had the villany [sic] to amuse

> the family by saying he caught it by inhaling after that leper Patterson, whom they saw, whose son I have described, when on my honor [sic] inhaled it out of different machines![136]

A comparison of chemical experience revealed the caution Thornton took in trials of airs. Of particular interest was his consultation with Adair Crawford, professor of chemistry at Woolwich and of St. Thomas's, who unfortunately died in the summer of 1795 after prolonged ill health. Crawford had been sufficiently interested in Beddoes' chemistry to try hydrocarbonate diluted with common air. He reported an effect of a "soothing tranquillity, such as opium is known sometimes to produce". At the same time, Thornton believed that he himself suffered from phthisis, which led him to try rooms with diminished oxygen. By comparison, he told Crawford that his chest pains disappeared "in the moist air of Cambridge, and that having not long back gone to Oxford to see some friends at that university, I was almost immediately affected with pain under the sternum and other marks of pulmonary affection". Crawford had meanwhile tried modified air but nothing could save him. He retreated to the country estate of Beddoes' patron Lord Lansdowne but there "this good man paid the debt to Nature."[137]

One possible answer to the long afflicted was to fit up closed rooms so that the atmosphere might be modified to suit the patient. This was Thornton's scheme for Bennet Street. He was motivated by earlier investigations conducted in France by Lavoisier in both the wards of the Hotel Dieu and in the pit of the theatre at the Tuileries. Chemical analysis revealed stark differences from the open air, but Thornton also had more dramatic examples he could call upon. There were the mephitic airs of the notorious Black Hole of Calcutta and shockingly high rates of infant mortality at the Dublin Lying-in Hospital.[138] If closed rooms could be catacombs of infection then, like the attempts in Dijon to fumigate the hospital, it was possible to conceive of spaces designed for therapeutic airs.

Thornton's claim that there was no one else in London engaged in the practice was highly suspect.[139] The evidence is otherwise. Thornton was especially attuned to the risks of the new inhalations, especially as public curiosity expanded. Public interest, of course, was a large part of the objective of Beddoes and Watt. As Priestley and Keir had earlier argued, the more hands in chemistry the better – so too with the democracy of factitious airs. The philosophical interest had a parallel in medical practise; little advance would be made if actors and apparatus were limited. This was uppermost in Beddoes' mind in 1794 when he had begged Watt to send a device to a Mr. Knight at Painswick, Gloucestershire, north of Bristol, who feared to wait another moment.[140] Without the nearby presence of a pneumatic physician who might have been found in an infirmary, neither the method nor the result could properly be recorded, notably so if the user was a desperate patient.

The application of airs remained problematic. After 1794, it also meant more players in a practice neither certain nor controlled. Indeed, there were many chemical reactions that produced airs the benefits of which were exceedingly unclear, if any. Thornton's letters to Watt and Beddoes were full of such warnings. Thus, he told Watt of one person who had engaged in self-medication as, for that matter, did many medics:

> a patient in the country poured warm vinegar on chalk, thinking to produce Fixed air, and inhaled the fumes. Soon after he felt most violent pains in his chest, inflammation augmented, and he became a speedy martyr to the trial.[141]

Here Thornton suspected impurities, but it was also clear that without controls, amateurs, as well as more adept experimenters, could get themselves into a great deal of difficulty. Watt's own workshop demonstrated such dangers frequently and revealed a deepening alarm. Watt's new mouthpiece, Thornton wrote in 1795, "will shew how much really depends on the modified airs, and my observations on forced inhalations . . . arose from seeing effects produced, beyond my expectations, but your invention will soon clear this up". As Thornton fully knew from patient's fears – and, as he put it, "from my own personal experience, there is always uneasiness, when I have inhaled the different factitious airs at forced inspirations."[142]

The failure to save the celebrated Crawford was one more case in the litany of disappointment. But Thornton refused to give up on factitious airs even in the face of professional resistance. He told Watt, if for no other reason, "Knowing the persecution that attends those branded as innovators, and the powerful combination that has formed against me, I was obliged to conduct my trials with the utmost verge of caution."[143] The "anti-pneumatists", as Thornton called them, "were neither liberal in their conceptions, nor conversant in that great branch of natural philosophy, which unfolds the properties of *permanently elastic fluids*". Like Beddoes, Thornton expected to be branded "enthusiast".[144] By 1796 Thornton had begun to collect reports in order to publish his own popular account of "Medical Extracts". At that point he thought the practice still in "its most infant state". While there is no evidence that he actually published any results before 1799, Thornton could commiserate with Watt and Beddoes. Of Watt at least, he wrote, "considering the numbers of great Geniuses, who have been bred up to Physic, none has been so unphilosophically treated." Indeed, his analysis of the resistance was virtually identical to that of Beddoes. He focussed on the dismal, "bureaucratic Trade" of the medics. He complained of the few patients who had been referred to him in London despite three years confronting incurable diseases. Even in the urban hospitals and infirmaries, where demand for apparatus grew daily, much obstinacy remained: "In Country towns it will be fairly tried, and even here opposition must be stifled, if the Airs, which I doubt not, aid Medicine so much, as to perform wonders with many turned out of hospitals as incurable, and these facts are recorded."[145] Mammon ruled medicine as much as did obstinacy.

The ideology of airs

Experimental atmospheres infused intense debates in the medical profession. But it would also seem new airs appealed more directly to those inclined to reform of many kinds. Medicine was inescapably part of the political spectrum. Such was explicitly understood by Thornton, who was not so cowed by the political campaigns of Pitt to avoid extolling Beddoes or his plans. In a remarkable passage in his *Philosophy of Medicine*, in 1799, Thornton pointed out the political obstruction by the "envious and interested against the *pneumatic remedies*". After wrapping himself in the reputation of Newton, rather than in the cowl of the monks, Thornton repeated the warning of Priestley's experience: "when philosophers forsake their pursuits of nature, which are every regular and uniform, to engage in the confusion of political contests".[146]

Since the middle of the 1790s, political agitation deeply troubled the tracks of aerial philosophers. This was especially true after rising unrest and unemployment, and the fears that led to arrests and trials for treason of the critics of Pitt's plutocracy. Despite the assault on virtually all forms of dissent, by 1795 Beddoes hung happily to the hope in airs and what he, like Cleghorn, believed "this grand expt. in behalf of humanity".[147] His commitment to experiment continued no matter his worries or those of his friends of possible prosecution. As we shall see, he was soon driven from the cover of medical therapy into political commentary. Yet, Beddoes was neither alone nor was he intimidated by Pittites to abandon the cause of a general reform. If he and his allies could show that new airs meant new and effective remedies across a range of diseases, affecting rich as well as poor, then he could claim a measure of credibility. While there was little clear proof of cures, there were many reports of them.

The search for cures implied social remedies as well. Indeed, when one reads the breadth of Beddoes' incessantly optimistic correspondence, the notion of a political reformation repeatedly surfaces, even to the fantasy of a Germany "republicanized". In medicine, there was oft times a reforming notion. Acknowledging that the benefits of airs were hardly clear, and hostile to any taint of radicalism, James Watt in industrial Birmingham remained certain that continued trials could be safely made. And in fashionable spa towns like Bath and Scarborough, the practise had caught on once the air apparatus arrived. Beddoes enthused about the "uncommon" accuracy of a Dr. (possibly William) Alexander in Scarborough who believed he had cured phthisis with fixed air. Throughout 1797 Beddoes was continually repeating these trials both upon himself and on patients.[148] In 1798 Thomas Creaser in Bath, who saw Beddoes often, was continuing with pneumatic medicine, partly with a theory of using "Oxygenous, Hydrogenous, & Azotic Gas" to reduce "the excitement" of the arterial system. Following upon efforts to explain the benefits of spa water, Creaser believed the azote thrown up by the Bath waters was the most effective means of lowering what he described as the agitation of the arteries in diseases like asthma. Watt, nevertheless, complained that the hydrocarbonate did more good to his rheumatism than did the Bath waters.[149]

In spa fashions a great irony lurked. According to the late Roy Porter, Beddoes despised the "glitterati" and those "obscene temples to Mammon as watering-places" as much he did the quacks. Even when Watt told the reactionary John Robison in Edinburgh that he considered "quack" an honour bestowed on him by "enemies of innovation," Beddoes had by then turned in a new direction at the Pneumatic Institution. He found a grateful set of patients, often the working poor, more numerous than the sedentary and indulgent rich entertained in spas. Like other practitioners, he struggled with the anxieties of his aerial subjects, much as had Thornton in London. But in Clifton, his patients were "decent people – small tradesmen & upper servants". Notably, Beddoes claimed, "they like better to come to us than go to the dispensary or a hospital – whither indeed they wd not go at all – We reject none."[150] Spa or no, poor clients were more reliable.

There was wisdom here, at least if you held the belief that in the breathing lay the prospect of a great medical and social transformation. As Porter once remarked of Beddoes, "Health had to be the touchstone of radicalism."[151] The poor might accept, while the wealthy might need more coddling. Beddoes saw himself apart, happily independent of a sorry profession – yet he was also a likely target of the growing suppression of innovation, in medicine just as in politics.[152] Beddoes tilled the endless tales of treatments by gasses, and Watt, alternatively certain of cures and wary of politics, kept up the harvest in his garret laboratory. Half a decade after Jessy's death, half a decade after the alarms of rebellion drove some into the arms of cynical Tory champions of a threadbare constitution, Watt and Beddoes were still probing Priestley's airs.

In early November 1799, Watt was at Clifton with Beddoes trying Davy's respirable nitrous air. It demonstrably made many giddy but, in a mirror held up to the palsied, appeared to cure paralytics.[153] Before the end of the month, Watt was back in Birmingham and reported to Black on the Clifton trials. Evidently one patient at Clifton was fully cured and had returned to his work as a tailor, while Beddoes continued with some seriously consumptive patients. With Carmichael in Birmingham, Watt was also preparing and trying a new air, the effects of which were a profound vertigo. There were more than a few willing volunteers, although some present were disappointed as the first air holder was quickly exhausted. A second bag put off the subjects as the air tasted strongly of nitric acid, and a third was thought unsafe because of problems in the preparation. Watt was confident that Beddoes' experiments demonstrated that no patient's lungs were injured in such trials, except in the case where the airs turned out to be "manifestly acid."[154] These experiences demonstrated the major hurdle of consistent preparation, quite apart from problems of dosage and the uncertain measurement of effects, even when in a single laboratory with the same operators.

Among experimenters of considerable experience, consistency still could not be assured. At the turn of the century, from Paris to Pennsylvania, the experience was much the same. Indeed, in his reports to Berthollet in Paris, then just having returned from the Napoleonic expedition to Egypt, Watt described the Clifton trials of Beddoes and Davy with a gaseous oxide of azote prepared from nitrate of ammonia. The effects, he told Berthollet, were remarkable despite the many obstacles to practice or to public comprehension. Highly stimulating were the results, in some "exciting a temporary but pleasing delirium", some paralytics patently cured, "others relieved in cases which were other ways desperate".[155] Berthollet was just then re-engaging with chemistry, on decomposition of sulphates by carbon, along with the production of carbonic acid (carbon dioxide) and a gas he believed composed of carbon, oxygen, and hydrogen, the results of which seemingly challenged reports of Cruickshank in London. Likewise, Priestley (in exile in Northumberland, Pennsylvania) received from Watt a furnace and apparatus, as well as manganese, which made it possible for him to produce fixed and inflammable airs. Like Berthollet, Priestley also addressed the work of Cruickshank, notably through the use of a Voltaic pile, which he believed helped to preserve the doctrine of phlogiston even in the decomposition of water.[156] The Watt-Beddoes network continued to be essential to the debates among pneumatic chemists. But for Beddoes there was a higher purpose than chemical adventures alone. With Watt, the intention was ultimately to ameliorate the health of the nation. While Watt may have carefully avoided any attacks on government, he acknowledged that a dispensary or institution could turn knowledge into preventive. Watt quietly skirted "Pitt's proto-Thatcherite policies of grinding the face of the poor".[157] The airs were nonetheless trapped in the currents of Pitt's counter-revolution.

In 1797, Beddoes had opened a lengthy pamphlet, *Alternatives Compared*, with a particularly telling point. In the crisis of the 1790s, he argued that the rich and the powerful

> must not forget there is a disposition to the discovery of political as well as physical truth. Let them be content to take themselves only for what they are: subjects and interpreters, not arbitrary controllers, of the law of nature. This doctrine of philosophic humility may not be very congenial to the temper of those accustomed to be obeyed.[158]

This, in my view, clearly established the link between a fraying constitution, a desperate sense of political siege, and stringent resistance to pneumatic visions. These matters, Beddoes believed, could not be reconciled unless there was a reform in politics and a revolution in medicine.

By the 1790s Beddoes felt what many others feared: history's relentless charge and a world turned upside down. To his own dismay, Watt believed the authorities of Birmingham had turned mob encouragers, while town democrats would make a new aristocracy of themselves.[159] To have sensed these divisions as ones not just of politics but also of class and economy is one of the keys to understanding Beddoes' objectives. E. P. Thompson once argued that the period between 1791 and 1795 was crucial to the formation of a particular "working-class consciousness".[160] In our view, this may actually be too narrow a compass. Surely there were those among entrepreneurs and manufacturers who recognized in new wealth their own distinction from the long-standing power of tradition, land, or of Church and King. If consciousness or self-definition was thus refined in the first industrial revolution, perhaps as an economic and as cultural experiment, it also drove new fears.[161] A few short months *before* the Priestley riots, new anxiety was produced by the widespread distribution of Paine's *Rights of Man*, in answer to the alarms of Burke, as though any written gospel could drive a revolution rather than reflect one. Paine's publisher was Joseph Johnson, to whom Beddoes would himself ultimately turn.[162] The riots had made *Rights* seem a more dangerous manifesto than might have truly been the case. But it is impossible to divorce the impact of the *Rights of Man*, especially after 1792 with the campaign to prosecute Paine, from the great plan for a broad reformation in political economy and social medicine. Diffusion of knowledge and dispersal of instruments were thus public initiatives, as later praised by the radical John Thelwall, writing about the assemblies of men.[163] Like many, Watt was highly suspicious of the ideology expressed in such campaigns.

From Bristol, Beddoes was compelled to address endemic disease in Britain's rapidly changing industrial landscape. But the chemical knowledge of industrialists obviously suggested the possibility of a solution to chemically charged airs and, therefore, implied a means of the absorption of aerial therapies. The very ways of irritants to the lungs also proposed the pathways to remedy. The turn to James Watt in March 1794 was driven by a need to find a way to promote the pneumatic chemistry across the entire medical spectrum of physicians, surgeons, and apothecaries.[164] This was, of course, before Beddoes was consulted in the sad affair of Jessy's health.

Beddoes could not avoid conflict with the very authority that Burke thought besieged. This was especially the case after 1792 when the attack on Paine had gathered steam and Paine was tried *in absentia*. Indeed, in that summer Beddoes had already been identified by the Home Office amongst a number of potentially disaffected persons that included Priestley, the attorney John Frost, and Thomas Cooper of Bolton.[165] It was in their wake that Watt and Beddoes found themselves enveloped in uncertainty over airs and turmoil in politics. Watt, characteristically, headed for cover. To the Genevan geologist Jean-André de Luc, he expressed nothing but hostility for the French and their Revolution:

> their history shews them always to have been, inconstant, insolent, cruel, and ambitious, as the indian said the Rum was made for indian & indian was made for Rum, so Despotism was made for the french & they for despotism.[166]

For Watt this was bad enough, especially given the enthusiasm the younger James Watt might have had for the early days of the Revolution. However, throughout 1792, with British alarm on the rise, Watt felt certain there were more local dangers. By then the "devil of republicanism" seemed to be gaining ground, with fresh rioting erupting in the Midlands. Watt made the very connection that Burke had eloquently anticipated:

> unfortunately a set of vile people also have their apostles in this country, have had but too much effect in unsettling the minds of many and I much dread the consequences if the french do not meet some signal check, either from themselves or somebody else.[167]

Everywhere he turned, there were Jacobins. There was little to dissuade him.

Unfortunately, the assault on Thomas Walker's house in Manchester in December 1792 looked all too familiar through Birmingham eyes. It seemed clear that the drunken mob had been put up to attack any promoters of political reform. Walker had been one of the founders of the Manchester Constitutional Society, made up, it was said, of "mechanicks of the lowest class" and in which young James cultivated his Jacobin sympathies.[168] The reaction of Watt, senior, was by then predictable. He wrote immediately to his son, then in Paris among real Jacobins, and to Walker in Manchester. As far as he was concerned, mob government was odious enough, except in comparison to "that Bigotry which can descend to employ them". Once again he saw the hand of the High Church harnessing mobs in its own cause.[169]

The alarms manipulated by Pitt and his gang had an ample and willing audience. Even so, among those inclined to innovation or reform there were limits. As Watt told the consumptive William Withering of the Lunar Society, upon Withering's return from Portugal:

> It is vain for any man to attempt to combat the opinions of a nation in the hour of their prejudices, & at the time when we have been seriously alarmed by the machinations of the Jacobins to disorganize England as they have done France. The dreadful disasters which mob government has brought upon that country, make all unprejudiced men here, tremble at the mention of reform, lest by pulling out an seemingly useless peg, the bands which unite us should be [?] . . . unloosed & the System fall to pieces.[170]

Interestingly enough, to disclaim any talents for pillage or confusion was very much the public position later taken by Beddoes and repeated by his first, and highly circumspect, biographer John Stock. There is a key here to Beddoes' own conduct. While writing his early missive, *A Guide to Self-Preservation and Parental Affection*, in 1793, he clearly had in mind advice to the poor who could not entirely be blamed for being victims of disease.[171] This was a question of his political sympathy without being drawn into the widening gyre of social distress.

Unfortunately for Beddoes, his timing was terrible. In an age of riot, his assertion that labourers paid the price simply for being low born could not be devoid of political consequence. This was, of course, before he and Watt engaged in the promotion of pneumatic remedies. But as he took on the decay of the Bill of Rights, and with his reputation in the Home Office as a seditious person well documented, Beddoes ultimately acknowledged to Watt:

> I know very well that my politics have been very injurious to the airs. . . . Yet as every stroke aimed at liberty equally threatens science, morals & humanity, it requires great self denial to look on patiently & silently, when such great interests are at stake.[172]

The middle years of the 1790s were a turning point. By the summer of 1794, the government's attack on so-called Jacobins had reached the point of show trials of reformers on charges of treason, although juries often refused to convict.[173] Political persecution clearly defined many sympathies. If indeed it reflected the emergence of a working-class consciousness, it also surely revealed the crisis of the labouring poor and of the necessity of measures to address public health. James Johnstone, M.D., of the Worcester Infirmary, in 1796 wrote of the severe difficulties of the needle grinders, in a line running from Bromsgrove to Alcester, especially at Redditch within 15 miles of Birmingham. The scene was too grave to ignore. The dry grinding of needles, Johnstone argued, was inevitably fatal so that workers "hardly ever attain the age of forty years." Johnstone was of the view that the workers' habits – notably of heavy drinking following relatively high wages intended to compensate for the dangers – certainly made matters worse. It was the "metalline and stony particles of dust which fly off" that induced the "continual irritation of the lungs by the dust of small particles of iron and stone, and their gradual concretion on the air cells of the lungs". This produced the "suppurative inflamation" and ulceration exhibited in pulmonary phthisis.[174] Working conditions were clearly hazardous, and living conditions were notorious. By 1792, John Ferriar of the Manchester Infirmary had complained of the way the hovels into which the poor were crowded and of how the conditions in the cotton mills furthered the spread of epidemics. In a note appended to the publication of a broadsheet, addressed to the *Committee for the Regulation of the Police*, Ferriar complained of work-houses, the last resort of the poor. He sensibly remarked that "they always appeared to me to be managed on the erroneous principle of treating the Poor as criminals, instead of receiving them as persons entitled to public support." He proposed, moreover, the establishment of fever wards in the larger towns, such as Manchester, with "at least considerable assistance from Government" as it was unlikely that private subscriptions would ever prove dependable.[175] The wealth of the well-disposed could not be relied upon – and, as Ferriar noted, it was hardly reasonable for the rich to assume they would not be infected by the poor.[176]

The industrialisation of Britain's towns brought about a stark recognition of conditions ripe for infectious disease, often exacerbated by the methods of manufacturers. Polluted waters and polluted airs were the bequest of manufactures. The inevitable debate exercised those provoked by political economy as well as medical airs. At that moment, Watt and Beddoes published their first collection of *Considerations on the Medicinal Use, and on the Production of Factitious Airs*. They were anticipated by Richard Pearson of the Birmingham General Hospital in his *Short Account of the Nature and Properties of Different Kinds of Airs, So far as related to their Medicinal Use* in January 1795. This was also the very year in which the Rev. Joseph Townsend, geologist, physician, and Rector of Pewsey, Wiltshire, chose to pen his *Guide to Health*. He accepted the possibilities then promoted by Beddoes and by the London botanist and medical practitioner Robert Thornton. By then both Beddoes and Thornton had reported beneficial results by use of factitious airs on numerous disorders, including pneumatic, paralytic, palsies, loss of sight and hearing, and even typhus.[177]

Consequently, in Manchester John Ferriar drew attention to newly discovered airs. These were the widely discussed oxygen, hydrogen, and hydrocarbonate (from carbon monoxide and water) employed in treating asthma and phthisis. While Ferriar refused to draw any definitive inference from the trials then being reported, he determined to "continue to use the pneumatic medicine, but only in those disorders which prove intractable to common remedies" – by which he likely meant the use of laudanum, or even the magical foxglove. Ferriar was not entirely discouraged. He acknowledged Beddoes had "opened a new and extensive train of observation, in which even disappointment may prove instructive".[178] Significantly, Ferriar, as physician to the Manchester Infirmary, was not solely concerned with the ailments of the rich. He took detailed notes of the spread of disease among those working poor confined to hovels where "most of the inhabitants are paralytic, in consequence of their situation in a blind alley, which excludes them from light and air. Consumptions, distortion, and idiocy, are common in such recesses." While workers' quarters were cheap, crowded, and disease infested, he also declared there was no reason why the wealth-generating cotton mills of Manchester should be as unhealthy as they then so obviously were. It was hardly novel to suggest that at least cleaning the mills periodically and providing proper ventilation could attack some root causes of contagion.[179] Toward the end of the century, increasing attention was paid to diseases prevalent in industrial towns. Hence, John Johnstone, M.D., physician to the Birmingham General Hospital, in 1795 printed a short account of a button-gilder who experienced severe headaches, impaired vision and, ultimately, tremors in his limbs to the point that he could barely walk.[180]

This was all too common a story. At that very moment, Richard Pearson, also of the General Hospital and initially highly sceptical, printed a superb and optimistic account of a few artificially produced airs – that is, oxygen, azote, three kinds of inflammable airs, and carbonic acid air – that had met with so much opposition.[181] Airs and political economy were a provocative combination.

Ideology and industry

To understand the objectives of Thomas Beddoes, we need to realize the force of the industrial transformation of Britain. In an era of political alarm, which trapped the Pittite administration in a wave of paranoia, Beddoes refused to be as circumspect as others might. An excellent example is the London pneumatic practitioner Robert Thornton. While an avowed monarchist, Thornton, as enthusiastic as he was about pneumatic medicine, was comparatively cautious about his political opinions.[182] It is interesting that in 1795, at the pinnacle of Pittite repression, Thornton nonetheless chose to address political subjects. However turgid his prose, despite his inherent conservatism, he was not averse to recognizing the problems of the lower orders. He was loath to tamper with a constitution that, however limited its popular representation, must be preserved: "we may be assured, before we venture upon a reformation, that the MAGNITUDE OF THE EVIL JUSTIFIES THE DANGER OF THE EXPERIMENT." Social discontent may have deep foundations but no ready reformation. Thornton argued:

> The common sort of people who have leisure to think always find fault with the times, and some must have reason, for the merchant gains by peace, and the soldier by war; the shepherd by wet seasons, and the ploughman by dry: when the city fills, the country grows empty; and while trade increases in one place, it decays in another.[183]

Thus, with the example across the Channel, the complaints even of well-meaning men could end in a trail of disaster. He was, in effect, much closer to Watt than Beddoes in his suspicion of those who might magnify complaints:

> They blow up sparks wherever they find the stubble is dry: they find out miscarriages wherever they are, and forge them often where they are not; they find fault first with the persons in office, and then with the prince or state; sometimes with the execution of laws, and at other times with the institutions, how ancient and sacred soever. – They make alarms pass for actual dangers, and appearance for truth; represent misfortunes for faults, and mole-hills for mountains; and by the persuasion of the vulgar, and pretences for patriots, or lovers of their country, at the same time that they undermine the credit and authority of the government.[184]

Like Watt, Thornton worried about the fury of the ungovernable mob. He referred directly to the Gordon Riots of 1780, a sanguinary example of the rage never far from Burke's mind. Thornton wrote about the obvious dangers, even if complaints were justifiable,

> when they are once put in motion, they soon get beyond all restraint and controul [sic]. – The rights of man, to life, liberty, and property, oppose but a feeble barrier to them; the beauteous face of nature, and the elegant refinements of art, the hoary head of wisdom, and enchanting smile of beauty, are all equally liable to become obnoxious to them; and as all their power consists of Destruction, whatever meets with their displeasure must be devoted to ruin.

They create, he argued, "nothing but prospects at which every friend of mankind must shudder."[185] Thornton offered caution, just as Beddoes offered reasons to despise the younger Pitt – and, between them, Watt confined his private fears to his network of correspondents. Watt was not one for print, if Beddoes was.

After the collapse of notorious prosecutions for sedition (notably of Thomas Walker, Thomas Hardy, the Rev. John Horne Tooke, and John Thelwall) a few may have been emboldened. Among philosophers and medics, the debate over reform in general came ever increasingly to focus on the desperate lives of the labouring poor.[186] They were, most clearly, the victims of polluted airs, indeed even in the very manufacturing towns where chemists saw remedies. While some of Beddoes' correspondence for this contentious period has disappeared, even possibly destroyed by his widow or by his first biographer, much remains in the traces of his recipients. It is here that we can see the gathering of Beddoes' determination. In the midst of seeking support for a Pneumatic Institute, in the spring of 1795 he wrote repeatedly to Giddy. But his main theme was the "quondam patriot" Pitt and corruption, especially in a draft letter to the persecuted shoemaker Thomas Hardy that he shared with Giddy. Here Beddoes mustered his indignation against Pitt, who had, he declared, "like Robespierre succeeded in securing his authority by terror". Pitt was the villain of the piece and Beddoes was warming to the cause.[187] By November, Beddoes launched into print with broadsheets about the abuse of the Bill of Rights. His fundamental principle was an informed populace. Indeed, in a postscript that he printed just days after *A Word in Defence of the Bill of Rights*, Beddoes proclaimed that rational people "cannot be dupes of wicked demagogues." The key to this lay in resisting any government infringement of rights. This was an essential enlightenment doctrine, which sounds very much like the assertions of Joseph Priestley and James Keir: "As men become more humanized by knowledge, they cannot indeed become less free. Instruction must now daily spread; and the majestic stream of human reason hold on its uninterrupted course."[188] For Beddoes, Pitt's crimes were not solely ones of bills against seditious meetings.

To Beddoes, Pitt was the perpetrator of the extreme want among the poor. His policies proved destructive of trade and employment, without regard to damage to a replaceable workforce and an unconnected, manipulated, rabble. It was self-evident that politics magnified want. This placed the new medicine, emanating from the pneumatic rooms of Bristol's aptly named Hope Square, on a new footing. The mission of medicine was the alleviation of pain and the improvement of the life of the many. In 1796, Beddoes published with Johnson in London a profoundly blunt confrontation with Pitt and his policies. This was even more remarkable because of the intense scrutiny to which any opponent of the regime was then subjected.[189] Far from being the authors of their own misfortune, Beddoes argued, the poor were sacrificed on the altars of privilege, wealth, and power.

His two-hundred-page assault, on the *Public Merits of Mr. Pitt*, revealed the trap from which inhabitants of cellar or cottage had little chance of escape. Beddoes described the conditions of the poor as the consequence of overwork and dismal food. In industry, matters were even worse:

> I shall say nothing of the pestilence that so frequently visits the cotton works; of the palsy contracted during the fabrication of certain metallic toys; of the extensive ravages which the natural small-pox is still suffered to commit, or of the fevers which, like the fires of Vesta, eternally rage among the poor of some manufacturing towns.[190]

If overcrowding in garrets and cellars magnified disease, so too did the new poisons that arose every day in towns. Citing Ferriar's *Medical Histories*, Beddoes made it a point to repeat that "it is hardly possible to prevent communication of the disease to the families of the rich, among whom it would never have been produced."[191] This was but one of a series of pamphlets that tied together the threads of medicine, politics, and public health. In 1797, in his *Alternatives Compared: Or, What Shall the Rich Do to Be Safe?*, Beddoes pursued Pitt's domestic and international policies. He challenged Westminster to "compare Great Britain as it is, with Great Britain as it was" and to demand a change in the ministry. War had ruined manufactures, and many of those once employed in them, but the rich had not reflected on the consequence: "Twenty thousand poor families starved, an unknown and countless rabble killed off, without interruption to their enjoyments, will have no systematic influence on the conduct of the mass of the affluent, however well they may be disposed."[192] His theme was the deliberate immiseration of the majority, and the self-destructive indulgence of the wealthy few.

From Birmingham, Manchester, and from Darwin in Derby, Beddoes had received repeated descriptions of the children of the poor "who are in general *ill fed, ill lodged*, and *ill clothed*". Much was the consequence of ruinous wages and uncertain employment, which induced an inevitable comparison of "*high life with low.*" Where was the morality in crowded manufactures made even worse when trade declined and employment ceased? Consumer gains from falling prices could hardly induce happiness amongst the artisans of buckles and of baubles for the rich. For Beddoes this was as much about a moral economy emerging from rapidly expanding manufactures, especially in the Midlands he knew so well. Thus, confronting Pittites, he argued that prosperity would be best achieved

> by enabling a whole people to purchase largely of the productions of industry. Were no man, woman, or child in Great Britain obliged to go in rags, the home consumption of our labourious fellow-citizens would be more profitable than the custom of all the nobility and gentry of Europe.[193]

In such conditions, could it be any wonder that mobs might readily be raised? But the solution was political. The clear dangers in 1797 demanded the rich see their best interest lay in joining with others in ridding Britain of the ministry of Pitt and his gang. Beddoes was brave enough, at such a juncture, to point out that "The disorders occasioned by ministers [were] equally to be dreaded with those occasioned by mobs."[194]

Beddoes' planned Pneumatic Institution in Bristol should not then be seen as just an infirmary or even a laboratory. It was surely both, but it was also more. By the turn of the century, even during a prolonged war with the French, he could not resist confronting orthodoxy on all fronts. He had no use for the "reptiles that plant themselves on the high road of improvement".[195] He did not merely mean opposition to pneumatic innovation. What he intended was a re-orientation that explicitly attacked the political economy that crippled any acknowledgment of social medicine. In 1800, he wrote to Giddy:

> I believe the condition of the poor, particularly the peasantry, to have been growing worse for many years – & that the war is only a great aggravation of a disease under which the country without that wd have certainly laboured – Indeed I think this is deducible from the firmest principles of Pol. economy.[196]

Thus, at the turn of the century, Beddoes relocated his venture from fashionable Clifton to the slums of the Bristol docks in a Preventive Institution for the Sick and Drooping Poor. Roy Porter's vision of Pitt "grinding the face of the poor" has had far too many political imitators since, as though labourers were an impediment to the rights of the well-born.

The key to Beddoes, since his earliest forays with many into pneumatic medicine, was the proposition of political responsibility for remedies to public health.[197] This, of course, required a recognition of the rapid change that threatened to engulf the Pitt regime. It is somewhat ironic that the very markers of industrial pollution were those very spaces that suggested the effect of airs on the general population of urban centres. This his first biographer recognized. Thus, by 1802, when Beddoes published the first volume of his *Hygëia*, he refused to ignore the obvious foundation of ill health that lay in economic disparity, comparing the conditions of labourers to those of slaves.[198] In a way that few but Beddoes dared, he challenged the priests of political economy who justified, in the name of political tranquillity, the grim exploitation of the labouring poor.

Herein lay the effect of the poisons of industry on the unsuspecting, and ultimately upon the doomed. Was there not then enough worry that acknowledgment of noxious airs could only make worse? To a sometime democratic world, the question may seem bizarre. Beddoes, in *Hygëia*, asserted that ubiquitous lead, then not clearly in "the catalogue of poisons", was dangerous to "activity and enjoyment, produces palsy too, of the worst species".[199] Thus, even commonly experienced effects were traceable to those sources not immediately or obviously alarming. If the notion of the spread of knowledge were to be a tenet of enlightened reformers, then understanding of chemical consequences went far beyond the critically dangerous to the long-term results of exposure.

The whole business of airs remained both a puzzle and a promise during the 1790s. In 1794, Beddoes had mused over the fact that

> Japanners are constantly breathing the vapours of resinous substances, but I never could observe that they were more or less subject to phthisis than others; casters of fine brass work very often die consumptive, much more so than any other set of artists in Birmingham. They dust their moulds with powdered rosin, the vapour of which rises copiously when the melted metal is poured in. But the mischief can hardly be attributed to this vapour, otherwise the Japanners would be affected; nor yet to the flowers of zinc [zinc oxide], which are copiously diffused through the work-shops, because the casters of large brass work are not peculiarly liable to become consumptive. I suppose the Phthisis in these instances to be caused by the mechanical action of the powdery matters which float in the air in great quantities in these fine casting shops, and are necessarily taken in with the breath. Whilst flints for the potteries are pounded in mortars, the people so employed universally died consumptive, and the grinders of needles now often experience the same fate.[200]

Vapours were inescapable, but did they cause diseases or could they cure them?

As we have seen, Beddoes was far from alone in concern over public hygiene in the manufactories and workshops of a rapidly industrializing regime. These effects clearly had consequence in what E. P. Thompson, following Marx, once described as the sites of "engines of immiseration".[201] For pneumatic practitioners, there was more than an occasional link between industrial airs and innovations in medicine. That link lay in the social awareness of some physicians. By the turn of the century, Thomas Beddoes was repeating the earlier observations of Ferrier in the Manchester manufactories. But, he reminded his readers, "the self-neglect of the workmen, and the unconcern of the masters, are not phaenomena confined to the needle manufactory. We shall find them universal, wherever there is danger of pulmonary consumption." He was adamant that "An immense list of artisans of different name, whose labours are carried on amid the floating particles of earth and metals, might be subjoined to the needle-grinders."[202] Here he included plaster and marble workers in Paris who suffered from "*maladie des grès, or maladie de St. Roch*" the tubercular symptoms of which were all too common. Similarly, among the flax dressers, as when "softer dust, that arises during the fabrication of various animal and vegetable substances, is equally capable of bringing on a fatal irritation of the membrane, investing the air-pipes." As he put it succinctly, in many such trades "Consumption is the reigning malady." Once in trades, there was no escape:

> The girls who come fresh from the high and airy district of the Cévennes, to perform the various operations in silk manufactory at Nimes, from the preparation of the cocoons to the spinning of the silk, suffer like our flax-dressers. Even in a few days, they are described as losing their bloom and all their vivacity. They are seized with a dry cough, by which they soon come to be continually harassed. They complain of oppressive pain in the chest. Fever often succeeds, and they die consumptive.[203]

Beddoes contemplated a world where a moral economy might hold sway over the "state-pestilence of polite luxury." The consequence was dire for all: "although it parade in open day, and by the light of a thousand torches, contrives, unsuspectedly by the multitude, to mangle and destroy whatever it meets, and on both sides of its path." Crucially, Beddoes was a democrat who demanded an address to the social determinants of diseases that were clearly not always confined to the first victims of contagion. Could we not, he asked, address those "whom the political oeconomist looks, as upon a race of two-legged cattle stalled in our manufactories, in order that in due season, they may be driven to glut the dogs of war"?[204] To judge from Beddoes' contemporaries, like Ferriar, it is not that such questions had not been asked before. It was rather an issue of what responses would be effective at a time when labourers were merely a means to an end, and yet who also became vectors of disease where social barriers obviously proved no quarantine. Surely, as we now know, even Watt's own household, amongst his family and servants, was riddled with disease. Broader medical visions were clearly necessary – and they did exist, even in the political economy of the late eighteenth century. As Beddoes clearly put it, "They suggest, at least, the reasonable expectation that the cultivation of truth, by experiment and observation, will finally accomplish what legislation would only have made itself ridiculous in attempting."[205]

The consequences were obvious. Enlightened industrialists understood that not all mills need be dark or satanic. In the Rev. Stebbing Shaw's Arcadian vision of Soho, written just as Beddoes penned his *Hygëia* on public health, we have a virtually bucolic image of Birmingham quite different from the cellars and choking mills of Manchester. Shaw wrote of Soho:

> The rules of this manufactory have certainly been productive of the most laudable and salutary effects. And, besides the great attention to cleanliness and wholesome air, &c. this manufactory has always been distinguished for its order and good behaviour, and particularly during the great riots at Birmingham.[206]

But the riots of Birmingham had also revealed a deeper rot: of labourers at risk of the scrap heap when trade turned down and of their manipulation by magistrates enamoured of Church and King and of their own self-worth. It was no wonder that the promotion of pneumatic medicine was caught up in political tides. This, of course, lay behind Joseph Banks' refusal to assist Beddoes in plans for a pneumatic institution. This too, as James Watt junior pointedly remarked, in a letter to Ferriar in Manchester, was the political trap in which Beddoes had caught his "cloven *Jacobin* foot".[207]

Notes

1 Matthew Boulton and James Watt to Sir John Scott, 25 April 1793, BCL, JWP copy books; John Johnstone, M.D., ed., *The Works of Samuel Parr, Prebendary of St. Paul's, with Memoirs of His Life* (London: Longman, Rees, Orme, Brown and Green, 1828), I, 336; cited in Kenneth R. Johnston, *Unusual Suspects. Pitt's Reign of Alarm and the Lost Generation of the 1790s* (Oxford: Oxford University Press, 2013), 47. See also Rev. William Field, *Memoirs of the Life, Writings, and Opinions of the Reverend Samuel Parr* (London: Henry Colbourn, 1828), I, 304–315; R. B. Rose, "The Priestley Riots of 1791", *Past and Present*, **18** (November, 1960), 68–84, esp. 70, 72; Albert Goodwin, *The Friends of Liberty: The English Democratic Movement in the Age of the French Revolution* (Cambridge, MA: Harvard University Press, 1979), 180 ff.

2 Matthew Boulton to unknown, 18 July 1791, BCL, MBP (see abbreviation list) 249/219.

3 Cf. Wil Verhoeven, *Americomania and the French Revolution Debate in Britain, 1789–1802* (Cambridge and New York: Cambridge University Press, 2013); E. P. Thompson, *The Making of the English Working Class* (New York: Vintage Books, 1963), 37; Johnston, *Unusual Suspects*, 315. On the radical context see Michael Andrew Zmolek, *Rethinking the Industrial Revolution: Five Centuries of Transition from Agrarian to Industrial Capitalism in England* (Leiden and Boston: Brill, 2013), 537 ff.

4 Cf. Goodwin, *The Friends of Liberty*, *passim*; T. H. Levere, "Dr Thomas Beddoes: Chemistry, Medicine, and the Perils of Democracy", *Notes & Records of the Royal Society*, **63** (September, 2009), 215–229; and Johnston, *Unusual Suspects*, *passim*.

5 Porter, 187–188.
6 Jan Golinski, *Science as Public Culture: Chemistry and Enlightenment in Britain, 1760–1820* (Cambridge and New York: Cambridge University Press, 1992), 158–161, 176 ff.
7 See the Introduction to this volume. Quoted in Trevor H. Levere, "Dr. Thomas Beddoes and the Establishment of his Pneumatic Institution: A Tale of Three Presidents", *Notes and Records of the Royal Society*, **32** (July, 1977), 42. See also Stock (1811); reprint edition (Bristol: Thoemmes Press, 2003), 36.
8 Beddoes to Davies Giddy (later Gilbert), 5 May 1791, CRO, GD 41/6. Here possibly reflecting John Thelwall. See Thompson, *Making*, 157–159.
9 Porter, 74–77, 169.
10 Thompson, *Making*, 466; Johnston, *Unusual Suspects*, 214.
11 J. Reynolds to Davies Giddy, 26 August 1791, CRO, DG 41/30. On Reynolds see Hugh Torrens, "The Reynolds-Anstice Shropshire Geological Collection, 1776–1981", *Archives of Natural History*, **10** (1982), 429–441; Stansfield, 61.
12 James Watt to Jean-André e de Luc, 19 July 1791; James Watt to [Gilbert?] Hamilton, 24 July 1791, BCL, JWP copy books. James Watt to Jean-André de Luc, 19 July 1791; James Watt to [Gilbert?] Hamilton, 24 July 1791, op. cit.
13 James Watt junior to James Watt, 14 August 1791. BCL, JWP W/6/14; See also "The Chemists of Paris to Doctor Priestley", in *A Scientific Autobiography of Joseph Priestley, 1733–1804: Selected Scientific Correspondence Edited with Commentary*, ed. Robert E. Schofield (Cambridge, MA: MIT Press, 1966), 257–259.
14 James Watt to Joseph Black, 23 November 1791. BCL, JWP 4/12/31; reprinted in CJB, II, 1143–1144.
15 Watt to Jean-André de Luc, 29 August 1791. BCL, JWP copy books.
16 Beddoes to Giddy, 1 December, [1791?], CRO, DG 41/44.
17 Cf. Johnston, *Unusual Suspects*, 153. On the reaction to Pitt, see Jay, esp. 132–134.
18 Beddoes to Giddy, 4 July 1792, CRO, DG 41/16; Stock (1811), 72. Nonetheless, Beddoes' views were virtually barometrical, changing with every rumour of better days. See e.g. Beddoes to Giddy, 19 November 1792, CRO, DG 41/38.
19 Beddoes to Giddy, 11 May [1793?], CRO, DG 41/3.
20 Watt to Black, 17 July 1793, BCL, JWP 4/12/19; see CJB, II, 1208. On the battle of metaphors, especially of Priestley and Burke, see Maurice Crosland, "The Image of Science as a Threat: Burke versus Priestley and the 'Philosophical Revolution' ", *British Journal for the History of Science*, **20** (July, 1987), 277–307, esp. 281ff; Golinski, *Science as Public Culture*, 176 ff.
21 Cf. Stock (1811), 117; Thompson, 126; Johnston, *Unusual Suspects* gives a broad picture of the climate of suspicion.
22 Beddoes to Giddy, probably late 1792 or very early 1793, CRO, DG 41/22. See also F. W. Gibbs and W. A. Smeaton, "Thomas Beddoes at Oxford", *Ambix*, **9** (1961), 47–49, and T. H. Levere, "Dr. Thomas Beddoes at Oxford: Radical Politics in 1788–1793 and the Fate of the Regius Chair in Chemistry", *Ambix*, **28** (1961), 61–69. Although the Crown was to have funded the chair, it is not clear that this was to have been a Regius Professorship. By 1792 Beddoes was to be found among a Home Office list of seditious persons. National Archives (Kew) [NA], Home Office Papers [HO] 42/21. 28 July 1792. By November, the Home Office was clearly contemplating prosecuting "Dr. Beddowes" for "sowing sedition" in the neighbourhood of Shifnall. Evan Nepean, Under-Secretary of State, to Isaac Hawkins Browne, 1 November, 1792. HO 42/22/233; Stansfield, 79.
23 Priestley to Boulton, n.d., BCL, MBP 249, no. 212; and no. 213. Priestley to Boulton, 25 November 1777; Robert E. Schofield, ed., *A Scientific Autobiography of Joseph Priestley, 1733–1804: Selected Scientific Correspondence Edited*

with Commentary (Cambridge, MA: MIT Press, 1966), 161–161. See also Simon Schaffer, "Measuring Virtue: Eudiometry, Enlightenment and Pneumatic Medicine", in *The Medical Enlightenment of the Eighteenth Century*, eds. Andrew Cunningham and Roger French (Cambridge and New York: Cambridge University Press, 1990), esp. 290–291; Victor D. Boantza, "Collecting Airs and Ideas: Priestley's Style of Experimental Reasoning", *Studies in History and Philosophy of Science*, **38** (2007), 506–522; and Boantza, "The Rise and Fall of Nitrous Air Eudiometry: Enlightenment Ideals, Embodied Skills, and the Conflicts of Experimental Philosophy", *History of Science*, **51** (2013), 377–412. Also, Levere, "Measuring Gases and Measuring Goodness", in *Holmes and Levere* (2000), 105–135.

24 Rev. Stebbing Shaw, *The History and Antiquities of Staffordshire*, Vol. II (London: J. Nichols and Son & Co., 1801), 117.

25 Shaw, *Staffordshire*, II, 117, 121.

26 See, for example, the observations in the seventeenth century of Bernardino Ramazzini of Modena, now often deemed the father of occupational health. J. S. Felton, "The Heritage of Bernardino Ramazzini", *Occupational Medicine*, **47** (3) (1997), 167–179; G. Franco, "Ramazzini and Workers' Health", *The Lancet*, **354** (4 September, 1999), 858–861, and Schaffer, "Measuring Virtue", 293. Take the problem of the diseases of chemists or apothecaries, or of lead manufacturers who in turn produced poisonous glazes for potters, or of glass or mirror makers mixing mercury on the island of Murano, near Venice. According to one letter to the Royal Society, published in the first volume of the *Philosophical Transactions* in 1667, workers on mirrors were highly affected by palsy or, in the view of Ramazzini, "oftentimes die Apoplectic", Bern. Ramazzini, *A Treatise of the Diseases of Tradesmen* (London: Printed for Andrew Bell, Ralph Smith, Daniel Midwinter, et al., 1705), 28, 30, 39; Extract of a Letter, Lately Written from Venice, by the Learned Doctor Walter Pope, to the Reverend Dean of Rippon, Doctor John Wilkins, Concerning the Mines of Mercury in Friuli, *Philosophical Transactions*, **1** (April, 1665), 21–26, esp. 24. Cf. Mark Jenner, "The politics of London air John Evelyn's Fumifugium and the Restoration", *The Historical Journal*, **38** (September, 1995), 536–551.

27 Thomas Percival to Watt, 16 September 1786, BCL, JWP 4/48/7.

28 James Watt to Mr. Gilbert Hamilton, 24 July, 1791. BCL, JWP copy books. Italics are underlined in original.

29 Thompson, *Making*, 113; Goodwin, *The Friends of Liberty*, 265. Cf. Frida Knight, *The Strange Case of Thomas Walker* (London: Lawrence & Wishart, 1957).

30 Beddoes to Giddy, 9 December 1792. CRO, DG 41/57.

31 Stock (1811), 196–197, 229–230; Porter, 164; Golinski, *Science as Public Culture*, 160 ff.

32 James Watt junior to his father, September, 1793, BCL, JWP W/6/30. The reference is to John Frost, attorney and former associate of Pitt, who was sentenced to the pillory for sedition. See Thompson, *Making*, 129. For an excellent account of the difficulties between James Watt and his son, see Peter M. Jones, "Living the Enlightenment and the French Revolution: James Watt, Matthew Boulton and Their Sons", *The Historical Journal*, **42** (1999), 157–182; cf. Eric Robinson, "An English Jacobin: James Watt, Junior, 1769–1848", *Cambridge Historical Journal*, **11** (1955), 349–355; Jones, *Industrial Enlightenment: Science, Technology and Culture in Birmingham and the West Midlands 1760–1820* (Manchester and New York: Manchester University Press, 2008), esp. 212. See also Knight, *The Strange Case of Thomas Walker*.

33 Thompson, *Making*, 172; Johnston, *Unusual Suspects*, 102 ff.

34 Beddoes to Giddy, 14 March 1795,CRO, DG 42/30; Beddoes to Giddy, 2 March 1795, DG 42/35. See notably Johnston, *Unusual Suspects.*
35 Thompson, *Making*, 177–178. On the economic problems behind increasing disturbance, see Chris Evans, *Debating the Revolution in Britain in the 1790s* (London: I. B. Tauris, 2006), 88 ff.
36 James Wat, junior to his father, September, 1793, BCL, JWP W/6/28; Beddoes to Giddy, March 14, 1795, CRO, DG. 42/20.
37 Beddoes to Giddy, 21 January 1802, CRO, DG 42/8.
38 Beddoes to Giddy, 23 April 1796, CRO, DG 42/23. On Beddoes' views of medical institutions and disease, see Porter, *Doctor of Society*, 43 ff.
39 Thomas Beddoes, *Extract of a Letter on Early Instruction, Particularly That of the Poor* (25 January, 1792), 17, 20; the printer remained anonymous. Beddoes was only one of many determined to expose the prejudices of Burke. See Smolek, *Rethinking*, 540 ff.
40 Beddoes to James Watt, 21 April [probably 1798], BCL, JWP W/9/7.
41 On Keir, see Stock (1811), 37–38; Stansfield, 62–63. James Keir, *The First Part of a Dictionary of Chemistry &c.* (Birmingham: 1789), preface, iii-iv.
42 On William Henry see W. V. Farrar, Kathleen R. Farrar, and E. L. Scott, "Thomas Henry's Sons: Thomas, Peter and William", in *Chemistry and the Chemical Industry in the 19th Century: The Henrys of Manchester and Other Studies*, eds. Richard L. Hills and W. H. Brock (Aldershot and Brookfied: Ashgate, 1997), II, 189; also William Henry to James Watt, 15 December 1790[?], BCL, JWP 6/35/32. On watching French chemists see also Thomas Cooper to James Watt, 14 September, 1790, BCL, JWP 4/48/2.
43 Watt to Joseph Black, 3 February 1783, BCL, JWP 4/12/46; reprinted in CJB, I, 611. On Watt's chemistry see especially David Philip Miller, *James Watt, Chemist: Understanding the Origins of the Steam Age* (London: Pickering and Chatto, 2009).
44 Porter, 17. See also David Philip Miller and Trevor H. Levere, "'Inhale it and See?' The Collaboration between Thomas Beddoes and James Watt in Pneumatic Medicine", *Ambix*, 55 (March, 2008), 5–28.
45 James Watt to [Gilbert] Hamilton, 16 May 1790; Watt to James Watt junior, 20 July,1790; Watt to Hamilton, 8 March, 1791, BCL, JWP copy books; William Withering, "Experiments and Observations on the Terra Ponderosa, &c; Communicated by Richard Kirwan, Esq. F. R. S.", *Philosophical Transactions*, 74 (1782), 293–311.
46 Watt to James Watt junior, 16 October 1790, BCL, JWP copy books.
47 Cf. F. F. Cartwright, "The Association of Thomas Beddoes, M.D. with James Watt, FRS", *Notes and Records of the Royal Society of London*, 22 (September, 1967), 131–143; Dorothy A. Stansfield and Ronald G. Stansfield, "Dr. Thomas Beddoes and James Watt; Preparatory Work 1794–96 for the Bristol Pneumatic Institute", *Medical History*, 30 (1986), 276–302, esp. 282 ff.; Miller and Levere, "'Inhale It and See?'", 1–24; Jay, 97–100.
48 James Watt to Joseph Black, 15 May 1794, BCL, JWP 4/12/28; Watt to Black, 9 June 1794, JWP 4/12/27; See CJB, II, 1236–1240.
49 Watt to Darwin, 30 June, 1794, BCL, JWP copy books. See also Darwin to Watt, 25 May, 11 June, 3 July 1794 in CED, 436–437, 440–441.
50 Ewart to Watt, 5 October 1794, BCL, JWP 4/23/45. On Ewart see Stock (1811), 98; L. Stewart, "His Majesty's Subjects: From Laboratory to Human Experiment in Pneumatic Chemistry", *Notes and Records of the Royal Society*, 63 (September, 2009), 231–245, esp. 231–232. Beddoes was particularly averse to seeking support of "certain fashionable Doctors & if I had, I should almost feel degraded." Beddoes to Thomas Wedgwood, 4 November 1794, WMB W/M, 35.

51 Stansfield, 33–34. On Guyton see William A. Smeaton, "Platinum and Ground Glass: Some Innovations in Chemical Apparatus by Guyton de Morveau and Others", in *Instruments and Experimentation in the History of Chemistry*, eds. Frederick L. Holmes and Trevor H. Levere (Cambridge and London: MIT Press, 2002), 211–237, esp. 225 ff.
52 Beddoes to Watt, 27 February 1795, BCL, JWP 4/65/18. See Chapter 1, this volume.
53 James Carmichael Smyth to Matthew Boulton, 25 October 1793, BCL, MBP 254/228. On Smyth see Stock (1811), 134; George Stronach, "Smyth, James Carmichael (1742–1821)", ODNB.
54 Beddoes to Watt, 21 August 1794, BCL, JWP 4/23/25.
55 Beddoes to Watt, 7 July 1794[?], BCL, JWP 4/23/28. On Saunders, see Susan C. Lawrence, *Charitable Knowledge. Hospital Pupils and Practitioners in Eighteenth-Century London* (Cambridge and London: Cambridge University Press, 1996), 310; Noel G. Coley, "Medical Chemists and the Origins of Clinical Chemistry in Britain (circa 1750–1850)", *Clinical Chemistry*, **50** (2004), 961–972.
56 James Carmichael Smyth to Matthew Boulton, 21 February 1795, BCL, MBP 254/229; Beddoes to Giddy, 12 February 1795, CRO, DG 42/36.
57 James Watt to Thomas Garnett, 9 February 1797, BCL, JWP copy books; reproduced in part, but misattributed to James Carmichael, in A. E. Musson and Eric Robinson, *Science and Technology in the Industrial Revolution* (Toronto: University of Toronto Press, 1969), 146.
58 James Watt, "Instructions for G. Watt", 15 November 1797; Watt to James junior, 16 November 1797, BCL, JWP copy books.
59 Partington to Watt, 24 January 1798[?], BCL, JWP W/9/9. Cf. Stuart Strickland, "The Ideology of Self-Knowledge and the Practice of Self-Experimentation", *Eighteenth-Century Studies*, **31** (Summer, 1998), 453–471.
60 Robinson Boulton to Beddoes, 17 November 1799, BCL, MBP 220/22.
61 Watt to Black, 8 December, 1794 BCL, JWP Letter Book 2, 188, reprinted in CJB, II, 1248.
62 Beddoes to Watt, 14 June 1794, BCL, JWP 2/22/31. See also Porter, *Doctor of Society*, 49–51. On Partington, see Paola Bertucci, "Revealing Sparks: John Wesley and the Religious Utility of Electrical Healing", *British Journal for the History of Science*, **39** (September, 2006), 341–362; Paul Elliott, " 'More Subtle than the Electrical Aura': Georgian Medical Electricity, the Spirit of Animation and the Development of Erasmus Darwin's Psychophysiology", *Medical History*, **52** (April, 2008), 195–220; also Miles Partington to Watt, 24 January, [possibly 1798], BCL, JWP W/9/9.
63 Beddoes to Watt, March 1795, BCL, JWP 4/65/23; Thomas Beddoes and James Watt, *Considerations on the Medicinal Use, and on the Production of Factitious Airs: Edition the Second: To Which Are Added Communications from Doctors Carmichael, Darwin, Ewart, Ferriar, Garnet, Johnstone, Pearson, Thornton, and Trotter; from Mr. Atwood, Mr. Barr, Surgeon to the Birmingham Dispensary, Mr. Walter William Capper, Mr. Gimbernat, Surgeon to the King of Spain, Mr. Sandford, Surgeon to the Worcester Infirmary, and others* (Briston: Bulgin and Rosser & London: J. Johnson, 1795), 146–147. See also, Miller and Levere, " 'Inhale it and See?' ", 19–20. On identifying airs, see Chapter 1, this volume.
64 On the frequency of such visits see John Key, seeking to improve windmills at Lisbon. Key to Watt, 3 November, 1791, BCL, JWP 4/13/6; also John F. Fulton, "The Place of William Withering in Scientific Medicine", *Journal of the History of Medicine & Allied Sciences*, 8 (January, 1953), 1–5; Jeffrey K. Aronson, "Withering, William (1741–1799)", ODNB; on Daer, Thompson, *Making*, 183;

and Larry Stewart, "Putting on Airs: Science, Medicine, and Polity in the Late Eighteenth-Century", in *Discussing Chemistry and Steam: The Minutes of a Coffee House Philosophical Society 1780–1787*, eds., Trevor Levere and Gerard L'E Turner (Oxford and New York: Oxford University Press, 2002), 237.

65 Beddoes to Giddy, 29 October 1793; CRO, DG 41/4.

66 Watt to Darwin, 30 June 1794; Watt to Beddoes, 15 July 1794, BCL, JWP copy books; Beddoes to Watt, 30 October 1794, JWP 4/23/8.; JWP Letter Book 2, 168; or Watt to Black, 31 August 1794, JWP 4/12/24., reprinted in CJB, II, 1243; Beddoes to Josiah Wedgwood, 12 August 1794?, WMB, W/M 35. See Chapter 1, this volume.

67 Beddoes to Watt, 5 January 1796, BCL, JWP 4/23/4.

68 Watt to Black, 7 January 1796, BCL, JWP, Letter Book 2, 221; reprinted in CJB, II, 1277–1279. Our italics.

69 Watt to Beddoes, 28 October 1797, BCL, JWP copy Books.

70 Watt to Beddoes, 11 April 1799, BCL, JWP copy books.

71 Watt to Black, 22 November 1799, BCL, JWP copy books; reprinted in CJB, II, 1382–1383. Watt was unaware of Black's death two days earlier.

72 Watt to Black, 8 December 1799, BCL, JWP copy books; reprinted in CJB, II, 1386–1388.

73 Beddoes to Watt, 12 January 1801, BCL, JWP, 6/33/38.

74 Watt to Beddoes, 20 June 1802; 7 August 1802; Watt to Franz Anton Schweidiauer, 2 January 1803, BCL, JWP copy books. Schwediauer or Swediauer (1748–1824) a Viennese physician and, according to Black a disreputable projector, known also to Beddoes, was a Dantonist in the Revolution. See Black to his brother Alexander, 20 February 1791, in CJB, II, 1116. On Gregory's travels as geologist, see Hugh S. Torrens, "The Geological Work of Gregory Watt, His Travels with William Maclure in Italy (1801–1802), and Watt's 'Proto-Geological' Map of Italy (1804)", in *The Origins of Geology in Italy, Geological Society of America Special Paper*, eds. G. B. Vai and W. G. E. Caldwell, **411** (2006), 179–197.

75 See Miller, *James Watt, Chemist.*

76 Watt to Black, 15 March 1789, BCL, JWP 4/12/33; Watt to Black, 5 May, 1789, JWP 4/12/35; Watt to Black, 5 December, 1790, JWP 4/12/32, Reproduced in CJB, II, 1005, 1014–15, 1111; Watt to Joseph Black, 5 May 1789, JWP 4/12/35. In CJB, II, 1014–1015. Crawford, FRS, physician at St. Thomas's and chemist at the Woolwich Military Academy was a member of the Chapter Coffee House Society. See Levere and Turner, *Discussing Chemistry and Steam*, *passim*.

77 Watt to Joseph Black, 5 December 1790, BCL, JWP 4/12/32; Watt to Black, 13 March 1789, JWP 4/12/33; Watt to Black, 17 July 1793, JWP 4/12/29. Reproduced in CJB, II, 1110–1111, 1005, 1207–1208.

78 Beddoes to Tom Wedgwood, 21 May 1795; also Tuesday, n.d., WMB, W/M 35.

79 Stock (1811), 165.

80 John Key to James Watt, 3 November 1791, BCL, JWP 4/13/6; Beddoes to Watt, 30 October 1794, JWP 4/23/8.

81 Watt to Thomas Percival, 24 November 1794, BCL, JWP 4/65/19. See John V. Pickstone, "Thomas Percival and the Production of Medical Ethics", in *The Codification of Medical Morality*, eds. R. Baker, Dorothy Porter and Roy Porter (Dordrecht: Kluwer Academic, 1993), 161–178.

82 Beddoes to Thomas Wedgwood, 7 November 1794, WMB,W/M 35; [Robert Thornton], *The Philosophy of Medicine: Or, Medical Extracts on the Nature of Health and Disease, Including the Laws of the Animal Oeconomy, and the Doctrines of Pneumatic Medicine* (London: Printed by C. Whittingham, Dean-street, Fetter-Lane, 1799), I, 433.

83 Beddoes to Watt, 4 March 1795, BCL, JWP 4/65/25.
84 Thomas Henry to James Watt, 11 March 1795, BCL, JWP 4/65/27. See Wilfred Vernon Farrar, "Thomas Henry (1734–1816)", in *Chemistry and the Chemical Industry in the 19th Century*, eds. Richard L. Hills and W. H. Brock (Aldershot: Variorum, 1997), I, esp. 202.
85 Beddoes to Boulton and Watt, 5 November 1794, BCL, JWP 4/65/30. On Lord Daer, see Thompson, *Making*, 183; Stewart, "Putting on Airs", 236–237; Johnston, *Unusual Suspects*, 293.
86 Beddoes to Watt, October 1794, BCL, JWP 4/65/32; Beddoes to Watt, 20 May 1794, JWP 4/65/1.
87 Beddoes to Watt, October 1794, BCL, JWP 4/65/32.
88 Beddoes to Watt, 24 October 1794, BCL, JWP 4/65/33; Beddoes to Watt, 14 October 1794. JWP 4/65/35.
89 Beddoes to Watt, 14 October 1794, BCL, JWP 4/65/35; Miller and Levere, " 'Inhale it and See?' ", 23–24.
90 Watt to Sir Joseph Banks, 7 December 1794, BCL, JWP 4/65/28. A slightly different version is reproduced in Neil Chambers, ed., *Scientific Correspondence of Sir Joseph Banks, 1765–1820* (London: Pickering and Chatto, 2007), IV, 340. On the failure to interest Banks, see Jay, 244–245.
91 Banks to Watt, 7 December 1797, BCL, JWP 4/65/37. This does not appear in the collected correspondence of Banks but the discussion was initiated, over a month earlier, from Birmingham by William Withering and Watt. See Withering to Banks, 31 October 1797 and Watt to Banks, 5 November 1797 in Chambers, ed., *Scientific Correspondence of Sir Joseph Banks*, IV, 506–508.
92 Stock (1811), 100; Levere, "Dr Thomas Beddoes: Chemistry, Medicine, and the Perils of Democracy", 224; Jay, 96–97, 101–102, 105–106.
93 WMB, W/M 35, Beddoes to Thomas Wedgwood, 25 June 1795.
94 Beddoes to Watt, 24 February 1796, BCL, JWP 4/65/36; Beddoes and Watt, *Considerations on the Medicinal Use, and on the Production of Factitious Airs*, Part I. By Thomas Beddoes, M.D. Part II. By James Watt, Engineer. Edition the Third. Corrected and Enlarged (Bristol: Printed by Bulgin and Rosser, For J. Johnson, in St. Paul's Church-Yard, London, 1796), Addenda, 1; Stansfield, 30; On Girtanner see Jay, 67, 136–137.
95 Beddoes to Watt, 26 May 1797, BCL, JWP W/9/29.
96 WMB, Wedgwood correspondence [WC], E 3317 and 3318, Watt to Tom Wedgwood, 16 and 23 December 1799.
97 On Davy's pneumatic experiments see June Z. Fullmer, *Young Humphry Davy: The Making of an Experimental Chemist* (Philadelphia: American Philosophical Society, 2000); and Jay, esp. chapters 5–6. On Davy, see Chapter 1, this volume.
98 Beddoes to Watt, 29 May 1799, BCL, JWP 6/35/4; Miller and Levere, " 'Inhale it and See?' ", 19–21; Stansfield, 162 ff.
99 Beddoes to Watt, 17 December 1799, BCL, JWP 6/35/1.
100 Beddoes to Watt, 12 January,1801, BCL, JWP 6/33/38; Beddoes to Watt, 23 December 1800, JWP 6/33/70. The declaration did not stop the *Anti-Jacobin* from immediately attacking Beddoes' hopes for restoring muscular energy. See *Anti-Jacobin Review and Magazine* (August, 1800), 425. http://babel.hathitrust.org/cgi/pt?id=inu.30000080768322;view=1up;seq=437
101 Beddoes to Watt, 20 January 1801, BCL, JWP 6/33/69.
102 Joseph Priestley to Watt, 17 October 1801, BCL, JWP 6/33/64.
103 William Henry to James Watt junior, 15 June 1804, BCL, JWP 6/41/1; Gregory Watt to William Henry, 4 August, 1804, JWP 6/41/11.

104 Beddoes to Watt, 13 January 1804, BCL, JWP 4/8/34; 5 March 1804, JWP 4/8/33.
105 Thomas Beddoes, M.D., *A Letter to Erasmus Darwin, M.D: On A New Method of Treating Pulmonary Consumption, and Some other Diseases hitherto found Incurable* (Bristol: Printed by Bulgin and Rosser, Broad-Street; Sold by J. Murray, No. 32, Fleet-street, and J. Johnson, No. 72, St. Paul's Church-yard, London; also by Bulgin and Sheppard, J. Norton, J. Cottle, W. Browne and T. Mills, Booksellers, Bristol, 1793), 3.
106 Beddoes to Watt, 25 June 1794, BCL, JWP 4/23/30; Watt to Darwin, 14 August 1794, JWP Copy books.
107 Lind to Watt, 20 February 1795, BCL, JWP 4/65/21; Stewart, "Putting on Airs", 220, 240. Underlined in original.
108 Watt to Lind, 25 February 1795, BCL, JWP C1/15; Lind to Watt, 7 August 1786, BCL JWP MS 3219/4/29 no. 6, W/9.
109 J. Haygarth to Thomas Wedgwood, 19 May 1795, WMB, WC, E20–17716; Christopher Booth, *John Haygarth, FRS (1740–1827): A Physician in the Enlightenment* (Philadelphia: American Philosophical Society, 2005), 37, 92, 101, 121.
110 *A Catalogue of the Very Valuable and Extensive Library of Thomas Beddoes, M.D.* (London: London: Leigh and S. Sotheby, 1809); British Library, Sotheby 60(2). See Chapter 4.
111 Beddoes to Tom Wedgwood, 3 August [1793?], WMB W/M 35, See, for example, Thomas Beddoes, *Reports Principally Concerning the Effects of the Nitrous Acid in the Venereal Disease, By the Surgeons of the Royal Hospital at Plymouth, and By Other Practitioners* (Bristol: Printed by N. Biggs, for J. Johnson, St. Paul's Church-yard, London, 1797); Thomas Beddoes, *Communications Respecting the External and Internal Use of Nitrous Acid; Demonstrating its Efficacy in Every Form of Venereal Disease, and Extending its Use to Other Complaints: With Original Facts, and a Preliminary Discourse, by the Editor* (London: Printed for J. Johnson, No. 72, St. Paul's Church-Yard, 1800), xxviii. [British Library, B399co. 1799]. See also Francis Geach to Beddoes, 26 July 1797 in Robert Thornton, *The Philosophy of Medicine: Or, Medical Extracts on the Nature of Health and Disease, Including the Laws of the Animal Oeconomy, and the Doctrines of Pneumatic Medicine* (London: Printed by C. Whittingham, 1799–1800), V, 405.
112 James Watt to Lind, 25 February 1795; BCL, JWP C1/15; Thomas Henry to Watt, 28 December, 1794, JWP C1/16. On the Henrys, see W. V. Farrar, Kathleen R. Farrar and E. L. Scott, "Thomas Henry's Sons: Thomas, Peter and William", in *Chemistry and the Chemical Industry in the 19th Century*, eds. Richard L. Hills and W. H. Brock (Aldershot: Variorum, 1997), II, *passim.*
113 See Thomas Beddoes, M.D., and James Watt, Engineer, *Considerations of the Medicinal Use and Production of Factitious Airs*, 2nd. edn., corrected and enlarged (London: printed for J. Johnson, St. Paul's Church-Yard, 1796).
114 Beddoes to Tom Wedgwood, 17 March 1795, WMB, W/M 35.
115 Beddoes to Tom Wedgwood, 27 March [1795?], WMB, W/M 35.
116 Beddoes to Tom Wedgwood, n.d. [poss. 1794; but not after January 1795], WMB, W/M 35.
117 Beddoes to Watt, 20 June 1796; BCL, JWP W/9/52. Beddoes printed the case in the *Considerations*, Part IV, 3. It was reprinted later by the London physician, Robert Thornton, in his *The Philosophy of Medicine: Or, Medical Extracts on the Nature of Health and Disease, Including the Laws of the Animal Oeconomy, and the Doctrines of Pneumatic Medicine* (London: Printed by C. Whittingham, Dean-street, Fetter-Lane, 1799), I, 509–513.

118 Cleghorn to Watt, 12 May 1796, BCL, JWP W/9/56. On Cleghorn, see Levere and Turner, *Discussing Chemistry and Steam*, 220.
119 Darwin to Watt, 17 August 1794 (note diagrams), BCL, JWP 4/23/35; see also Darwin to Watt, 1 June 1796, JWP W/9/55, and CED, 448–450, 498–499. On Derby see Paul A. Elliott, *The Derby Philosophers: Science and Culture in British Urban Society, 1700–1850* (Manchester: Manchester University Press, 2009). For Darwin, Beddoes and geology or mineralogy, see Chapter 2, this volume.
120 Darwin to Watt, 28 April, 1795, BCL, JWP 4/65/6. See CED, 478.
121 Beddoes to James Watt, n.d., BCL, JWP, W/9/48.
122 WMB, WC, E20–17716, J. Haygarth to Thomas Wedgwood, 19 May 1795.
123 http://www.medicalheritage.co.uk/Bath/Bath%20Medics.html; Creaser to James Watt, 1797, BCL, JWP W/9/24; Creaser to Watt, 23 October 1798, W/9/1; Thomas Beddoes, M.D., *Contributions to Physical and Medical Knowledge, Principally from the West of England* (Bristol: Printed by Biggs & Cottle, for T. N. Longman and O. Rees, Paternoster-Row, London, 1799), 335–338.
124 Humphry Davy to Watt. 19 December, 1800? BCL, JWP 6/33/66. See also David Knight, *Humphry Davy: Science & Power* (Oxford: Blackwell Publishing, 1992), 26–41. June Z. Fullmer, *Young Humphry Davy: The Making of an Experimental Chemist* (Philadelphia: American Philosophical Society, 2000), 310–314; BCL, JWP 6/20/6. Henry Sully to Watt, 20 February 1805; http://www.wiveliscombe.com/a_short_history. On the adoption of galvanic medicine, see Jan Golinski, “From Calcutta to London: James Dinwiddie's Galvanic Circuits”, in *The Circulation of Knowledge between Britain, India and China: The Early-Modern World to the Twentieth Century*, eds. Bernard Lightman, Gordon McOuat and Larry Stewart (Leiden and Boston: Brill, 2013), 75–94.
125 J. Seward to Watt, 24 August 1796, BCL, JWP W/9/42; Larry Stewart, “His Majesty's Subjects: From Laboratory to Human Experiment in Pneumatic Chemistry”, *Notes & Records of the Royal Society*, **63** (2009), 231–245, esp. 239.
126 Beddoes to Watt, January 1804, BCL, JWP 4/8/35. See also Richard Pearson, *A Short Account of the Nature and Properties of Different Kinds of Airs, So Far as Relates to Their Medicinal Use; Intended as an Introduction to the Pneumatic Method of Treating Diseases, with Miscellaneous Observations on Certain Remedies Used in Consumptions* (Birmingham: Printed by Thomas Pearson. Sold by R. Baldwin, No. 47, Pater-Noster Row, London, 1795).
127 http://www.haven.u-net.com/BIOBO-2.htm. We owe this reference to Mr. Marc Macdonald.
128 Cf. Miranda Millendorf, *The World in a Book: Robert John Thornton's Temple of Flora*. Ph.D. thesis, Harvard University, 2013; https://www.academia.edu/400046.
129 Beddoes to Watt, 7 July 1794[?], BCL, JWP 4/23/28. On Saunders see also Cherry L. E. Lewis, “Doctoring Geology: The Medical Origins of the Geological Society”, *Geological Society, London of London, Special Publications*, **317** (2009), 49–92, esp. 52 ff. It was to the Physical Society at Guy's that the radical and materialist John Thelwall gave a controversial lecture in January, 1793. See Thelwall, *An Essay towards a Definition of Animal Vitality; Read at the Theatre, Guy's Hospital, January 26, 1793; in Which Several of the Opinions of the Celebrated John Hunter are Examined and Controverted* (London: T. Rickaby, 1793).
130 Thomas Beddoes, M.D., *Letters from Dr. Withering, of Birmingham, Dr. Ewart, of Bath, Dr. Thornton, of London, and Dr. Biggs, Late of the Isle of Santa-Cruz; Together with Some Other Papers, Supplementary to Two Publications*

on Asthma, Consumption, Fever, and other Diseases (Bristol: Printed by Bulgin and Rosser, Broad-Street. Sold by J. Johnson, No. 72, St. Paul's Church-Yard; and H. Murray No. 32, Fleet-street, London; also by Bulgin and Sheppard, J. Norton, J. Cottle, W. Browne, and T. Mills, Bristol, 1794), 22, Thornton to Beddoes, 7 December 1793; and 36, Thornton to Beddoes, 4 January 1794.

131 Beddoes to Watt, 5 November 1794, BCL, JWP 4/65/30.

132 Watt to Black, 8 December 1794, BCL, JWP 4/12/23. Reproduced in CJB, II, 1248–1249.

133 His comments to Tom Wedgwood on the lack of expertise in London were more than misleading but may also reflect his suspicions of Thornton's extensive medical practise. See Beddoes to Wedgwood, 9 June 1795, WMB, W/M 35. Cf. Jay, 111.

134 Beddoes to Watt, 12 December 1794, BCL, JWP 4/65/7; quoting from Thornton to Beddoes, 15 November 1794. Like Saunders, Richard Pearson, Darwin, and many others we have encountered, Daniel Hill was a member of the Medical Society at Guy's and also author of *Practical Observations on the Use of Oxygen, or Vital Air, in the Cure of Diseases* (London, 1820; See *A List of the Members of the Medical Society of London for the year MDCCLXXXIX: By the Medical Society of London* (London: Cicero Press, 1789).

135 Beddoes to Watt, 4 March 1795, BLC, JWP 4/65/25.

136 Thornton to Watt, 22 July 1795, BCL, JWP 4/65/2. By the end of the decade, Thornton claimed to have simplified the Watt apparatus for making airs. See the anonymously published, but properly attributed to Robert Thornton, *The Philosophy of Medicine: Or, Medical Extracts on the Nature of Health and Disease, Including the Laws of the Animal Oeconomy, and the Doctrines of Pneumatic Medicine* (London: Printed by C. Whittingham, 1799–1800), I, 346. This was the work "of Dr. Thornton's in which Dr. Beddoes is extolled & placed upon the Pinnacle of Glory". See Richard Lovell Edgeworth to Sophy Ruxton, continued by Maria Edgeworth, n.d. [1795] (MS has October 1796 deleted), Edgeworth Papers, National Library of Ireland, Pos. 9027, manuscript 140. On the attempts to devise a new inhaler see Paul Luter, "John Wilkinson and the medicinal use of Ether," *Brosley Local History Society*, **29** (2007), 11–14. See Chapter 5, this volume, for William Clayfield, who handled the production of gases in the Pneumatic Institution.

137 Thornton to Beddoes, 20 August 1795 and to Watt, 27 August 1795, BCL, JWP 4/23/15.

138 Thornton, *The Philosophy of Medicine: Or, Medical Extracts* (1799), I, 339–343, 426–427; cf. John Edmonds Stock (1811); reprint edition, Bristol: Thoemmes Press, 2003), 98.

139 Thornton to Watt, 22 July 1795, BCL, JWP 4/65/2.

140 Beddoes to Watt, 4 September 1794, BCL, JWP 4/23/22.

141 Thornton to Watt, 27 September 1795, BCL, JWP 4/23/14.

142 Thornton to Watt, 27 September 1795, BCL, JWP 4/23/14. On the purity of airs, see Chapter 1, this volume.

143 Thornton to Watt, 27 September 1795, BCL, JWP 4/23/14.

144 Thornton, *The Philosophy of Medicine: Or, Medical Extracts* (1799), I, 428, 434. See also Beddoes' comment to Darwin as being declared a "silly projector" or a "rapacious empiric". Beddoes, *A Letter to Erasmus Darwin*, 3.

145 Thornton to Watt, 31 December 1796, BCL, JWP 4/9/33.

146 Thornton, *The Philosophy of Medicine: Or, Medical Extracts*, I, 433–434.

147 Beddoes to Watt, 27 February, 1795, BCL, JWP, 4/65/18.

148 Watt to Charles Campbell, April 1797, BCL, JWP copy books; Beddoes to Watt, 24 October, 1797, BCL, JWP W/0/19.

149 Creaser to Watt, 23 October 1798, BCL, JWP W/9/1; Watt to Creaser. 6 November, 1799, JWP copy books; cf. Noel G. Coley, "The Preparation and Uses of Artificial Mineral Waters, (ca. 1660–1825)", *Ambix*, **31** (March, 1984), 32–48; Simon Schaffer, "Measuring Virtue: Eudiometry, Enlightenment and Pneumatic Medicine", in *The Medical Enlightenment of the Eighteenth Century*, eds. Andrew Cunningham and Roger French (Cambridge and New York: Cambridge University Press, 1990), 281–318.
150 Watt to John Robison, 30 June 1798, BCL, JWP copy books; Beddoes to Watt, July, 1799, JWP 6/35/6: Porter, 77–78. On Clifton see Jay, 76–77, 175.
151 Porter, 188; Beddoes to James Watt, 27 February 1795, BCL, JWP 4/65/18.
152 See, for example, *The Anti-Jacobin Review and Magazine* (May–August, 1800), 424–428.
153 Watt to Black, 6 November 1799, BCL, JWP copy books; reproduced in CJB, II, 1376–1377.
154 Watt to Black, 22 November 1799; Watt to Black, 8 December 1799, BCL, JWP copy books, reproduced in CJB, II, 1382–1383, and 1386–1388 (the word "acid" is missing in the transcription).
155 Watt to Berthollet, 1 January 1800, BCL, JWP copy books.
156 Berthollet to Watt, 23 September 1801? BCL, JWP 6/33/65; Priestley to Watt, 17 October 1801, JWP 6/35/64. See Priestley, "Observations and Experiments Relating to the Pile of Volta", *Nicholson's Journal*, **1** (1802), 198–204, and "On the Air from Finery Cinder and Charcoal, with Other Remarks on the Experiments and Observations of Mr. Cruickshank", *Nicholson's Journal*, **3** (September, 1802), 52–54; Jan Golinski, "Davy and the Lever of Experiment", in *Experimental Inquiries: Historical, Philosophical and Social Studies of Experimentation in Science*, ed. H. E. Le Grand (Dordrecht, Boston, London: Kluwer Academic Publishers, 1990), 99–136, esp. 104; Giuliano Pancaldi, "On Hybrid Objects and Their Trajectories: Beddoes, Davy and the Battery", *Notes and Records of the Royal Society*, **63** (September, 2009), 247–262, esp. 258. On Cruickshank's chemistry see Guy H. Neild, "William Cruickshank (FRS — 1802): Clinical chemist", *Nephrology Dialysis Transplantation*, **11** (1996), 1885–1889.
157 Porter, 169. On the benefit to the poor of the Pneumatic Institution see Watt to Beddoes, 3 March, 1802, BCL, JWP copy books.
158 Thomas Beddoes, *Alternatives Compared: Or, What Shall the Rich do to Be Safe?* (London: Debrett, 1797), 1–2.
159 Watt to Joseph Black, 23 November 1791, BCL, JWP copy books; reproduced in CJB, II, 1144.
160 Thompson, *Making*, 181.
161 Note Peter M. Jones, *Industrial Enlightenment: Science, Technology and Culture in Birmingham and the West Midlands 1760–1820* (Manchester and New York: Manchester University Press, 2008), esp. 199. On the confusions of a rising manufacturing class, see Neil Davidson, *How Revolutionary Were the Bourgeois Revolutions?* (Chicago: Haymarket Books, 2012), esp. 79–92.
162 On Joseph Johnson prosecuted for sedition, see Helen Braithwaite, *Romanticism, Publishing and Dissent: Joseph Johnson and the Cause of Liberty* (New York: Palgrave Macmillan, 2003); Jay, *passim*, esp. 133–134, 167; Johnston, *Unusual Suspects, passim*, esp. 104; and Chapter 4, this volume.
163 Thompson, *Making*, 185; citing John Thelwall, *The Rights of Nature, against the Usurpations of Establishments: A Series of Letters to the People of Britain, on the State of Public Affairs, and the Recent Effusions of the Right Honourable Edmund Burke: Letter the First* (London: H. D. Symonds, 1796), 21.
164 Jay, 97.
165 "Abstract of Papers relative to seditious Persons", 28 July 1792, NA, HO 42.21, f. 214; Stansfield, 77–79; Johnston, *Unusual Suspects*, 99–100.

166 Watt to De Luc, 11 November 1792, BCL, JWP copy books.
167 Watt to Gilbert Hamilton, 30 May 1792.
168 Eric Robinson, "An English Jacobin: James Watt, Junior, 1769–1848", *Cambridge Historical Journal*, **11** (1955), 349–355; Thompson, *Making*, 112, 120; Goodwin, *Friends of Liberty*, 265, 337.
169 Watt to Thomas Walker, 8 January 1793; Watt to James junior, 8 January 1793, BCL, JWP copy books.
170 Watt to Withering, 29 April 1793, BCL, JWP copy books.
171 In Beddoes, *A Word in Defence of the Bill of Rights against Gagging Bills* (Bristol: N. Biggs, and London: J. Johnson, 1795), 8; quoted in Stock, 117. Stock clearly had access to much of Beddoes' letters, many of which seem to have disappeared apart from those still in the collections of his correspondents. If Stock or Beddoes' widow was responsible, Stock not have been entirely unsympathetic to Beddoes' politics. Johnston, *Unusual Suspects*, 87–88, 102–103.
172 Beddoes to Watt, 26 December 1795, BCL, JWP 4/23/6.
173 Thompson, *Making*, 108–137; Goodwin, *Friends of Liberty*, chapters 7–9. Johnston, *Unusual Suspects*, Appendix 1, 329–330; see the Introduction to this volume. Beddoes' first biographer, Stock, fled to America to avoid trial and possible execution; see Appendix 1 on Stock, this volume.
174 James Johnstone, M.D., "Some Account of a Species of Phthisis Pulmonalis, Peculiar to Persons Employed in Pointing Needles in the Needle Manufacture" (read 1 February, 1796), in *Memoirs of the Medical Society of London* (London: J. Johnson, 1799), V, 89–93.
175 John Ferriar, *To the Committee for the Regulation of the Police, in the Towns of Manchester and Salford* (Manchester, 1792); Bodleian copy, http://find.galegroup.com.cyber.usask.ca/ecco/infomark.do?&source=gale&prodId=ECCO&userGroupName=usaskmain&tabID=T001&docId=CB130204814&type=multipage&contentSet=ECCOArticles&version=1.0&docLevel=FASCIMILE
176 John Ferriar, *Medical Histories and Reflections* (Warrington: W. Eyres, and London: T. Cadell, 1792), I, 244. Significantly, this was later quoted, although from the wrong volume, in Thomas Beddoes, *Essay on the Public Merits of Mr. Pitt* (London, J. Johnson, 1796), 163.
177 Cf. Rev. Joseph Townsend, *A Guide to Health; Being Cautions and Directions in the Treatment of Diseases, Directed Chiefly to the Use of Students*, 2nd edn. (London: Cox, Dilly, Murray and Owen, 1795), 38–44.
178 John Ferriar, *Medical Histories and Reflections* (London: Cadell and Davies, 1795), II, esp. 241. On Ferriar and Pearson, see Golinski, *Science as Public Culture*, 160–161.
179 Ferriar, *Medical Histories and Reflections*, 181, 196–197.
180 John Johnstone, *An Essay on Mineral Poisons* (Evesham: J. Agg, 1795), 166.
181 Richard Pearson, *A Short Account of the Nature and Properties of Different Kinds of Airs, So Far as Relates to Their Medicinal Use; Intended as an Introduction to the Pneumatic Method of Treating Diseases* (Birmingham: Thomas Pearson, 1795), "Advertisement",7–8.
182 Cf. Golinski, *Science as Public Culture*, 160; Martin Kemp, "Thornton, Robert John (1768–1837)", ODNB (2004). http://www.oxforddnb.com/view/article/27361
183 [R. J. Thornton], *The Politician's Creed: Being the Great Outline of Political Science: From the Writings of Montesquieu, Hume, Gibbon, Paley, Townsend, &c. &c.* By an Independent, 1. (London: Printed for T. Cox, St. Thomas's-Street, Borough; and sold by Johnson, St. Paul's Church-yard; Robinson's, Paternoster-Row; Owen, Piccadilly; and Manson, Pall, Mall, 1795), 124, 161–162.
184 [Thornton], *The Politician's Creed*, 167–168.
185 [Thornton], *The Politician's Creed*, 172, 174.

186 See Goodwin, *Friends of Liberty*, 353–358.
187 Beddoes to Giddy, January (or 7 February), 1795; CRO, DG 42/4; Beddoes to Giddy, 12 February 1795 DG 42/36; Beddoes to Giddy, 2 March 1795, DG 42/35. See also Stock (1811), 106; cf. Tim Nyborg, *Radical Chemist: The Politics and Natural Philosophy of Thomas Beddoes*. M.A. thesis, University of Saskatchewan, 2011.
188 [Thomas Beddoes], *Postscript to the Defence of the Bill of Rights* (Bristol: 20 November 1795), 3–4. See also Beddoes, *A Word in Defence of the Bill of Rights against Gagging Bills* (Bristol: N. Biggs, and London: J. Johnson and T. Chapman, 17 November 1795). Copies in CRO, DG 42/17 and 42/19; Beddoes, *Essay on the Public Merits of Mr. Pitt* (London: J. Johnson, 1796), esp. 156; Stock, 114 ff.
189 Golinski, *Science as Public Culture*, 159; Johnston, *Unusual Suspects*, 104; cf. Roy Porter, "Taking Histories, Medical Lives: Thomas Beddoes and Biography", in *Telling Lives in Science. Essays on Scientific Biography*, eds. Michael Shortland and Richard Yeo (Cambridge and New York: Cambridge University Press, 1996), 215–242, esp. 222–223.
190 Beddoes, *Essay on the Public Merits of Mr. Pitt*, 160.
191 Beddoes, *Essay on the Public Merits of Mr. Pitt*, 162–163. Beddoes' citation is from the wrong volume of the *Medical Histories*. The correct reference is in volume I (1792), 243.
192 Beddoes, *Alternatives Compared: Or, What Shall the Rich Do to Be Safe?* (London: J. Debrett, 1797), 11, 17. On the public disputes over the consequences of war, see J. E. Cookson, *The Friends of Peace: Anti-war Liberalism in England, 1793–1815* (Cambridge: Cambridge University Press, 1982.)
193 Beddoes, *Essay on the Public Merits of Mr. Pitt*, 166, 169–170. Italics in original. Cf. Stock, 209.
194 Beddoes, *Alternatives Compared*, 73–74.
195 Beddoes, *Notice of Observations Made at the Medical Pneumatic Institution* (Bristol: Biggs and Cottle; London: Longman and Rees, 1799), 3–4.
196 Beddoes to Giddy, 11 October 1800, CRO, DG 42/35.
197 Porter, 164, 169; Jay, 201; Johnston, *Unusual Suspects*, 107, 109.
198 Thomas Beddoes, *Hygëia: or Essays Moral and Medical, On the Causes Affecting the Personal State of our Middling and Affluent Classes* (Bristol: J. Mills, and London: R. Phillips, 1802), I, Essay Third, 11.
199 Stock (1811), 196–197.
200 Beddoes, 12, 15; See also Tim Carter, "British Occupational Hygiene Practice 1720–1920", *Annals of Occupational Hygiene*, **48** (4) (2004), 299–307, esp. 301.
201 Thompson, *Making*, 203–205.
202 Thomas Beddoes, M.D., *Hygëia: or Essays Moral and Medical, on the Causes Affecting the Personal State of our Middling and Affluent Classes* (Bristol: Printed by J. Mills, St. Augustine's Back, For R. Phillips, No. 71, St. Paul's Church-Yard, London, 1802), II, Essay Seventh, 28.
203 Beddoes, *Hygëia*, II, 29–30; Stock (1811), 230–232.
204 Beddoes, *Hygëia*, I, 79.
205 Beddoes, *Hygëia*, II, 80.
206 Rev. Stebbing Shaw, *The History and Antiquities of Staffordshire*, Vol. II (London: J. Nichols and Son, &c., 1801), 121.
207 James Watt junior to Thomas Ferriar, 19 December 1794. Quoted in Trevor Levere, "Dr. Thomas Beddoes and the Establishment of His Pneumatic Institution: A Tale of Three Presidents", *Notes and Records of the Royal Society*, **32** (July, 1977), 41–49, at 44. See also Jay, 107.

4 Book collector, library omnivore, and critic

Trevor Levere

Introduction

Beddoes was a physician, a chemist, a mineralogist and geologist, and a bibliophile. He read Latin like many if not most eighteenth-century physicians but, unlike most, read several living European languages. He was an eclectic translator and reviewer of English, Latin, and foreign books, especially German ones that he introduced to English readers. He advised the Edinburgh-born John Murray (1737–1793), founder of the eponymous publishing house, about scientific publications, and for some years Murray was his principal publisher. Beddoes then moved to the radical Unitarian publisher Joseph Johnson, who published not only his books, but also his political and moralizing tracts and broadsheets. Beddoes was at pains to acquire and study an extraordinary range of medical, scientific, philosophical, and literary works, especially those of German authors. Humphry Davy wrote of Beddoes that "[h]e had great talent & much reading";[1] "much" is an understatement. His networks included physicians, chemists, industrialists, politicians, and political radicals.

Beddoes used his friends and acquaintances at home and abroad, as well as booksellers in England and Germany, the latter mostly through the offices of Matthew Robinson Boulton, to borrow and to buy books. He built an extraordinary personal library, sold after his death in more than 2,000 lots, more than half of them German. He was a severe critic of Bodley's Librarian in Oxford; an admirer of the university library in Göttingen; a devourer of books, journals, and periodicals. Samuel Taylor Coleridge described himself as a library cormorant; Beddoes was an omnivore when it came to books. All this is patent in what remains of his correspondence and in the sale catalogue of his library. The use that Beddoes made of books is not immediately obvious. Unlike Coleridge, when he returned borrowed books they were not enriched by marginalia. He did not engage in the pedantry of academic scholarship, and references to other works in his own publications are scarce. There is little discussion of books in his correspondence, although there are frequent references to periodicals from France. When the French wars started in 1793, it became illegal to import publications from

France. For an avowed democrat, this may explain the paucity of French books in his library.

Author, translator, reviewer, medical student: foreign interests, Oxford, Edinburgh, and the Murray years

Books were always an important part of Beddoes' life, although he began modestly. When he wrote to his father in 1776, itemising his expenses for the previous year, he reckoned that he could live in Oxford for about 80 pounds a year, with books accounting for rather more than 10% of the total.[2] The costs would have been higher had he not developed great skill in borrowing books and manuscripts of lectures from fellow students and teachers alike.

He also, as a typically impoverished student, took every opportunity to review books. The earliest surviving correspondence between the publisher John Murray and Beddoes dates from 1782, and is about book reviews.[3] It isn't clear who first approached whom, but Murray was eager for copy, and was soon regularly soliciting Beddoes. Beddoes in turn was eager for books; his relationship with Murray was symbiotic.

Murray had enrolled at Edinburgh University for just one term, then entered the marines and served in the Seven Year's War. For several years thereafter, he sought employment but failed to hit on a career until 1768, when he bought the business of William Sandby, a London bookseller; he knew nothing about the book trade, but charged in, observing that "Many blockheads in the trade are making fortunes."[4] Murray, no blockhead, made his fortune, partly from reprinting. Over the next twenty-five years, he published about a thousand books, around one-quarter in science and medicine. Beddoes was to become a kind of scientific advisor and editor for Murray.[5] Murray, in his first exchanges with Beddoes, wasn't sure if he should address him as Mr. Beddoes or Dr. Beddoes; at this stage Beddoes was plain Mr., having spent five years as a student at Oxford, and then studying with Sheldon in London. It was Sheldon's course in physiology that had led Beddoes to make notes on Spallanzani. It may have been at one of Sheldon's lectures that Murray and Beddoes first met, although it could equally have been at Murray's London business. Who first proposed that Murray publish Beddoes' translation of Spallanzani is not clear, but Beddoes did the necessary work at about that time, and Murray published it in 1784.[6]

Murray rapidly gained confidence in Beddoes' judgment, as well as his flexibility as an author. Beddoes gave Murray "the liberty to alter or even to reject", which convinced him "of your good sense, & wishes to assist me".[7] Murray asked Beddoes to write reviews for his new journal, *The English Review*,[8] before the journal was publicly announced.[9] Beddoes obliged, and Murray came to depend upon Beddoes, and to flatter him, as when he asked him to review the third volume of William Cullen's latest volume on "the practice of physic": "I really do not know to whom to apply, to perform this; and you will much oblige by your compliance. Secrecy shall be inviolable."[10] By then, Beddoes had completed his studies at Oxford[11] and his work under Sheldon in London. In the following year, Beddoes would go to Edinburgh to study medicine; Cullen was an eminent professor in the University of Edinburgh, and First Physician to his Majesty for Scotland. Reviews were, of course, anonymous, and guessing who their authors might be was a popular sport. Murray regularly sent Beddoes review copies of books on medicine, chemistry, and natural philosophy.[12] In the summer of 1783, when Beddoes was back in Shropshire and helping to deal with an outbreak of fever, Murray kept writing to him: "I was disappointed not to receive an account of the foreign Book I sent you . . . – my Review suffers by your assistance being withdrawn, I will nevertheless expect some copy for it in your next cover."[13] In September he was flattering and complaining:

> the Articles you favourd me with will appear in the E.R. for September, your distance is a great disadvantage to me for with your assistance the E.R. was superior to any other – What becomes of you in the winter . . . – the treatise on the Plague is too good to be lost and I shall Certainly print it whatever may be its success – you still have a foreign Book which I sent you I only remember you of it, for I dare not press you for assistance, however anxious I may be to receive it.[14]

In October, he asked Beddoes when he meant to visit Oxford "because if you will not come to London from thence, I shall go forward to Oxford".[15] Clearly, Beddoes had become worth cultivating as reviewer and even as friend. In November, Murray encouraged Beddoes to commission him for sundry purposes, sending him a lottery ticket, along with "4 India Bandana Handkerchiefs" procured by Mrs. Murray, who had also hemmed them.

> Any articles you could furnish for the E. Review, I should be doubly grateful for, as the dearth of books at this Instant is so great that I really know not where to provide for that publication. Pray do for me what you can. Do you go at all to Edinburgh[?] They have got prodigious classes at that City this winter. When shall we see you in London . . . ?[16]

He also began regaling Beddoes with literary gossip: "Dr Cullens method of Issuing his new Edition is truly mean, and pitiful and brings his liberality as a Gentleman much in question. His conduct in this business is much talked of here."[17]

Meanwhile, Murray had commissioned a different Dr. Cullen, one Edmund Cullen, a leading Irish physician, to translate Bergman's *Opuscula*[18] into English, but he was unhappy with Cullen's work, so he sent it to Beddoes:

> Pray tell me how you like it, and touch it without Ceremony for improvement as you go over it. Mark it also in the order it should be printed, for

> if I am not mistaken the author has neglected to do this. Were you to go soon to Oxford to Stay any time I would print it there. I shall indeed be at a great loss to put it to press without your assistance.[19]

Then Murray's shopman reported that he had seen a printed sheet of a rival translation of the work then under way.

> In this situation I offer you thirty guineas either to come to London or Oxford immediately in order to carry this work through, and I mean it merely for *this* work. I know this consideration of itself will not have weight with you. But your love of Bergman may prevail, provided reasons of great moment do not keep you at Shifnal. You will observe that no time is to be lost; and this evening I forward franks to Dr Cullen to forward the remaining part of his translation to me.
>
> It is needless to add more till I hear from you. If you prefer London, advise, and knowing your taste I shall take Lodgings for you; and two presses shall be immediately employed. This would also answer well on account of Spallanzani – I shall expect your answer without delay. But were you to bring it yourself I should like it Better.[20]

Before he set off in the autumn of 1784 for further medical and chemical studies in Edinburgh, Beddoes shepherded the preparation of volumes one and two through the press, and once in Edinburgh was urged by Murray to undertake the translation of volume three.[21]

Edinburgh University in the late eighteenth century was a leading European university in both medicine and chemistry, far outshining the more introverted English universities; Joseph Black's students came from all Europe and beyond.[22] His foreign students brought with them lecture notes from home, and Beddoes was soon borrowing them, and getting a firsthand introduction to the world of chemists and physicians.[23] He was clearly operating some kind of barter, for he wrote to his friend Charles Brandon Trye in 1785 that he wanted to hang on to the first volume of Hunter's lectures and asks for a loan of the others "as I shd. then be certain of procuring the use of Van Doeveren's lectures, which contain I am told many good things never published & in particular long speculations on the vital principle."[24] In 1784, William Hunter's first two introductory lectures in his course on anatomy were published posthumously;[25] Beddoes' reference to several volumes of Hunter's lectures suggests that he was asking about unpublished manuscript notes, which he could temporarily exchange for manuscript notes of the Leiden physician's lectures. Van Doeveren had written a dissertation on animal life in 1758,[26] when Leiden was the leading medical school of Europe, although by the 1780s the Edinburgh school of medicine was fast catching up, and – thanks especially to William Cullen, James Gregory, and Joseph Black – could be seen as having overtaken Leiden.

Christoph Girtanner, having studied both chemistry and medicine, was one visitor to Edinburgh because of Black, and would have been able to tell Beddoes about the splendid library in Göttingen, a university town in Lower Saxony. The University of Edinburgh had its own impressive collections in medicine, and there were besides the libraries of the Edinburgh Infirmary, the Royal College of Physicians of Edinburgh, and the Royal Medical Society of Edinburgh.[27]

Beddoes had arrived in Edinburgh with letters of introduction to Murray's brother-in-law James Gilliland, and to his friend Dr. Duncan.[28] Beddoes asked Murray to send him goods from London, including lengths of silk and bottles of port, and Murray replied: "Make no apologies for your commissions, as Mrs. Murray and Myself have great Pleasure in being of the least use to you."[29] Murray wrote separately to Gilliland about Beddoes: "You will be pleased with his Manners & I know not a worthier person. I am much obliged to him."[30]

Beddoes had ben busy. With his translations of Spallanzani and Bergman's *Opuscula* behind him, other translations for Murray followed: Bergman's *Disquisitio de attractionibus electivis (Elective Affinities)*, and Scheele's essays published by the Academy of Science in Stockholm (*Chemical Essays)*.[31] The translation of the former came to Beddoes by a roundabout route. William Withering – physician, naturalist, and member of the Lunar Society of Birmingham, best known for his work on the action of digitalis as a medicine – wanted to translate Bergman for Murray, and Beddoes, aware of this even though he had already undertaken to make the translation himself, relinquished it in Withering's favour. So far, so good. But then, Murray reported to Beddoes, Withering handed the task to "a Journeyman of whom he is ashamed, on which account he throws it back upon you with disagreeable expences" – behaviour that Murray condemned as "not entirely honourable". But if Beddoes was willing to complete the task, Murray would support him. "If the project goes forward the tract of Bergmans which I have just published with some corrections may I suppose be rendered useful."[32] Beddoes did undertake the translation, and it was published in 1785 by Murray in London and by Charles Elliot in Edinburgh, without Beddoes' name appearing, but described as *Translated from the Latin by the Translator of Spallanzani's Dissertations*.

Beddoes was not so much the translator as the editor of Scheele's *Chemical Essays*. Carl Wilhelm Scheele is now chiefly remembered for his discovery of "fire air", an independent discovery of oxygen, although interpreted in a very different theoretical framework. Beddoes had proposed as translator the physician F. X. Schwediauer (or Swediauer),[33] whom Murray distrusted, as did Joseph Black;[34] but Murray was willing to accept Schwediauer as long as Beddoes supervised the translation. The translation was made, Beddoes added some material and corrected Schwediauer's English, and Murray published the volume in 1786.[35]

Beddoes, Oxford libraries, and politicization

His studies in Edinburgh completed, Beddoes visited Oxford to collect his M.D., an Oxford degree being necessary for the title of Dr., and an Edinburgh education being essential for the informed practice of medicine. In 1787, after a trip to France and perhaps also to Germany, Beddoes returned to give chemical lectures in Oxford. Murray kept in touch with him, and found him helpful from time to time, but Beddoes was no longer treating Murray as his sole publisher, although Murray was not keen to cut the apron strings. Once in Oxford, Beddoes set about writing his lectures, which entailed spending a lot of time in the Bodleian Library. A library, after all, was the "institution which chiefly distinguishes a seat of learning from other places".[36] And learning, like the sciences, should be truly international. Beddoes needed access to the full range of continental as well as British texts on science and medicine. After his experience with the Library of the University of Edinburgh, Beddoes was singularly unimpressed with the state of the Bodleian. Given his own success in obtaining the books he needed, he had little sympathy with librarians who failed to perform their jobs properly.

He addressed a swingeing list of criticisms to Rev. John Price, Bodley's Librarian.[37] The criticisms filled a nineteen-page booklet,[38] a superb polemic, and Beddoes' first printed work as Reader[39] in Chemistry. Beddoes noted that

> freedom of enquiry into the state of the University seems to have been too much discountenanced. Had such discussions been encouraged, the close of the eighteenth century had never found us with so many wants, and so patiently acquiescing under them; I mean, such wants as that of an opportunity to acquire a competent knowledge of Natural History, of an institution for instructing the youth of a great commercial State in the principles of commerce and manufactures. . .. But my present business is with an institution of still greater importance, since, if it be neglected, or suffered to fall into decay, no other can flourish, nor any literary pursuits be effectively carried on.[40]

Beddoes claimed that he had some experience of public libraries, and some information about them, and "I think it not rash to affirm, that all Europe scarce affords an example of one so little calculated to answer the purposes for which a public library is designed." He criticized the behaviour of the Librarian, the incompetence or laziness of those responsible for buying books, and the fact that the money raised for the library was not spent on acquisitions.[41] Because of his chemical lectures, Beddoes was free to use the library only on Saturdays and Mondays, but was unable to consult the books he needed, because the Librarian, who alone could accommodate him, was absent on those days because of his curacy.

Beddoes complained, accurately, that serials were not maintained and sets were incomplete:

> We have the Berlin Memoirs only to the year 1779, although several volumes have been since published, and the neglect has been noted in the list of Libri Desiderati. The same inattention is observable with respect to the Memoirs of the French Society of Medicine, and the Petersburgh Transactions.[42]

Again,

> Haller's Elementa Physiologiae were purchased in 1784; but an imperfect copy was mistaken for perfect [lacking the last three volumes]. . .. The other works of Haller are likewise incomplete. I find only three volumes of his Disp. Anatomicae, and only one of his Bibliotheca Anatomica: the Bibliotheca Med. Practicae is wanting, though a part of the work.[43]

The Bodleian Library holdings today generally have the key works, but some of them are a mixture of volumes from the first and second editions, indicating that Beddoes' strictures were accurate, and that missing volumes in first editions were made up from second editions.[44] Not only were many such important works missing, but the library spent too much on works of little or no use or authority; as an example of such a work and such an expenditure, Beddoes gives Sir John Hill's *Vegetable System*, "for which one hundred and forty pounds were given" – and today's catalogue indicates that that work was purchased incomplete, and missing volumes were made up from the second edition. Beddoes writes with relish that "Fabricius, the great Entomologist, indignant at his total want of good faith, styles him Joannem Hilol, damnandae memoriae."[45] Then there was the problem of buying poor English translations instead of the French originals: "no person, who knows how such translations are fabricated, would choose to rely on their authority."[46] Beddoes' experience as a translator and improver of others' translations for Murray gave him an informed vantage point.

The Bodleian Library, according to Beddoes, acquired too few foreign books:

> I have heard it said, that many foreign books cannot be procured; but there is no difficulty, when once the proper channels are known. To wait for the most valuable foreign books, till they appear in the shops of our booksellers, is to wait till the rivulet has ceased to flow; but I would myself undertake to procure almost any one of modern date.[47]

We have already seen some of the ways Beddoes obtained foreign books; we shall soon encounter others. Beddoes was particularly incensed by the Bodleian Librarian's deficiencies when it came to buying German books. No care was taken

> to supply us with the authors of a country, who may justly contest the palm of Science and Literature with those of any other nation. It may be said, indeed, that if we consider the small number of readers of German, that the use of books written in that language would be very limited. But in a place of education, I think it is rather to be expected, that the means of making literary acquisitions should be provided, than that persons should come already furnished with them.[48]

It would help if the library had the latest German and French dictionaries instead of ones long out of date, since language – and especially the language of the sciences – saw a constant stream of new words. And surely the effort of learning to read German could not be wasted,

> unless we can suppose that the powers of Haller, Heyne, Meiners, and Michaelis, desert them, when they write in their mother tongue. . . . [H]ow can such writers, as Jerusalem, Doederlein, Michaelis, Mendelssohn, or Lessing, be searched for new arguments on either side, while our high-priests of learning take no care to introduce their offerings into her temples?[49]

Beddoes' list of great German writers is a remarkably eclectic one, including an historian; Jewish, protestant, and enlightenment philosophers; theologians; a poet; biblical scholars; and a classicist.[50]The list is impressive for its range: few Englishmen in the 1780s would have been familiar with more than one or two of these writers. But the list does not include all the German books that Beddoes reviewed, and it represents only a very small part of the astonishing range of German authors in Beddoes' own library. As we shall see later, the sale catalogue of that library following Beddoes' death dramatically avoids the inadequacies of the Bodleian Library: roughly half of the two thousand lots in that sale were either German texts or Latin texts written by German authors, ranging over the whole gamut of scientific and humanistic learning.

Beddoes had a slew of criticisms about Bodley's Librarian; and he ended by making unfavourable comparisons of the Bodleian Library and its budget for acquisitions with the libraries of Edinburgh and Göttingen. The Edinburgh library he chose was not the university library, but the Advocates Library, founded in 1682, partly funded by the fee paid by every advocate on his admission, and, since the reign of Queen Anne,[51] serving as a copyright library for every book entered in Stationers-hall. The other copyright libraries were those of the universities of Oxford and Cambridge, and the

Royal Library. In 1788, Hugo Arnot reported that the collection of the Advocates library amounted to

> upwards of thirty thousand volumes, in all sciences, and in many languages. . .. The books are lent out to the members of faculty in such a manner that the institution is more useful than that of any library we know.[52]

Beddoes reported, in his memorial, that

> [t]he Advocates library at Edinburgh, the most useful in Great Britain, has an annual income of £700; that at Gottingen, of £1100. Now, as these sums are not found too large for the purchase of books, by what kind of arithmetic can it be made to appear, that a smaller sum . . . [will suffice]. . .. I can discover no reason why an English should be inferior to a Scotch or an Hanoverian University, in any respect; nor why the nation should be without as ample a repository of all kinds of knowledge.[53]

Beddoes would have been delighted to know that Göttingen University has fifty-four of his publications in its library today, several in German translation but most in the original English. Oxford has sixty; and Edinburgh University has but twenty-eight.

It was not just the state of the Bodleian Library that frustrated Beddoes in the preparation for his lectures; he could not find a good elementary textbook for his course, although, as he wrote to Black, he had collected all the modern elementary books, especially the German ones, "which are not a few".[54]

Beddoes' criticism of Bodley's Librarian was printed, without the name of publisher or printer, and without date. The same is true of another and longer item, a poem of fifty-four pages in length, accompanied by forty pages of notes, the whole sold by Murray and by James Phillips at their London shops. The title of the work was *Alexander's Expedition Down the Hydaspes & the Indus to the Indian Ocean.*[55] The author's name does not appear anywhere in the work; nor does the printer's. The origin of the work, according to the transcription of a note of 9 October 1819 by William Anstice that accompanied a copy sent to Dr. Parr after Beddoes' death, was

> a conversation which took place at the table of the late Mr. William Reynolds, in which some men of taste and genius contended that the poetic effusions of Darwin were inimitable. Dr Beddoes maintained a contrary opinion, and to try the point, produced to the same party a short time afterwards, a manuscript of the present piece, as from his friend Darwin, and sent to him for his inspection, previous to publication: the advocates for Darwin's style were deceived, and the Doctor

> triumphed. Mr. Reynolds had it printed at his own expense, but for obvious reasons it was not published.[56]

The epic poem, in the style of Darwin's *Zoonomia* and *The Botanic Garden*, was radical in its politics. Beddoes' political sympathies were far from universal in England, and were indeed officially regarded as subversive of established order and authority. He became the target of spying by the government.[57]

Beddoes radicalised: Bristol and the Johnson era

In 1793, Beddoes, advocate for democracy and defender of the French Revolution, thought it best to leave Tory Oxford following correspondence between the Vice-Chancellor and the Home Office.[58] He moved to Clifton, near Bristol, to practice medicine, to write, and to experiment with pneumatic therapies. In the next fifteen years, he published a string of books and pamphlets (more than thirty titles). Bristol was a thriving city for the book trade, where at least one bookseller, John Brice Beckett,[59] wrote on scientific subjects. In Bristol, Beddoes dealt with the Baptist bookseller, printer, and publisher Joseph Cottle, who printed or published several titles, including *Contributions to Physical and Medical Knowledge, Principally from the West of England and Wales*, and *Notice of Some Observations Made at the Medical Pneumatic Institution* (both printed in Bristol by Biggs and Cottle for Longman and Rees, 1799); Longman and Rees were Bristol publishers and book merchants who set up a business in London. Cottle, who befriended Coleridge and was for a while his publisher, also obliged Beddoes by receiving packages for him, including mouthpieces and silk bags for his breathing apparatus, and at least one box of German books, presumably sent through Boulton and Watt.[60] Beddoes also dealt with Bulgin and Rosser of Bristol, who printed and published parts of various editions of Beddoes' and Watt's *Considerations*, as well as *A Letter to Erasmus Darwin* (1793) and several other items between 1793 and 1796.

Beddoes' principal publisher in his Bristol period was, however, Joseph Johnson.[61] From the 1760s, Johnson published medical works[62] and dissenting religious tracts; he became the principal publisher for Dissenters. He developed a friendship with Joseph Priestley, and became his exclusive publisher. This, along with his support for the French Revolution and the publication of many political pamphlets, gave him a reputation as a radical and the publisher of radical authors.[63] He published key rebuttals of Burke's *Reflections on the Revolution in France*. He was clearly sympathetic to Beddoes. In 1793 he published Beddoes' works jointly with John Murray, with whom he cooperated, at least for a while. Murray wrote to Johnson in 1787:

> A Coalition is what I have cordially wished for, if it can be established upon principles of reciprocity. For it is no matter whose scheme here is

> the best; a public competition will infallibly hurt both. Obstinacy therefore should be avoided on both sides, & all of us keep steadily in our Eye, what will tend most to the general interest. These are my genuine sentiments upon the business & I remain &c.[64]

The coalition included joint publishing by Murray and Johnson of several of Beddoes' works.[65] But Beddoes moved gradually to Johnson, starting with *Observations on the nature of demonstrative evidence; with an explanation of certain difficulties occurring in the elements of geometry: and reflections on language* (London: printed for J. Johnson, 1793). Beddoes produced a series of anti-war pamphlets and protests against the restriction of free speech in the 1790s, which Johnson first sold and then had printed in London.[66] These tracts included *A Word in Defence of the Bill of Rights, Against Gagging* Bills (1795), *Where Would Be the Harm of a Speedy Peace?* (1795), *Essay on the Public Merits of Mr. Pitt* (1796), and *A Letter to the Right Hon. William Pitt, on the Means of Relieving the Present Scarcity, and Preventing the Diseases that Arise from Meagre Food* (1796). In 1799, Johnson was prosecuted for seditious libel for publishing a pamphlet by the Unitarian minister Gilbert Wakefield and was sentenced to six months in prison.[67] Gradually, gagging bills and treason trials had the result that Dundas and Pitt intended, and radical protest was increasingly stifled.[68] After Johnson's imprisonment, he was more cautious, but he continued to publish medical and scientific work, including Humphry Davy on nitrous oxide, and Priestley's *Memoirs*.[69] His efforts met with an acid response by the lawyer, Whig politician, and acidulous contributor to the *Edinburgh Review*, Francis Jeffrey:

> It has often occurred to us . . . that there is universally something presumptuous in provincial genius, and that it is a very rare felicity to meet with a man of talents out of the metropolis, who does not overrate himself and his *coterie* prodigiously. . . . We need only run over the names of Darwin, Day, Beddoes, Southey, Coleridge, and Priestley, to make ourselves perfectly intelligible.[70]

Although Jeffrey, as a Whig, wanted to curb the power of the Crown, he was even more eager to curb the excesses of radicalism.

Beddoes remained active as a reviewer, writing more than one hundred and sixty reviews for the *Monthly Review*.[71] Seventy-two of them were for the Foreign Supplement. Sixteen of the books were in French, among them eleven on chemical subjects; twenty-eight were in German; and two were in Italian.[72] Most of the French reviews are of journals. Beddoes' reviews were by no means all on medical or chemical topics. In 1796 he published what may be the first English review of Immanuel Kant's *Zum ewigen Frieden*.[73] His library contained first editions of both of Kant's *Critiques*, along with his works on geography, anthropology, and other subjects. Kant's metaphysical and epistemological concerns seriously interested Beddoes, although Kant's notion of *a priori* knowledge entirely failed to convince him. In his *Observations on the Nature of Demonstrative Evidence* (1793), Beddoes argued that Locke, in his *Essay on Human Understanding*, rightly understood that all knowledge derived from experience; but Locke, as a philosopher of language, was lacking in Beddoes' eyes. Beddoes asserted that John Horne Tooke, in *The Diversions of Purley* (1786), a conversational essay on philology and the structure of language, corrected and developed Locke's insights, and helped to reveal the faults in Kant's *Critique of Pure Reason.* In making his case, Beddoes gave readable and accurate translations of several pages of Kant, there being then no English translation available.[74]

Beddoes' was truly a lifelong European endeavour, in literature and philosophy as well as in science and medicine.

Private collections

Beddoes' book collecting began no later than his undergraduate days. Studies in Oxford and more importantly in Edinburgh; travel on the Continent; extensive book reviewing; and a wide network in science, medicine, and literature enabled him to keep building his library. The Bristol Library Society, for all its strengths, was wholly inadequate for his purposes, and Bristol itself was far from the libraries of London, Oxford, and Edinburgh. Beddoes needed to build his own library, professionally and culturally. He satisfied his need for books with spectacular success and with a truly European range. One window on that range is provided by the sale catalogue of his library. In November 1809, almost one year after his death on 23 December 1808, an auction was announced: *A Catalogue of the very valuable and extensive library of Thomas Beddoes, M.D. of Clifton, near Bristol, lately deceased: containing a very capital collection of modern publications in all the departments of surgery and medicine; voyages and travels, antiquities, natural history and belles lettres: likewise all the late best German writers on the above subjects, which will be sold by auction by Leigh and S. Sotheby, Booksellers, at their House, No. 145, Strand, opposite Catherine Street, On Friday, November 10, 1809, and Nine following Days, (Sundays excepted) at 12 o'Clock.*[75] The auctioneer's copy of the catalogue, identifying purchasers for each lot, and giving prices, was in the British Library.[76] The list of books is impressive, larger than that of many subscription libraries.[77] Beddoes' library was remarkable for a West Country physician, and would have been remarkable even for a metropolitan one. Its range is astonishingly wide: the only major field of Beddoes' interests not prominent in the catalogue is that of politics, and it is tempting to surmise that absence may have arisen because of weeding or sanitation by his widow or by Beddoes' biographer John Stock, whose own experience with the dangers of radicalism had been overwhelming. It is also worth remarking that only two of Beddoes' numerous publications are in the catalogue; some, especially the medical ones, may have gone to his son, Thomas Lovell Beddoes.[78]

The library is striking for its size (more than 2,000 lots), and for the very high proportion (more than half) of foreign titles, especially German ones. Stock illustrates this in an anecdote concerning Beddoes and a visitor, a Dr. Frank from Vienna. Frank was enjoying the freedom to travel in the brief window of peace that followed the Treaty of Amiens in 1802. He met Richard Lovell Edgeworth, Beddoes' father-in-law, and, armed with an introduction from Edgeworth, arrived in Clifton, the suburb of Bristol where Beddoes lived and had his pneumatic practice.[79] There were at that time at least three Dr. Franks from Vienna who had published one or more medical books. When Beddoes greeted his visitor, having kept him waiting, he did so from behind an armful of books, arranging them in piles, each by a different Dr. Frank, and asked "Which Dr. Frank are you?"[80] The sale catalogue lists twenty-six volumes by one or another Austrian or German Dr. Frank.

There were 2,131 lots. The first five days of the sale disposed of 1,071 lots, one in Dutch, the rest in German or in Latin by German authors, mainly devoted to chemistry and medicine, but with a very considerable number of works of literature, history, and philosophy. Beddoes' interest in German culture – medical, philosophical, and literary – is clear from this catalogue, and also appears in his published reviews, in his correspondence, in his animadversions directed at Bodley's Librarian, and in his encouragement of the young Samuel Taylor Coleridge to study in Germany. The remaining five days of the sale saw the sale of books in English, French, German, Greek, Portuguese, Latin, Spanish, Italian, and Dutch; it is worth noting that Beddoes' circle of friends and correspondents mirrored his library, being equally international, and included chemists, naturalists, philosophers, and physicians from many countries. His English books are also comprehensive in chemistry, as one would expect. The auctioneer's copy includes the names of purchasers, of whom there were more than seventy-five, including Joseph Banks,[81] for whom Beddoes dead was more profitable than Beddoes living. The library was thus comprehensively fragmented.

The first lot in the sale was "RICHTERS Chirurgische Bibliothek, 9 vol.", sold for nine shillings. August Gottlieb Richter was a German surgeon who edited his surgical journal from 1791 until 1797 and was the author of a seven-volume work on surgery, *Anfangsgründe der Wundarzneykunst* (Göttingen, 1782–1784); the sale catalogue lists this (lot 84) as a two-volume set dated 1782, as well as his *Abhandlung von den Bruchen*, in two volumes (Linz, 1788). This is the Richter whom Beddoes, when a medical student in Edinburgh, reported was "as highly spoken of as any man I ever heard of".[82] Girtanner, who had told Beddoes about the library at the Georg-August University of Göttingen,[83] was represented by his *Abhandlung über die Krankheiten der Kinder und über die physische Erziehung derselben* (Berlin, 1794), and by his *Ausführliche Darstellung des Darwinschen Systemes der praktischen Heilkunde, nebst einer Kritik desselben* (Göttingen, 1799).

The great physiologist and physician Albrecht von Haller was represented by six titles, including all eight volumes of his *Elementa physiologiae corporis humani* (Lausannae, 1757) and all eleven volumes of his *Artis medicae principes. Recensuit praefatus est Albertus de Haller* (Lausannae, 1769): Beddoes was not going to repeat the sins of the Bodleian Library. Among the best-represented physicians and surgeons in the catalogue were Christoph Wilhelm Friedrich Hufeland (twelve lots), Johann Christian Reil (eight lots), Kurt Polycarp Joachim Sprengel (eight lots), Christian August Struve (ten lots), Georg Christian Gottlieb von Wedekind (six lots), and Melchior Adam Weikard (seven lots).

There were many hundreds of volumes dealing with different areas of medicine, including pneumatic medicine (there is a German translation of one of Beddoes' own books on consumption and its treatment).[84] There are also chemical monographs and textbooks by Scheele, Ritter, Volta, Gren, Klaproth, and a host of others; and a run of Crell's *Chemische Annalen* starting in 1784. The library sale catalogue covers far more than medicine and chemistry. Eleven of Kant's books are there, including not only the treatises on pure reason and on judgment, but also works on anthropology, on the metaphysical foundations of natural philosophy, and more besides. Also in the sale were what became recognized as canonical texts in *Naturphilosophie* and the Romantic movement, including Schelling's *Erste Entwurf eines Systems der Naturphilosophie* (Leipzig, 1799); two volumes of the works of Novalis (Georg Philipp Friedrich Freiherr von Hardenberg); seventeen lots of the writings of Jean Paul (born Johann Paul Friedrich Richter); four works by Schiller, including two volumes of his poems and the verse-drama *Die Räuber*, which Coleridge briefly planned to translate into English; and four works by Johann Friedrich Wilhelm Ritter, *Naturphilosoph*, electrochemist, and disciple of Schelling.

The French books in the sale covered the standard up-to-date range in chemistry, including Berthollet, Chaptal, Fourcroy, Guyton, and Kirwan (in the French translation made by Mme. Lavoisier, with detailed point-by-point rebuttals by Lavoisier and others).[85] Lavoisier himself appears in Robert Kerr's translation, *The Elements of Chemistry* (1790), lot 1576, and in Thomas Henry's translation of the *Essays Physical and Chemical* (London, 1776), lot 1577; since Beddoes had no trouble reading French, and since his dealings with Murray had revealed some of the problems with translations, it is curious that he only had the English versions of these two works. Joseph Priestley, the leading pneumatic chemist of his day, and one therefore of particular interest to Beddoes, is represented by his metaphysical and materialistic *Disquisitions relating to Matter and Spirit* (1777), as well as by his *Experiments and Observations on Different Kinds of Air* (1790), the most up-to-date of the editions, and his last-ditch defence of the phlogiston theory, *The Doctrine of Phlogiston Established* (1800), long after it was generally vanquished.[86] Sir John Pringle, President of the Royal Society before Joseph Bank's reign, was among Priestley's patrons,[87] and it is pleasing to see the seventh edition of his *Observations on the Diseases of the Army* (London, W. Strahan, 1775) immediately after Priestley's works in the sale catalogue. Edmund Goodwyn, to whom Beddoes dedicated his translation of Mayow, is represented by *The Connexion of Life with Respiration; Or, an Experimental Inquiry into the Effects of Submersion, Strangulation, and Several Kinds of Noxious Airs, on Living Animals: With an Account of the Nature of the Disease they Produce...* (London, 1788). Beddoes had the first part (all that was published) of his Birmingham friend James Keir's *Dictionary of Chemistry* (1789); Keir had vainly asked Beddoes to work with him on the dictionary.[88] Beddoes did not, however, have Keir's *A Treatise on the Various Kinds of Permanently Elastic Fluids or Gases* (London, Cadell and Elmsley, 1779), a work that was already dated when Beddoes studied in Edinburgh. Besides science, he had a smaller collection of French literature, including a ninety-two-volume set of the complete works of Voltaire (1785), but lacking volumes seven and twenty-three.

What of the circumstances in which Beddoes built his library? Most of Beddoes' professional career took place in wartime, from the American Revolution (or War of Independence), which broke out while he was an undergraduate and which took on the dimensions of a world war, to the Napoleonic Wars, which continued beyond his death.

Some civilities were maintained during these conflicts. Joseph Banks negotiated the liberation or exchange of British and French scientists.[89] Interested M.P.s managed to get hold of French periodicals and political reports. The regular international mail, however, largely dried up. Trading vessels had to negotiate the blockade of Napoleonic France. In 1797, Louis Odier, Fellow of the Royal Society of Edinburgh, and physician and professor of medicine in Geneva, wrote to Beddoes: "I have just read your considerations on the medical use of factititous air. I had heard of them long ago, but in this part of the world we cannot easily get a speedy sight of English books."[90] Since Beddoes had published the first part of the book to which Odier referred in 1794, and the other parts in 1795,[91] we are looking at a delay of two to three years. In an admittedly extreme case, it took six years to ship one set of philosophical and chemical apparatus from the Low Countries to the new university in Russian-controlled Dorpat (now Tartu, Estonia).[92] Beddoes worked hard to get continental books. We shall see later the role of Matthew Robinson Boulton and his agents in Germany, as well as DeBoffe and Bohte, who specialized in the sale of foreign books in London.

Shipments to America were generally easier and safer, but not always. In December 1794 the industrial chemist Thomas Henry wrote with news sent by his son. Thomas junior had taken out US citizenship and attached himself to the household of Joseph Priestley, who had left England in the wake of the Birmingham riots. After spending some time in the "back settlement" of Northumberland, Pennsylvania, he returned to Philadelphia, "quite cured of the ideas he had strongly entertained of the advantages & happiness attendant on first settlers. He intends reading Chemical Lectures, &

pursuing the regular line of his Profession."[93] Henry senior had been asking for a breathing apparatus and chemical apparatus from Watt and Wedgwood,[94] some for himself, some for his son, who had been

> on the point of delivering Lectures on Chemistry. . .. But in addition to the misfortunes he has hitherto been involved in, I fear a number of his books, & among the rest a Copy of Dr. Black's Lectures, are lost in the Hope, the ship in which . . . many . . . passengers sailed, the beginning of August, & had not arrived the first of November.[95]

Such losses were also reflected in shipments to American libraries in the late eighteenth century.[96]

Acquiring books had been easier and safer before the French Revolution, when neither the sciences nor the nations of Europe were at war. Beddoes continued to obtain French books and periodicals from 1789 to 1792. Lavoisier, who met him in Dijon in 1787, had sent him a copy of the antiphlogistonists' response to Kirwan[97] in 1788. After 1793, with Britain and France at war, Beddoes obtained few French works. Unsurprisingly, not much was published by French chemists in the years of the Terror and its immediate aftermath; the Republic had scant regard for its men of science. Lavoisier was guillotined in 1794, along with the other tax farmers of the ancien régime. Only Antoine-François Fourcroy, who was a born survivor if ever there was one, and had a somewhat jaundiced view of Lavoisier's supreme role in the Chemical Revolution, managed to publish in the early 1790s. Even he did not publish at the height of the Terror, and resumed with a steady stream of books only in 1798.[98]

One of Beddoes' sources for French books and papers before 1793 was Joseph DeBoffe,[99] whose trade in French newspapers was to lead to his being called as a witness in the high-profile, hotly contested and highly controversial treason trials in 1794. In his cross-examination by the defendant John Horne Tooke, who was acquitted on the charge of high treason, DeBoffe stated that he had sold "some hundreds" of *Le Moniteur*[100] (the principal French newspaper during the French Revolution and the Napoleonic Wars) to those noblemen and gentlemen who were regular subscribers. Beddoes was among the readers of this newspaper.[101] DeBoffe viewed the sale of *Le Moniteur* as entirely legitimate, "down to the time when the communication was totally stopped".[102] In January 1793, Louis XVI was guillotined, after which the National Convention declared war on England. England banned the import of French books and papers. We have already seen[103] that French chemistry was identified, most notably but not uniquely by Burke, with French revolutionary politics. The exchange of chemical books and papers suffered accordingly.

We can see something more of the tension associated with scientific and foreign books in the history of circulating libraries in late eighteenth-century Leeds. A clear instance is the Leeds Foreign Library, founded in 1778 and incorporated into the Leeds Library in 1814. The Leeds Library, the oldest surviving subscription library in Britain, had a turbulent early history. The library's first Secretary and second President was Joseph Priestley, increasingly visible as a radical in politics and religion. The library's subscribers included many medical men. In 1793, the New Subscription Library was established, with a section dealing with "Arts, Chemistry, Philosophy and Optics"; one of the founders of the new library was William Wood, Priestley's successor as minister of Mill Hill Chapel. Perhaps because of its books, perhaps because of its founders, as late as 1801 the artist Joseph Harrington wrote in his diary: "There is also a Library which probably from having been established by certain people is called the Jacobin Library. – The other, the Old Library, on the contrary is called the Anti-Jacobin Library."[104]

War or no war, Beddoes managed to keep a stream of German books coming.[105] He seems not to have had dealings with London booksellers specializing in German books, and although he used Bristol printers and publishers, including Biggs and Cottle,[106] there seem not to have been Bristol booksellers specializing in French and German publications.[107] That is hardly surprising. Although Carl Heydinger had operated as printer and bookseller at various locations in Soho until about 1784, there was no resident German bookseller in London from 1784 to around 1793. A German circulating library, the Deutsche Lese-Bibliothek in the Strand, started business in 1793. James Remnant opened a German bookshop in West Smithfield in the same year, and Henry Escher opened another in Bloomsbury in 1794. Graham Jefcoate notes that "[d]espite a vogue for the German language and its literature late in the century, the German-speaking world remained almost completely unknown in England." Few German works were translated into English, whereas thousands were translated from English into German.[108]

> [T]he English, until the end of the [eighteenth] century at least, were largely excluded from intellectual and literary developments in the German-speaking world. German books were largely absent in British libraries, but the library of the University of Göttingen, then the premier research library in Europe, acquired very many English books.[109]

The dearth of English knowledge about German writers ended in the 1790s, when German literature began to be treated more favourably than that of the hostile French Republic. As the *Monthly Magazine* reported in February 1797, German literature was on the upswing, and "[I]n mathematics, natural history, physic, experimental philosophy, they are second to none."[110] Beddoes' reviews for the *Monthly Review* played their part in introducing German thought to England.[111] But booksellers in Britain paid little part in his acquisition of German books, and there are no references in his surviving correspondence to British booksellers specializing in German books or periodicals.

Beddoes did, however, have his own ways of obtaining foreign books and periodicals, principally German with a very small sprinkling of French.

The members of the Lunar Society of Birmingham, with their international business and industrial connections, were a prime source. Even during wartime, they had their agents and spies active in Europe[112] – and if it was not quite business as usual, it was still business. They were well supplied with foreign books. In March 1798, Beddoes asked James Watt junior to tell Matthew Boulton that he had received the latter's recent volume of the *Magasin encyclopédique*,[113] published in Paris. It was "the best book for a physician's chaise I have seen." In the same month, he told James Watt junior that "I yesterday received from Dr. Thomson at Naples 3 little mineralogical tracts," and he asked for the loan of Crell's journal, the *Chemische Annalen*,[114] for 1793, 1794, and 1795, "as they must contain the translation of academical papers which I have not seen".[115] Years later, in 1804, Watt wrote to Beddoes that Reimarus had written to the London house of Boulton and Watt, and would forward books to that house.[116]

For all Watt's helpfulness, it was Matthew Robinson Boulton, son of Matthew Boulton, James Watt's partner in the business of building pumps and steam engines, who was to provide Beddoes' main conduit for German books. In August 1798, Beddoes wrote to him, hoping that he had safely received his French books back again (he had, and in good condition), and then asked him about German books: "Perhaps it might be in your power to render me an essential service in the line of German literature."[117] Beddoes had been in touch with one A. Deriabin, a much-travelled Russian who had attended lectures by Joseph Black, corresponded with Gregory Watt about minerals and casting brass and steel, had financial dealings in several countries, and acted as an agent for Boulton and Watt, supervising, among other enterprises, sales of James Watt's copying press in St. Petersburg.[118] In 1793 Beddoes reported to Boulton that Deriabin had advised him

> to give a discretionary commission to one Martini of Leipsic;[119] but says I must have a correspondent at Hamburgh, who may pay Martini ready money, receive & forward the books. If your house at Hamburgh wd. do this, I shd. be very glad. The only precaution they need observe is not to pay above £25 in one year without especially directions from me.

Boulton did not reply instantly, and a mere five days later, Beddoes wrote more insistently: "I wish to know soon whether you can do me the favour I ask – otherwise it will be too late to get books from the Leipsic [Michaelmas] fair."[120] Boulton replied three days later that he would be very happy to oblige Beddoes, and gave one piece of advice. "[F]oreign Books when bound pay a higher Duty, than if imported in sheets",[121] so Beddoes should give appropriate instructions. Boulton maintained a German branch office throughout the wars. He arranged for the Leipzig printer and bookseller Gottfried Martini, a successor to Johann Christian Martini, to buy books, with Busch, an employee at the firm's office in Hamburg, paying him, and receiving and forwarding the books to London or Birmingham and thence to Beddoes in Clifton; Beddoes was given no invoices until the books reached him, which meant that

Boulton and Watt were acting as his bankers.[122] The system was soon running smoothly. Early in 1794, Beddoes was planning a series of lectures in Bristol, "on the properties of animal nature, & particularly that of man":

> To make the course more distinct & lively I find it necessary to procure some expensive works of anatomy from Germany. I have sent an order to Leipsic which will exceed the sum I mentioned as the utmost Mr Busch should pay Martini – viz £25 in one year – & shd you write to Hamburgh I request you to desire Mr Busch to pay him his bill which may be £40 or 50. . . .
>
> I regret troubling you again. By putting me in the way of procuring books from Leipsic, you have done me a most acceptable service.[123]

Beddoes' books came in steadily from Germany, thanks to Boulton's well-run business, although there were occasional losses and delays. The tiresome business of clearing customs was handled by the firm.[124] Beddoes' requests did not greatly inconvenience Boulton and Watt; their employees did the work. Sending books by the mail coach or goods wagon within England was generally but not always reliable. While Beddoes was lecturing on chemistry and mineralogy at Oxford, Mrs. Reynolds, wife of the Quaker ironmaster William Reynolds, wrote to him from Ketley in Shropshire that she was "surprized to hear thy Books are not come to hand they were sent from hence by the waggon near a month since exceptg. one wh. was omitted but will be sent in one of the boxes".[125]

Borrowing books

Beddoes, in addition to creating his own library, was a frequent borrower of books and journals from friends, and, to a lesser extent, from lending libraries, although in 1794 the Bristol Library Committee "doubted whether he was a fit person to be entrusted with books".[126] It appears that the Committee later changed its mind, since the Bristol Library Society's records tell us that Beddoes borrowed books from 1798 to 1803.[127] His borrowings included issues of the *Philosophical Transactions of the Royal Society of London*, which he did not own since Banks's hostility guaranteed that he would never become FRS. Beddoes nonetheless published four papers in the *Philosophical Transactions* between 1787, when he began lecturing at Oxford, and 1792, the last year of peace with France.

He sought the loan of books from his friends with an insistent energy that made it hard to refuse him. In 1791 he borrowed "a number of a journal in which [the mining geologist and mineralogist Abraham Gottlob Werner] acknowledges the absolute identity [of basalt and granite] in the most express terms". This was important not only for Beddoes' own studies of minerals, but also for his lectures at Oxford.[128]

He later made especially good use of the services of Davies Giddy, his close friend and former student at Oxford, and books, newspapers, and journals were a staple of their correspondence. In October 1791 he wrote to Giddy with his usual mix of politics, science, and medicine:

> If Beurmonville is Ajax, Dumourier is Hercules. He is the greatest of the subduers of monsters. I have had the Moniteur of the 3rd of Octr. since last nights post. You shall hear soon of Mr Volta's &c expt. – I think to be able to apply them so as to create a new system of medicine.

Being in receipt of the latest French news was politically sensitive, and it was this, rather than any ideas about "a new system of medicine", which moved Beddoes to add: "[B]urn the letter."[129] Beddoes was, however, soon to be entirely public and explicit in his castigation of Pitt and his gang.

His correspondence with Giddy also showed awareness of sciences remote from medicine, chemistry, and geology. Thus, for example, he wrote to Giddy in 1798: "I see by the Journal Encyclopedique [ou universel] that Lagrange has been doing a good deal respectg. methods – e.g. throwing out infinites – he has published Euler's Algebra with great additions."[130]

Giddy became especially useful to Beddoes in borrowing and purchasing books in 1806. In 1804, Giddy, who had been high sheriff and then deputy lieutenant of Cornwall, was elected as the M.P. for Helston, and then in 1806 became M.P. for Bodmin. Giddy had a London residence, and as a M.P. had access to diplomatic bags from abroad, and free postage within Britain. Beddoes was not going to miss such an opportunity. He was constantly asking Giddy to drop into this or that bookshop[131] – Lunn's[132] in Soho Square or Joseph DeBoffe's nearby – and have books sent to him. That was fine for books purchased, but booksellers were understandably less keen to arrange the shipment of borrowed books.

Matthew Robinson Boulton, through whose good offices Beddoes obtained many of his German books, also lent books and journals to Beddoes. In 1798, Beddoes wrote to him: "When the Magasin Encyclop accumulates to a vol. or two, if you will lend them to me, I will take care to return them with 3 weeks of their arrival."[133] In the following year, he wrote to Boulton that he had returned a set of numbers of the *Magasin Encyclopédique* long since, and was chagrined that they had not reached him.[134] He used his connection with Boulton to borrow as well as purchase books. Later on, Beddoes obtained his own run of the *Magasin* from 1795 to 1806, although the sale catalogue notes that "some numbers" were wanting. On another occasion, Beddoes wanted to borrow the first twenty-five volumes of the *Journal de médecine* belonging to Dr. Maxwell Garthsore, physician accoucheur and Fellow of the Royal Societies of Edinburgh and London. Garthsore was willing, but DeBoffe was not happy with the request, since it meant taking responsibility for the shipment of books that he had not supplied. Giddy acted as Beddoes' intermediary and overcame DeBoffe's reluctance. Beddoes wrote translated abstracts of all "of those [volumes} which Dr. G. had to spare", and was uncharacteristically nervous about asking DeBoffe and Garthshore for another twenty-five volumes, which would bring him up to his own abstract; but, he told Giddy, "I dare not ask for 25 more, I fear." He returned the first batch by coach, having had them packed under his eye. He wanted to purchase a complete set, if he could; in the end, he bought an incomplete set.[135]

Another important source for Beddoes' borrowing was the antiquary and collector Francis Douce from the British Museum, who sent Beddoes packages of old books on medicine.[136] Beddoes' interest in such books was by no means antiquarian. He had a long-standing interest in public health and preventive medicine, being concerned to track changes and discover remedies where he could. Like French physicians in the ancient regime, his ideas were in line with the Hippocratic medicine of "airs, waters, and places".[137] Cures were slippery things, and the best hope for medicine was for prevention.[138]

Beddoes, although he was not the first, was among the few in Britain who saw the compilation of medical statistics as an essential prerequisite to sound policy, and his library was a key tool in this enterprise. Political arithmetic, the development of comprehensive statistics based on the analysis of bills of mortality, was one hundred and forty years old in England when Beddoes first attempted to arrive at comprehensive statistics tabulating mortality and its causes.[139] In 1793 he had published *The History of Isaac Jenkins and Sarah his Wife, and their Three Children*, as a moral tale instructing the poor in management for health, and instructing the gentry in philanthropy directed to the same end. In 1794 he published *A Guide for Self-Preservation, and Parental Affection; or Plain Directions for Enabling People to Keep Themselves and their Children Free from Several Common Disorders*. Then came *Hygëia: or Essays Moral and Medical on the Causes Affecting the Personal State of our Middling and Affluent Classes*.[140] In *Hygëia*, Beddoes expressed his central thesis: "One might engage at once to reduce the tribute of lives we render. . . . It must first, however, be generally believed with SYDENHAM, *that our chronic maladies are of our own creating*."

Prevention was the key, and it depended on knowledge. What was needed was a national organization for preventive medicine, in which information about public health would be collected by local and regional boards, and disseminated by the metropolitan board. "But the *expence! The danger of a job!* – what expence? The expence of a few thousand pounds a year for the greatest benefit that can be conferred on a people, from whom so many millions are drawn."[141] In one of the essays in *Hygëia*, he publishes this table:

Years	*Buried*	*Stated to have died of Consumption*	*Of Convulsions**	*Fevers*	*Small-Pox*
1790	18,038	4,852	4,003	2,185	1,617
1791	18,760	5,090	4,386	2,013	1,747
1792	20,213	5,255	4,646	2,236	1,568
1793	21,749	5,474	4,783	2,426	2,382

(*Continued*)

Years	*Buried*	*Stated to have died of Consumption*	*Of Convulsions**	*Fevers*	*Small-Pox*
1794	19,241	4,781	4,368	1,935	1,913
1795	21,179	5,733	4,758	1,947	1,040
1796	19,288	4,265	3,798	1,547	3,548
1797	17,014	4,776	3,804	1,526	512
1798	18,155	4,533	3,663	1.754	2,237
1799	23,068	5,721	4,512	2,712	2,400

* Convulsions in this table is a word that often stood for a variety of disorders, totally unlike each other.] Children commonly die convulsed whatever may have been their complaint; and almost all such deaths are referred to this head.[142]

The table shows

> [t]he whole number of deaths for several years together, at two different periods, . . . taken from the bills of mortality; and the four largest items subjoined. It will be seen what a proportion consumption bears to the whole.[143]

Beddoes recognized that some deaths attributed to consumption were in all likelihood brought about by other diseases; but some other deaths, attributed to such other causes as "asthma, phthisic, and cough" were in fact caused by consumption. No wonder that consumption was the target for his experiments in pneumatic medicine. But Beddoes wanted to know not only the nature of different diseases, but also the social and material conditions of their victims (including diet, sanitation, and heating), the role of environment (including climate), and the variation of these factors over time. Beddoes wanted data from cities, towns, and country parishes. The books that he borrowed from Francis Douce were mined for such information. In April 1808 he told Douce:

> That more substantial breakfasts were in use seems confirmed by every document – & butcher's meat was proportionally cheaper & more used – salted, baked & fresh. Many curious proofs of this in Dr. Whitaker's Deanery of Craven. Yr. seal had destroyed the name of the old author whose pthisic you say is not our consumption – Medical language is only growing by by [sic] degrees more analytic – Many disorders as catarrhis senilis – asthma – probably mesenteric atrophy were all lately confounded with consn. of the lungs[.][144]

In June, he wrote again:

> We suffer during the short continuance of our hot weather more from heat than [the] E. Indians – & in winter more from cold than the Russians &

Canadians – It appears clearly that the high temperature of ye rooms in which these people live, does not dispose to take colds – The Bp. of Quebec, a sensible observing man, was very subject to colds before he went to Canada – was totally free there, & for 2 years that he has been in England, he & his family have had nothing but colds, from which all had been exempt for 12 years before. Our island for dampness of air is a cellar in comparison with the interior of continents – I had, as you see, concluded that we are more cold-catching than our forefathers – You seem to think the same? – One specific question I wd. put to you is – where is your evidence for the fact? – a second. can you point any circumstance in the mode of living, which may have made us more tender? – Another point wd. be; are there indications of a change of climate? –

I shd. extremely like to see what A. Borde says of his phthisis – as medicine has grown more & more discrimination I had conceived that old authors have confounded ulcerative consumption of the lungs – humoural asthma – catarrhis senilis – under the common name of [pthisic] – though every one up to the level of the current information can now in most cases easily distinguish between these –

I shd. be very thankful for any old medical books bearing on my subject – & I wd be careful to return them safe & soon[.][145]

A week later, Beddoes wrote again to Douce, comparing modern medical knowledge with that of the ancients, urging again the importance of preventive medicine, reviewing continental literature about consumption, talking about books, and referring to the incidence of consumption in the military:

In some important departments of medicine I agree with you that we are not advanced beyond Galen. Indeed in fevers I believe Cullen & the Scotch school went retrograde. But in our knowledge & treatment of inflammatory & some other disorders, a degree of certainty, almost mathematical, has been obtained – & if you consider that in very many cases we can write down beforehand the changes which will be discovered on dissection, you will see that our practice as a much surer aim than that of the antients, who were so little versed in anatomy – But the unprofessional world is a very ass or buzzard not to see how much preventive medicine is their affair. . . .

I am obliged by yr. quotations from [Stubbes] & if you will send Borde &c. by any Bristol coach (– I except only the mail –), particularly by Fromonts' coach, I shall be more obliged still & will return them whenever you fix or in a reasonable time – . . .

I am sorry I did not know of Dr. Coombe's inclination to sell his old medical books. I wd. have bought them if I cd. . . .

I have paid attention to foreign books on consumption & have made some personal observations. In Germany, France, Poland &c it is a very common disease, almost as much so in some parts as with us – What

> is remarkable is that in the army & navy it has <of late> amazingly increased & almost taken the place of fever[.][146]

As Beddoes' correspondence with Douce suggests, and as his efforts to borrow, mine, and buy books shows, books were important in so many ways to Beddoes. The one category that was the exception was that of novels, which like many men of his time he found unsettling and which he regarded as positively unhealthy, especially for women.[147] His library and the books he borrowed were his pass-key to the international world of learning. He persevered in his polyglot reading. In 1800, he wrote to a correspondent that he was "engaged in the composition of a work on popular physiology & preventive medicine, in which I wd. derive all the aid I can from the whole literature of Europe".[148] He did not allow revolutions or wars or delinquent librarians to block his access to them.

Notes

1 Quoted in June Z. Fullmer, *Young Humphry Davy: The Making of an Experimental Chemist* (*Memoirs of the American Philosophical Society*), **CCXXXVII** (2000), 110.

2 Thomas Beddoes to Richard Beddoes, 1776, Bodley MS Dep. C. 135/2. The tenor of his letter to his father matches that of a letter to his mother shortly before his marriage to Anna Maria Edgeworth:

> I beg you to put from your mind all anxiety about my taking too much of my father's property. It cannot be better laid out than in establishing me. . .. And you may take it for granted that as soon as I can afford it, I will cease to take any thing or at most about £100 a year from you.

3 Murray to Beddoes, 24 December, 1782, NLS JMA (see abbreviation list) MS 41904, 92–93.

4 John Murray Letter-Book, 16 October 1768, quoted in William Zachs, Peter Isaac, Angus Fraser, and William Lister, "Murray family (per. 1768–1967)", ODNB. William Zachs, *The First John Murray and the Late Eighteenth-Century London Book Trade; with a Checklist of His Publications* (New York: Oxford University Press, for the British Academy, 1998). There is also a brief account of Murray in Richard Sher, *The Enlightenment and the Book: Scottish Authors and Their Publishers in Eighteenth-Century Britain, Ireland & America* (Chicago: University of Chicago Press, 2006), which has a brief account of Murray as publisher, and a briefer mention (187–188) of Beddoes' translation of John Brown: *The Elements of Medicine of John Brown, M.D. Translated from the Latin, with Comments and Illustrations, by the Author: A New Edition, Revised and Corrected, with a Biographical Preface by Thomas Beddoes, M.D. and a Head of the Author*, 2 vols. (London: Printed for J. Johnson, 1795).

5 Murray to Beddoes, 18 November 1782, NLS JMA MS 41904, 77; Murray to Beddoes 24 December 1782, ibid., 92–93; Murray to Beddoes 14 July 1783, ibid., 192.

6 Lazzaro Spallanzani, *Dissertations Relative to the Natural History of Animals and Vegetables*, trans. Thomas Beddoes, 2 vols. (London: Murray, and Edinburgh: Elliot, 1784), trans. from *Opuscoli di fisica animale e vegetabile, aggiuntevi alcune lettere relative ad essi opuscoli dal Signor Bonnet e da altri scritte all'autore*, 3 vols. (Venice: Presso Giammaria Bassaglia, 1782), with other texts.

7 Murray to Beddoes, 24 December 1782, NLS JMA MS 41904, 92–93.
8 *The English Review* was the most successful of Murray's periodicals. The first issue appeared in January 1783 and "continued monthly until 1797, when it merged with Joseph Johnson's *Analytical Review*". See William Zachs, Peter Isaac, Angus Fraser, and William Lister, "Murray family (*per.* 1768–1967)", ODNB.
9 Murray to Beddoes, 23 November 1783, NLS JMA MS 41904, 80–81.
10 Murray to Beddoes 17 June 1783, ibid., 173. William Cullen, *First Lines of the Practice of Physic, for the Use of Students in the University of Edinburgh*, 3 vols. (Edinburgh: Printed for William Creech, 1783), III.
11 Beddoes returned to Oxford in 1783 to take his M.A.
12 E.g. Murray to Beddoes, 24 December 1782, NLS JMA MS 41904, 92–3: "I have also on the 21st sent you/ 12 Observations/ Millman on the Scurvy/ Reed on Consumptions . . ./ Nicholson's introduction to natural Philosophy 8vo 2 vols. / The System of Vegetables of Linnaeus."
13 Murray to Beddoes, 14 August 1783, NLS JMA MS 41904, 209.
14 Murray to Beddoes, 25 September 1783, NLS JMA MS 41904, 220.
15 Murray to Beddoes, 2 October 1783, NLS JMA MS 41904, 224–225.
16 Murray to Beddoes, 25 November 1783, NLS JMA MS 41904, 254–255.
17 Murray to Beddoes, 12 November 1783, NLS JMA MS 41904, 257–258, concerning Dr. William Cullen.
18 *Physical and Chemical Essays: Translated from the Original Latin of Sir Torbern Bergman . . . By Edmund Cullen*, 2 vols. (London: John Murray, 1788). The title page for vol. III (Edinburgh: Murray, 1791) omits all reference to Cullen.
19 Murray to Beddoes, 1 March 1784, NLS JMA (see abbreviation list) MS 41904, 287.
20 Murray to Beddoes, 9 March 1784, NLS JMA MS 41904, 292. Murray also wrote to Cullen on 21 April 1784:

> I have greatest reason to believe there is another translation going forward in London. and this being the case it will be extremely unpleasant if it is published before us merely because the Copy on your part is not ready. I am extremely anxious about this Matter and you will no doubt see the propriety of your giving as much expedition as is possible for you to do. I mean to publish the 1st Volume whenever it is ready and it will be your fault solely if it is not done in 14 days from this date. I entreat you to attend to this.

NLS JMA MS 41904 313. Murray corresponded frequently with Cullen over the next three months, concluded that his conduct had been shameful, and relieved his indignation in publishing a vigorous denunciation, mostly in the form of reprinting his correspondence with Cullen: John Murray, *An Author's Conduct to the Public, Stated in the Behaviour of Dr. William Cullen, His Majesty's Physician at Edinburgh* (London: J. Murray, 1784).
21 Murray to Beddoes, 3 December 1784, NLS JMA MS 41904, 399–400; 31 January 1785; ibid., 417; 14 February 1785; ibid., 434–435.
22 Joseph Black to James Watt, 15 March 1780, BCL JWP 4/44/55, in CJB, I, 414–415.
23 Beddoes to Trye, n.d. [winter 1794–1795], Gloucestershire Record Office D303 C1/61. This letter is published in T. H. Levere and P. B. Wood, "Thomas Beddoes and the Edinburgh Medical School: A Letter to Charles Brandon Trye, c.1785", *University of Edinburgh Journal*, 32 (1986), 36–39. For information about Trye, see the entry Charles Brandon Trye by D'A. Power, revised by Michael Bevan, in the ODNB.
24 Beddoes to Trye, n.d. [winter 1794-1795], Gloucestershire Record Office D303 C1/61.
25 *Two Introductory Lectures Delivered by Dr. William Hunter, to His Last Course of Anatomical Lectures, at His Theatre in Windmill-Street: As They Were Left*

Corrected for the Press by Himself: To Which Are Added, Some Papers Relating to Dr. Hunter's Intended Plan, for Establishing a Museum in London: For the Improvement of Anatomy, Surgery, and Physic (London: J. Johnson, 1784).

26 Wouter van Douveren, Præs. *Disquisitio physiologica de eo quod vitam constituit in corpore animali, quam . . . examinandum proposuit auctor M. van Geuns* (1758), in Sandifort (E.) *Thesaurus dissertationum, etc., III* (Leiden: S. and J. Luchtmans, 1768).

27 The library of the Royal Medical Society was sold at auction in three sales at Sotheby's in 1969. The Royal College of Physicians of Edinburgh still has its own fine library, much of it from the eighteenth century. Hugo Arnot, *The History of Edinburgh, from the Earliest Accounts to the Present Time . . .* (Edinburgh: Printed for William Creech, and sold by Messrs Robinson & Co., Paternoster Row, London, 1788), 323–324 has a brief account of the library of the RCPE, which "contains many valuable books, chiefly in natural history, the greatest part of which belonged to Dr Wright of Kersie; and, upon his death, a donation was made of them by his heir, to the college." (324)

28 Murray to Beddoes, 11 December 1783, NLS JMA MS 41904, 260.

29 Ibid., 281–282.

30 Murray to Mr. Jas. Gilliland 25 December 1783, NLS JMA MS 41904, 268–269.

31 Torbern Bergmann, A *Dissertation on Elective Attractions*, translated from the Latin by the translator of Spallanzani's dissertations (London: John Murray, and Edinburgh: Edward Elliot, 1785), trans. from "Disquisitio de attractionibus electivis", which first appeared in *Nova acta Regiæ societatis scientiarum upsaliensis*, 2 (1775), 161–250 (from which Beddoes made his translation), then in Bergman, *Opuscula Physica et Chemica* (Lipsiae: Jo. Godof. Mülleriana, 1786), III, 291–470. *The Chemical Essays of Charles-William Scheele: Translated from the Transactions of the Academy of Science at Stockholm, with Additions by T. Beddoes* (London: John Murray, and Edinburgh: Charles Elliot, 1786).

32 Murray to Beddoes, 2 October 1783, NLS JMA MS 41904, 224.

33 Murray published Bergman's, *An Essay on the Usefulness of Chemistry, and Its Application to the Various Occasions of Life* (London: Jo. Godof. Mülleriana, 1783); the text was translated from the German (itself a translation from the original Swedish) by the Austrian physician and mineralogist Franz Xavier Schwediauer with the assistance of Jeremy Bentham, lawyer, polymath, and utilitarian philosopher. The translator of Bergman's *Essay* is not named on the title page, but is identified, along with bibliographic details, in William A. Cole, *Chemical Literature, 1700–1860: A Bibliography with Annotations, Detailed Descriptions, Comparisons, and Locations* (London and New York: Mansell, 1988). Schwediauer spent more than a decade in London and Edinburgh before removing to Paris in 1789, and he was a useful go-between in transmitting books to Joseph Black and others: see e.g. Schwediauer to Black, 30 January 1782, EUL MS Gen 875/II/41, in CJB I, 481. Schwediauer is mentioned in several entries in Neil Chambers, ed., *Scientific Correspondence of Joseph Banks, 1765–1820*, 6 vols. (London: Pickering and Chatto, 2007).

34 Black to his brother Alexander, 18 July 1789, in CJB II, 1034:

> Dr. Schwediauer has a good deal of knowledge in Chemistry & the Arts but he is a great Projector & pryer into Secrets and from all that I can learn he has succeeded very ill in his projects here & pushed on his schemes with a good deal of the Spirit of an Aventurier.John Murray to Dr. Garthshore, 23 June 1784, NLS JMA MS 41904, 338: "Mr. M. must again decline to meet with Dr. S. upon this subject because he knows from dear bought experience that *his* heart is unsusceptible of one generous impression."

35 Murray to Beddoes, NLS JMA MS 41904, 449–481 *passim*, several letters in the spring of 1785. *The Chemical Essays of Charles-William Scheele* (1786).

36 Memorial, 2.
37 David Vaisey observes in the ODNB, s.v. Price: "The advances made by the Bodleian Library during Price's forty-five years as its librarian owed little to his administrative skills." Price held ecclesiastical appointments elsewhere, while holding the Librarianship from 1768 to his death in 1813, whereupon the position passed to his godson Bulkeley Bandinel.
38 Memorial.
39 Beddoes' function at Oxford, to offer lectures in chemistry, accompanied by demonstration experiments, was clear; his title was less so, and at different times he styled himself lecturer, reader, and professor, the last title being retroactive and inaccurate. The Vice-Chancellor wanted to obtain a chair for Beddoes, with royal support, but politics and political espionage intervened, and Beddoes resigned from Oxford without a chair. See T. H. Levere, "Dr. Thomas Beddoes at Oxford: Radical Politics in 1788–1793 and the Fate of Regius Chair in Chemistry", *Ambix*, **28** (1981), 61–69, and F. W. Gibbs and W. A. Smeaton, "Thomas Beddoes at Oxford", *Ambix*, **9** (1961), 47–49. We are no longer convinced that the chair was to be a Regius Chair, but the Vice-Chancellor was certainly hoping that it would be sanctioned and financed by the crown.
40 Memorial, 4.
41 Memorial, 5.
42 Memorial, 9.
43 Memorial, 10.
44 See e.g. Albrecht von Haller, *Elementa physiologiæ corporis humani* (Lausanne, 1757–1778); in the Bodleian Library's copy, vols. VI–VIII are of the 2nd ed.
45 Memorial, 11. John Hill, *The Vegetable System, or, the Internal Structure and the Life of Plants*, 26 vols. (London: printed for the author, sold by R. Baldwin, 1762–1765); the Bodleian Library copy has volumes I and XXI of the 2nd ed.
46 *Memorial*, 11.
47 Ibid., 15.
48 Ibid.
49 Ibid., 15–16.
50 Albrecht von Haller, Swiss anatomist and physiologist, for seventeen years from 1736 a professor at Göttingen; Christian Gottlob Heyne, classicist and archaeologist; Christoph Meiners, philosopher and historian; Johann David Michaelis, biblical scholar; Johann Friedrich Wilhelm Jerusalem (1709–1789), protestant theologian; Johann Christoph Döderlein, theologian; Moses Mendelssohn, Jewish philosopher; Gotthold Ephraim Lessing, poet, dramatist, and Enlightenment writer on aesthetics, critical theology, and philosophy.
51 Statute of Anne 1710, *An Act for the Encouragement of Learning, by Vesting the Copies of Printed Books in the Authors or Purchasers of such Copies, during the Times Therein Mentioned.*
52 Hugo Arnot, *The History of Edinburgh . . .* (1788), 295–297. There were eighteenth-century catalogues of the library: T. Ruddiman, *A Catalogue of the Library of the Faculty of Advocates*, Pt. 1 (Edinburgh: Printed by Thomas, Walter, and Thomas Ruddiman, 1742); *Appendix to the Catalogue of the Advocates Library* (Edinburgh: Printed by William Smellie, 1787).
53 Ibid., 18–19. Samuel Taylor Coleridge, encouraged by Beddoes, studied in Germany, principally in Göttingen, where he matriculated at the university, and visited the library, which he enthusiastically reported was "without doubt . . . the very first in the World both in itself, & in the management of it." Coleridge to Mrs. [Sara] Coleridge, 10 March 1799, in *Collected Letters of Samuel Taylor Coleridge*, ed. Earl Leslie Griggs, 6 vols. (Oxford: Clarendon Press, 1956–1971), I, 475. See Trevor H. Levere, *Poetry Realized in Nature: Samuel Taylor Coleridge and Early Nineteenth-Century Science* (Cambridge: Cambridge University Press, 1981), 16–20.

54 Beddoes to Black, 6 November 1787, EUL MS Gen 875/III/52, 53, in CJB I 414–415.
55 [Beddoes], *Alexander's Expedition Down the Hydaspes & the Indus to the Indian Ocean* (1992). There is some discussion of this in Simon Schaffer, "The Asiatic Enlightenments of British Astronomy", in *The Brokered World: Go-Betweens and Global Intelligence, 1770–1820*, eds. Simon Schaffer, Lissa Roberts, and Kapil Raj (2009), 100–101.
56 This note and transcription, dated [18] June 1901, accompanies the copy in the Stirling Library, Yale University, and is dated Madeley Wood, 9 October 1819. See also Martin Priestman, *The Poetry of Erasmus Darwin: Enlightened Spaces, Romantic Times* (Ashgate: Farnham, Surrey and Burlington, VT: 2013), and Patricia Fara, *Erasmus Darwin: Sex, Science, & Serendipity* (Oxford: Oxford University Press, 2012).
57 An anonymous pamphlet published in 1790, and bearing the title *The Propriety of Sending Lord Sheffield to Coventry, Peace! Plenty! Bill of Rights! The Constitution! Yes! The British, Not Pitt's Constitution!* has been tentatively attributed to Beddoes by Stephen Holt, *Times Literary Supplement* (22 August 2003), 11. Whether Beddoes wrote it or not, the sentiments appear in later pamphlets that he wrote, and in his correspondence with Davies Giddy.
58 See Levere, "Dr. Thomas Beddoes at Oxford. Radical politics in 1788-1793 and the fate of the regius chair in chemistry," *Ambix* 28 (1981) 61–69.
59 Active from 1766 until his death in 1806.
60 Beddoes to Watt, 26 May 1797, and September 1796, JWP MS 3219/4/29 nos. 29 and 9.
61 Helen Braithwaite, *Romanticism, Publishing and Dissent: Joseph Johnson and the Cause of Liberty* (New York: Palgrave Macmillan, 2003). Carol Hall, art. "Joseph Johnson", ODNB. G. Tyson, *Joseph Johnson: An Eighteenth-Century Bookseller* (1979).
62 Joseph Johnson, *Books in Anatomy, Medicine, Surgery, and Natural Philosophy* (London: Printed for J. Johnson, No. 72, St. Paul's Church-Yard, 1785?).
63 A. A. Engstrom, "Joseph Johnson's Circle and the *Analytical Review*: A Study of English Radicals in the Late Eighteenth Century", Ph.D. dissertation, University of Southern California, 1986.
64 Murray to Johnson, 8 October 1787, NLS JMA MS 41905, 198.
65 Beddoes' works published jointly by Johnson and Murray included *A Guide for Self-Preservation, and Parental Affection or Plain Directions for Enabling People to Keep Themselves and Their Children Free from Several Common Disorders* (London: Printed by Bulgin and Rosser, for J. Murray, and for J. Johnson, 1793); Beddoes, *A Letter to Erasmus Darwin, M.D. on a New Method of Treating Pulmonary Consumption, and Some Other Diseases Hitherto Found Incurable* (London and Bristol: Printed by Bulgin and Rosser, for J. Murray, and J. Johnson, London; also by Bulgin and Sheppard, J. Norton, J. Cottle, W. Browne, and T. Mills, Booksellers, Bristol, 1793); and, in part, *Considerations on the Medicinal Use, and on the Production of Factitious Airs Part I: By Thomas Beddoes, M.D Part II: Description of an Air Apparatus; with Hints Respecting the Use and Properties of Different Elastic Fluids: By James Watt, Esq. Pt. III Considerations on the Medicinal Use, and on the Production of Factitious Airs: Pts. IV & V Medical Cases and Speculations* (Pts. I & II Printed by Bulgin and Rosser, for J. Johnson, No. 72, St Paul's Church-Yard, and J. Murray, No. 32, Fleet Street, London; Pt. III Printed by Bulgin and Rosser, for J. Johnson, in St Paul's Church-Yard, London, 1795; parts IV and V, 1796).
66 Braithwaite, *Romanticism, Publishing and Dissent*, 148.
67 J. W. Symser, "The Trial and Imprisonment of Joseph Johnson, Bookseller", *Bulletin of the New York Public Library*, 77 (1974), 418–435. See also Kenneth

R. Johnston, *Unusual Suspects: Pitt's Reign of Alarm and the Lost Generation of the 1790s* (Oxford: Oxford University Press, 2013), which also has major discussions of Priestley and Beddoes, and has a useful appendix on the Treason Trials of 1792–1798.

68 Goodwin, *The Friends of Liberty*; Johnston, *Unusual Suspects*.

69 Humphry Davy, *Researches, Chemical and Philosophical, Chiefly Concerning Nitrous Air, and Its Respiration* (London: Printed by Biggs & Cottle, Bristol, for J. Johnson, 1800). Johnson offered Davy the substantial sum of 60 pounds for this work: see Braithwaite (2003), 212, n. 49. *Memoirs of the Rev. Dr. Joseph Priestley to the Year 1795, Written by Himself: With a Continuation, to the Time of his Decease, by His Son, Joseph Priestley* (London: Reprinted from the American edition, by the several Unitarian Societies in England: and sold by Joseph Johnson, St. Paul's Church-Yard, 1809).

70 Quoted from the *Edinburgh Review* in Braithwaite (2003), 176.

71 For a list of books and their reviewers, see Benjamin Christie Nangle, *The Monthly Review, Second Series, 1790–1815; Indexes of Contributors and Articles* (Oxford: Clarendon Press, 1955).

72 Murray to Beddoes, NLS JMA MS 41904, 384–385. Murray told Beddoes that he obtained Italian books from Moline, presumably the London bookseller and publisher Pietro Molini (1729 or 1730–1806).

73 Thomas Beddoes, review of *Zum ewigen Frieden* (i.e. *To Perpetual Peace*), Emmanuel Kant, Koenigsburg 1795, in *The Monthly Review, or, Literary Journal*, **20** (1796), Foreign Supp. Article 2. For a list of Beddoes' reviews in this journal, see Stansfield, 254–260, taken from B. C. Nangle, *The Monthly Review, First Series, 1749–1789; Indexes of Contributors and Articles* (Oxford: Clarendon Press, 1934), and *The Monthly Review, Second Series, 1790–1815* (1955). See also René Wellek, *Immanuel Kant in England, 1793–1838* (Princeton: Princeton University Press, 1931), and Giuseppe Micheli, "The Early Reception of Kant's Thought in England, 1785–1805", in *Kant and his Influence*, eds. George MacDonald Ross and Tony McWalter (Bristol: Thoemmes Antiquarian Books, 1990), 202–314.

74 Beddoes, *Observations on the Nature of Demonstrative Evidence; With an Explanation of Certain Difficulties Occurring in the Elements of Geometry: And Reflections on Language* (London: J. Johnson, 1793), 3, 5, 89–101. John Horne Tooke, *Epea pteroenta: Or, the Diversions of Purley: Part I* (London: J. Johnson, 1786).

75 An edited alphabetic listing and interpretation of the books in Beddoes' library, derived from the sale catalogue, is available online at http://hdl.handle.net/1807/71096.

76 CBL. This item appears to have been misplaced, but a microfilm survives.

77 Rebecca Bowd, "Useful Knowledge or Polite Learning? A Reappraisal of Approaches to Subscription Library History", *Library and Information History*, **29** (2013), 182–195. The main subscription library in Bristol was that of the Bristol Library Society. There is much information about that library in G. Munro Smith, *A History of the Bristol Royal Infirmary* (Bristol: J. W. Arrowsmith Ltd., and London: Simpkin, Marshall, Hamilton, Kent & Co. Ltd., 1917), and in Paul Kaufman, *Borrowings from the Bristol Library, I773-I784: A Unique Record of Reading Vogues* (Charlottesville: University of Virginia Press, 1960).

78 Thomas Lovell Beddoes was a poet and physician. In 1824 he went to study medicine at the University of Göttingen, from which he was expelled in 1829, for "riotous drinking bouts". He completed his medical studies at Würzburg; thereafter, he practised medicine, and wrote verse and prose, some mournful, some satirical and democratic. The latter led to his deportation from Bavaria, and to an enforced move from Zürich. Thus far, the apple didn't fall far from the

tree. But unlike his father, Thomas Lovell Beddoes committed suicide in 1849. See Alan Halsey, "Beddoes, Thomas Lovell (1803–1849)", ODNB.

79 Neil Vickers, *Coleridge and the Doctors 1795–1806* (Oxford: Clarendon Press, 2004), 37–78, gives an admirable account of Beddoes' medical theories and practice.

80 Stock (1811), 300.

81 Banks purchased just one book: Gomes, Bernardino Antonio (the Elder), *Memoria sobre a ipecacuanha fusca do Brasil, ou cipo das nossas boticas, etc.* (Lisboa: Arco do Ceco, 1801).

82 Beddoes to Charles Brandon Trye, n.d. [1784 or 1785], Gloucestershire Record Office, MS D303 C1/61.

83 Beddoes' knowledge of the Göttingen Library is impressive, and we find it hard to believe that he had never visited it. All we can say for certain is that we have looked for evidence in Göttingen (including the register of borrowings from the university library, and correspondence by Girtanner) and in Beddoes' correspondence, but failed to find it. If the visit occurred, it would have been in 1787, before he went to Oxford to teach chemistry.

84 Thomas Beddoes, *Über die neuesten Methoden die Schwindsucht zu heilen, übersetzt von G.K. Kühn* (Leipzig: Sommer, 1803).

85 *Essai sur le phlogistique, et sur la constitution des acides, traduit de l'anglois de M. Kirwan* [by Mme. Lavoisier]; *avec des notes de MM. de Morveau, Lavoisier, de la Place, Monge, Berthollet, & de Fourcroy* (Paris: Rue et Hôtel Serpente, 1788). This edition is discussed in Seymour Mauskopf, "Richard Kirwan's Phlogiston Theory: Its Success and Fate", *Ambix*, **49** (2002), 185–205.

86 Joseph Priestley, *Disquisitions relating to Matter and Spirit: To Which is Added, the History of the Philosophical Doctrine Concerning the Origin of the Soul, and the Nature of Matter; with Its Influence on Christianity, etc.* (London: J. Johnson, 1777). *Experiments and Observations on Different Kinds of Air, and Other Branches of Natural Philosophy, Connected with the Subject. In Three Volumes; Being the Former Six Volumes Abridged and Methodized, with Many Additions* (Birmingham: Printed by Thomas Pearson; and sold by J. Johnson, London, 1790). *The Doctrine of Phlogiston Established, and That of the Composition of Water Refuted* (Northumberland, PA: Printed for the author by A. Kennedy, 1800).

87 Erich Weidenhammer, *Air, Disease, and Improvement in Eighteenth-Century Britain: Sir John Pringle (1707–1782)*, Ph.D. dissertation (University of Toronto, 2014).

88 See Chapter 1, this volume.

89 Gavin de Beer, *The Sciences Were Never at War* (New York: Thomas Nelson, 1960). H. B. Carter, *Sir Joseph Banks 1743–1820* (London: British Museum (Natural History), 1988).

90 Odier to Beddoes, 3 September 1798, Birmingham Record Office, MS 3219/4/29:20. For an account of Odier, see G. de Morsier, "La vie et l'oeuvre de Louis Odier(1748–1817)", *Gesnerus*, **32** (1975), 248–270. By comparison, Wedgwood, Beddoes, and others made a concerted effort to obtain foreign books, as did the Georg August Library in Göttingen. For Wedgwood's access to continental sources, see Larry Stewart, "Assistants to Enlightenment: William Lewis, Alexander Chisholm and Invisible Technicians in the Industrial Revolution", *Notes and Records of the Royal Society*, **62** (2008), 17–29.

91 Thomas Beddoes and James Watt, *Considerations on the Medicinal Use, and on the Production of Factitious Airs*, Part I by Beddoes, Part II, *Description of an Air Apparatus; With Hints Respecting the Use and Properties of Different Elastic Fluids*, by James Watt; Pt. III by Thomas Beddoes and James Watt; Pt. IV *Medical Cases and Speculations*; Part V, *Supplement to the Description of a Pneumatic Apparatus for Preparing Factitious Airs; Containing a Description of a Simplified Apparatus, and of a Portable Apparatus, by James Watt.*

92 T. H. Levere, "Spreading the Chemical Revolution: The Dutch Connection", *Bulletin of the Scientific Instrument Society*, **49** (June, 1996), 14–16.

93 Thomas Henry to Joseph Watt junior, 2 December, 1794, quoted in William Vernon Farrar, *Chemistry and the Chemical Industry in the 19th Century: The Henrys of Manchester and Other Studies*, eds. Richard L. Hills and W. H. Brock (Variorum: Ashgate, Aldershot & Brookfield, VT, 1997), II, 184.

94 Thomas Henry to James Watt 11 March 1794, JWP MS 3219/4/27.25. Thomas Henry to Dr. [?] 6 December 1794, BCL JWP MS 3219/4/27.24.

95 Thomas Henry to James Watt junior, 6 December 1794, op. cit.

96 See James Raven, *London Booksellers and Their American Customers: Transatlantic Literary Community and The Charleston Library Society, 1748–1811* (Columbia: University of South Carolina Press, 2001); G. M. Abbott, *A Short History of the Library Company of Philadelphia; Compiled from the Minutes, Together with Some Personal Reminiscences* (Philadelphia: Published by order of the Board of Directors, 1913); D. F. Grimm, *A History of the Library Company of Philadelphia, 1731–1835*, Ph. D. Dissertation, University of Pennsylvania (1955).

97 See W. A. Smeaton, "Louis Bernard Guyton de Morveau, FRS (1737–1813) and His Relations with British Scientists", *Notes and Records of the Royal Society of London*, **22** (1967), 113–130. Richard Kirwan, *Essai sur le Phlogistique* (1788) (see note 85).

98 William A. Smeaton, *Fourcroy, Chemist and Revolutionary, 1755–1809* (Cambridge: Printed for the author by W. Heffer, 1962). See also W. A. Smeaton, "French Scientists in the Shadow of the Guillotine: The Death Roll of 1792–1794", *Endeavour*, NS **17** (1993), 60–63. Jonathan R. Topham, "Science, Print, and Crossing Borders: Importing French Science Books into Britain, 1789–1815", in *Geographies of Nineteenth-Century Science*, eds. D. N. Livingstone and Charles W. Withers (Chicago: University of Chicago Press, 2011), 311.

99 Joseph DeBoffe, French bookseller, was at 7 Gerrard Street, Soho, London, from 1792–1807; from 1808–1815 the firm was at 10 Nassau Street, Soho, trading as J. C. DeBoffe. DeBoffe bought *Mémoires de l'Institut National des Sciences et Arts, pour l'an IV de la République : Littérature et Beaux-Arts*, avec planches, 16 tom., from Beddoes' library sale.

100

> *Le Moniteur Universel* was founded in Paris on 24 November 1789 under the title *Gazette Nationale ou Le Moniteur Universal* by Charles-Joseph Pancoucke [a publisher, and one of a family of booksellers and writers]. . . . It was the main French newspaper during the French Revolution and was for a long time the official journal of the French government and at times a propaganda publication, especially under the Napoleonic regime. *Le Moniteur* had a large circulation in France and Europe, and also in America during the French Revolution. . . . [It] ceased publication on 31 December 1868.

Wikipedia, art. "Le Moniteur Universel". See also *Encylopaedia Britannica*, online, art. "History of Publishing: Commercial newsletters in continental Europe"; Memorial.

101 See e.g. Beddoes to Giddy 8 October 1791, CRO MS DG 41/4.

102 The trial of John Horne Tooke occurred 17–22 November 1794. DeBoffe was also a witness at the trial of Thomas Hardy, founder and secretary of the London Corresponding Society, who was also acquitted on charge of treasonable practices on 5 November 1794.

> [JHT] Did you sell a great many? – [JDB] I have sold as many as I had regular subscribers for: I was not in the habit of selling papers loosely, but such noblemen and gentlemen as subscribed regularly, I had the honour of serving them with regularity. I have sold some hundreds in regular connexions, from the beginning of the Moniteur, down to the time when the communication was

totally stopped. . . . [JHT] You thought it a safe thing to sell these papers? – [JDB] Yes, while the post office sold them, and several of the shops sold them, as well as me, having paid the regular duty at the Custom-house.*A Complete Collection of State Trials and Proceedings for High Treason and other Crimes and Misdemeanors from the Earliest Period to the Year 1783, with Notes and Other Illustrations: Compiled by T.B. Howell . . . and Continued from the Year 1783 to the Present Time: by Thomas Jones Howell . . .*, vol. xxv (vol. iv of the continuation) 35 & 36 George III . . . 1794–1796 (London: Printed for Longman, Hurst, Rees, Orme, and Brown . . . and T. C. Hansard, 1818), 191.

103 See Chapter 1.

104 Rebecca Boyd "Useful Knowledge or Polite Learning? A Reappraisal of Approaches to Subscription Library History", *Library and Information History*, 29 (2013), 182–195, at 180 181, 190–191, quoting from J. Farington, *The Farington Diary*, ed. J. Greig (London: Hutchinson & Co., 1923–1928), I, 315.

105 Graham Jefcoate, "'Hier ist nichts zu machen': Zum deutschen Buchhandel in London 1790–1806", in *Literatur und Erfahrungswandel 1789–1830*, eds. Rainer Schöwerling, Hartmut Steinecke and Günter Tiggesbäumker (Munich: Fink, 1996), 47–59.

J. H. Bohte, 4 Yorke St., Covent Garden, Foreign Bookseller to the King, became active in the book trade and was a key source of German books in London after Beddoes died. Bohte subsequently had dealings with Beddoes' friend Coleridge, offering him books and receiving an invitation to the latter's 1818–1819 lectures: Kathleen Coburn, ed., *The Notebooks of Samuel Taylor Coleridge: Vol. 3, 1808–1819* (Princeton University Press, 1973), 4464. Sellers of German books in London in Beddoes' lifetime included Henry Escher (14 Broadstreet, St. Giles in 1799, 9 Gerard Street, Soho in 1802); Vaughan Griffiths (1 Paternoster Row 1797–1812); James Remnant (St. John's Lane, West Smithfield 1793; High Holborn [1795–1799]; 22 High St., St. Giles [1800–1801]). See Ian Maxted, *Exeter Working Papers in British Book Trade History*, "The London Book Trades 1775–1800: A Preliminary Checklist of Members", accessible online at http://bookhistory.blogspot.ca/2013/01/gloucestershire.html.

106 They printed *Contributions to Physical and Medical Knowledge, Principally from the West of England and Wales* (Bristol: Printed by Biggs and Cottle, for T. N. Longman and O. Rees, 1799).

107 See the list in Ian Maxted, op. cit.

108 Graham Jefcoate, "German printing and bookselling in eighteenth-century London: evidence and interpretation", in Barry Taylor, ed., *Foreign-language printing in London 1500-1900* (Boston Spa & London: The British Library, 2002), 1–36 at 7, 23–25.

109 Graham Jefcoate and Karen Kloth, *A Catalogue of English Books Printed before 1801 Held by the University Library at Göttingen* (Hildesheim: Olms-Weidman, 1987). Pt. 1, Books printed before 1700, I–II; Pt. 2, Books printed between 1701 and 1800, I–IV; Pt. 3 indices.

110 Graham Jefcoate, "German Printing and Bookselling in Eighteenth-Century London: Evidence and Interpretation", in *Foreign-Language Printing in London 1500–1900*, ed. Barry Taylor (Boston Spa & London: The British Library, 2002), 1–36 at 7, 23–25. *Monthly Magazine* (February, 1796), 35.

111 German thought first made significant inroads in England in the 1790s and gained steadily during the nineteenth century. For the process through which German thought was introduced in Britain in the nineteenth century, see Rosemary Ashton, *The German Idea: Four English Writers and the Reception of German Thought, 1800–1860* (Cambridge: Cambridge University Press, 1980).

112 J. R. Harris, *Industrial Espionage and Technology Transfer: Britain and France in the Eighteenth Century* (Ashgate: Aldershot, 1998) is the best treatment of this topic.
113 A. L. Millin, *Magasin encyclopédique, ou journal des sciences, des lettres et des arts* . . .(Paris: J. B. Sajou, printer, 1797).
114 As we have seen earlier, Beddoes subsequently acquired an almost complete run of the journal.
115 Beddoes to James Watt junior, [March, 1798], BCL BWP MS M IV B.
116 James Watt to Thomas Beddoes 10 January 1804, JWP MS 3219/4/119.57. Johann Albert Heinrich Reimarus, Hamburg physician, economist, and natural philosopher.
117 Beddoes to Matthew R. Boulton, 13 August 1798, microprint, *Industrial Revolution: A Documentary History*, Series I, part I, From Box B2 (220) – Item 14 (Marlborough: Adam Matthew, 1993–1999).
118 A. Deriabin, letters to Gregory Watt, 1798–1800, BCL JWP Box B2 (220) 14.
119 Gottfried Martini, publisher and bookseller, Leipzig, from 1799 through the early 1800s.
120 Thomas Beddoes to Matthew Robinson Boulton, 13 August 1793, BCL BWP Box 2 (220) 15.
121 M. R. Boulton to Beddoes, 16 August 1798, BCL BWP Box 2 (220) 15.
122 Beddoes to M. R. Boulton, 18 August 1798, BCL BWP Box B2 (220) 15.
123 Beddoes to M. R. Boulton, 18 August 1798, BCL BWP Box B2 (220) 18.
124 Beddoes to Davies Giddy, 11 October 1800, CRO MS DG 42:25.
125 Mrs. Ketley Reynolds to Beddoes, 4 March 1789, Bodley MS Dep. C.134/2.
126 Stansfield, 101 n.2.
127 Bristol Library Society, borrowing register, in Bristol Public Library. The complete list of Beddoes' borrowings is given in Appendix 2, this volume.
128 Beddoes to Giddy, 16 May 1791, CRO MS DG 41/15. For Beddoes' lectures, see Chapter 2, this volume.
129 Beddoes to Giddy, 8 October 179[1] or CRO MS DG 41/54 (or MS DG 41/55 179[1]; the MS is between two dockets, and could bear either number). Pierre de Riel, Marquis de Beurnonville, was a French general in the Revolution and subsequently in the Empire, who served under Charles-François du Périer Dumouriez. Dumouriez called him the "French Ajax" because of his stature and courage. The Moniteur is *Gazette nationale ou Le moniteur Universel.*
130 Beddoes to Giddy, 14 April 1798, CRO MS DG 42/2, referring to Joseph Louis Lagrange, *Théorie des fonctions analytiques contenant les principes de calcul differential, dégagés de toutes considerations d'infiniment petits, d'évanouissants, de limites et de fluxious et réduits à l'analyse algébraique des quantités finies* (1797).
131 For a wide-ranging account, see James Raven, *The Business of Books: Booksellers and the English Book Trade 1450–1850* (New Haven: Yale University Press, 2007).
132 William Henry Lunn was a bookseller at 332 Oxford Street, 1797–1801, sold his stock through Leigh and Sotheby in January 1802, and established the Classical Library at 30 Soho Square in 1802. His business did not succeed, and he committed suicide in June 1815.
133 Beddoes to M. R. Boulton, 18 August 1798, BWP Box B2 (220) 15.
134 Beddoes to M. R. Boulton, n.d. [1799], BCL BWP box B2 (220) 24.
135 DeBoffe to Giddy for Beddoes, 8 July 1807, CRO DD DG 43/49; Beddoes to Giddy, 29 July 1807, CRO DD DG 43/57; Beddoes to Giddy, 6 August 1807, CRO MS DD DG 43/56; Beddoes to Giddy, August 1807, CRO DD DG 43/38.

Beddoes to Giddy, [July 1807], CRO DD DG 43/38. The sale catalogue of Beddoes' library includes the *Journal de Médecine, Chirurgie, Pharmacie, etc., par MM: Corvisart, Leroux et Boyer*, 26 vols. (lacking vols. 12–17).

136 Beddoes correspondence with Douce, 22 June, 29 June, 6 July 1808, Bodley MSS Douce d.21/176–77, d.21/178f–g, d.21/178a–178e.

137 C. C. Hannaway, "The Société Royale de Médicine and epidemics in the Ancien Régime", *Bulletin of the History of Medicine*, **46** (1972), 257–273. J. P. Frank, "The Civil Administrator, Most Successful of Politicians", *Bulletin of the History of Medicine*, **16** (1944), 289–318. This and the next three paragraphs are taken from T. H. Levere, "Dr. Thomas Beddoes: The Interaction of Pneumatic and Preventive Medicine with Chemistry", *Interdisciplinary Science Reviews*, 7 (1982), 137–147.

138 This is introduced in Chapter 1, this volume, and becomes a major theme in Chapter 3.

139 John Graunt, *Natural and Political Observations, Mentioned in a Following Index, and Made upon the Bills of Mortality* (London: Printed by Tho. Rycroft, for John Martin, James Allestry, and Tho. Dicas . . ., 1662) is perhaps the first work of modern demography in England. In the early eighteenth century, numbers began to be used to assess the results of medical treatments. Francis Clifton (M.D., Leiden, 1724) was among the first to pursue this approach: Clifton, *The State of Physick, Ancient and Modern, Briefly Considered: With a Plan for the Improvement of It* (London: Printed by W. Bowyer, for John Nourse, 1732). For an overview, see Major Greenwood, *Medical Statistics from Graunt to Farr: The Fitzpatrick Lectures of the Years 1941 and 1943, Delivered at the Royal College of Physicians of London in February 1943* (Cambridge: Cambridge University Press, 2014); originally published 1941–1943.

140 Beddoes, *Hygëia* . . ., 3 vols. (Bristol: Printed by J. Mills, St. Augustine's Back, for R. Phillips . . ., London, 1802–1803).

141 Beddoes, *Hygëia*, II, 98 and III, 86.

142 Beddoes, *Hygëia*, II, essay vii, 5. Convulsions were seen most often in the young, and were accompanied by high fever.

143 Ibid.

144 Beddoes to Douce, 30 April 1808, Bodley MS Douce d.21/164–165. Thomas Dunham Whitaker, *The History and Antiquities of the Deanery of Craven, in the County of York* (London: W. Edwards and Son, 1805).

145 Beddoes to Douce, 22 June 1808, Bodley MS Douce d.21/176–177. Andrew Boorde, *The Breviary of Healthe, for all Manner of Sicknesses and Diseases the Which May Be in Man or Woman* . . . (London: Printed by Wyllyam Powell, 1552).

146 Beddoes to Douce, 29 June 1808, Bodley MS Douce d.21/178a–178e. Dr. Coombe was perhaps the Rev. Dr. Thomas Coombe, a man of wide and deep learning. As late as 12 November 1808, just over a month before he died, and writing "with the hand of a sick man", Beddoes wrote to the Secretary at War (CRO MS DD DG 43/69), asking for post-mortem records from military records. These were not forthcoming.

147 Beddoes to Giddy, 8 November 1792, CRO MS DD DG 41/5; Beddoes, *Hygëia*, I, 77:

> As to the sort of reading, most injurious to young females, I cordially assent to the opinion of almost all men of reflection. NOVELS, undoubtedly, are the sort most injurious. Novels render the sensibility more diseased. And they increase indolence, the imaginary world indisposing those, who inhabit it in thought, to go abroad into the real."H. J. Jackson, *Romantic Readers:*

The Evidence of Marginalia (New Haven and London: Yale University Press, 2005), 307, n.4, notes that "[t]he prolonged attack on women's reading has been the subject of many studies: a comprehensive recent example is Jacqueline Pearson's *Women's Reading in Britain, 1750–1835": A Dangerous Recreation* (Cambridge: Cambridge University Press, 1999).

148 Beddoes to Matthew Robinson Boulton, 15 July 1800, BCL BWP Box B2 (220) 26.

5 Models, toys, and Beddoes' struggle for educational reform, 1790–1800

Hugh Torrens and Joseph Wachelder

"The inventor of a child's toy goes out of the world a greater benefactor than the most splendid judicial orator."

– Tom Wedgwood[1]

Introduction

Beddoes always managed, through his many commitments and enthusiasms, to raise high expectations but, too often, as Roy Porter put it, he "suffered from trying to do too much too fast."[2] But his underlying philanthropy was never in doubt. This is abundantly true of his long-forgotten educational plans, which involved the promotion and then the manufacture of what were then known as "rational toys" (i.e. toys for use by the general public and for the educational advancement of their children).[3]

For the development of rational toys, the last decade of the eighteenth century turned out to be pivotal. The period between 1792 and 1800 entailed drastic changes in Beddoes' private and public life. As we know, he moved from the University of Oxford to Bristol,[4] to become a practising physician and to establish his Pneumatic Institution. From Bristol he got in touch with the Edgeworth family, with whom he entered into warm relationships. In 1794 he married Anna Edgeworth. In 1798 her sister Maria, partly inspired by Beddoes, joined forces with her father Richard Lovell Edgeworth and her stepmother Honora Sneyd to write *Practical Education*; Sneyd died before the writing was completed.[5] The book was a landmark in the history of education, and proved seminal for the development of educational toys in the first decades of the nineteenth century. The Edgeworths' theory of education stressed the formative role of a child's early experiences, and the value of practical experience and experiments.

In the 1790s, Beddoes was definitely trying to do too much too fast. However, his engagement with education, educational reform, primary education, the teaching of reading to children, mathematical instruction, and the use of models and toys did not result in distracted or disjointed efforts. On the contrary, his endeavours are highly coherent. Starting from Locke's sensationalist epistemology, in which knowledge arose from experience derived from sense perception, he connected David Hartley's theories about the nervous system[6] and mind with proposals for educational reform that fitted into the British educational establishment. He built upon Locke's own writings on education.[7] Yet Beddoes' interest in models and toys did add a playful element that matched the influence of Rousseauian pedagogy[8] in Great Britain. In the 1820s, when educational toys where becoming widespread, Beddoes was remembered as its pioneer.

In this chapter we will follow Beddoes' interest in models, toys, and educational reform in the last decade of the eighteenth century. Central to his educational ideas are opinions regarding the relationship between the senses, especially sight, hearing, and touch.

Beddoes' *Extract of a Letter on early Instruction, particularly that of the poor* (1792)

The *Extract of a Letter on early Instruction* is addressed to an unnamed lady and dated 25 January 1792. In this pamphlet of twenty-two pages, Beddoes unequivocally demonstrated his commitment to engage in major societal issues. He deplored slavery and cruelty to animals, condemned the power of the bayonet, and called for tolerance between men, irrespective of their religion or irreligion. He saw education as an effective tool to bring about societal change.[9] At the end of his *Extract*, he elaborated on what he called the "romantic idea" of establishing in England schools where young children would learn to read and write, and would also be taught about the arts and industry, in combination with a more general education, following a pattern set for at least ten years in Germany.[10]

The first part of the *Extract* compared different ways to teach reading and writing. Key to his approach was that education should combine different senses and be connected to the child's everyday environment and experience. This made him dispute the moral use of narrative fables or religiously inspired text books, like those available in "READING-MADE-EASY" books.[11] Before being able to understand metaphorical language and moral reasoning, children should be acquainted with clear-cut ideas, connected to their own observations. This had consequences for the topics Beddoes considered suitable for introducing children to reading. He recommended words that

> quicken his curiosity and induce him to look with a more steady attention on the objects around him. This idea should be carefully carried on through every stage of instruction. In the Fourth part of Mrs. Barbauld's Lessons,[12] from page 72, to the end it is executed with great success in the explanation of such terms as "web-footed, cloven-hoofed, quadrupeds," and others expressive of natural appearances.[13]

Yet Beddoes not only approved and disapproved of certain topics in the introduction of reading, but also held strong opinions about didactic methods, based on an assessment of the particular senses involved. The teaching of reading, he complained, was too often solely based on reading aloud: "I conceive that the habit in the CHILD of hearing and pronouncing certain sounds without any movement of the mind, will dispose the MAN to hear the same sounds with the same vacancy of thought."[14] Beddoes suggested an alternative to the READING-MADE-EASY books, based on two principles: First, one should go from the simple and concrete to the complex. Second, one should not rely exclusively on sight and hearing in teaching reading; children should, from the start, learn writing alongside reading, which would also involve the sense of touch, or, in Stansfield's terms, "what we would call kinetic experience".[15]

Beddoes developed his alternative method for teaching reading by highlighting the logical order of the different constituents of language. Letters were the basic units, which were composed into words, which in their turn made up sentences. Beddoes followed a suggestion by "a very ingenious French writer", who "proposes to make use of printed copies of an easy round writing hand, and he thinks that the action of imitating the letters as they become successively acquainted with them, would prove a source of amusement and very soon confer the power of distinguishing them quickly".[16]

Beddoes argued that using specimens borrowed from natural history and thereby engaging children's sense of touch would help them to learn. Hence,

> To compare the real object with the description would alleviate the labour of learning and improve the habit of observation. In many cases the relation between the structure and functions of animals is easily understood, as in the strong muscles that move the wings of birds. Our first acquaintance with such facts is always attended with the most lively sense of delight, and they are ever afterwards remembered and communicated with pleasure. Schools might easily be provided with some obvious specimens and simple preparations for this purpose: and as nothing what is extremely palpable should now be offered to the notice of children, the teachers might easily qualify themselves to give sufficient explanations.[17]

Beddoes considered that the senses, and especially the sense of touch, were indispensable in teaching reading and natural history. Even the most abstract of the sciences – mathematics – should be taught through the senses.

Beddoes' *Observations on the Nature of Demonstrative Evidence* (1793)

Beddoes' wrote his *Observations on the Nature of Demonstrative Evidence* when still residing in Oxford.[18] In the foreword, addressed to his friend

Davies Giddy, he argued, in opposition to current pedagogy, that models employing the sense of touch were useful in teaching mathematics:

> The more I consider the subject, the more I am inclined . . . to believe not only in the possibility, but the utility of rendering the elements of geometry palpable. If they be taught at an early age – a plan in which I think I see many advantages – models would make the study infinitely more engaging: From the mere slate and pencil most beginners experience a repulsive sensation. But if a child has something to handle and to place in various postures, he might learn many properties of geometrical figures without any constraint upon his inclinations. He would have no difficulty in transferring the properties of palpable to merely visible figures, nor in generalizing the inferences.[19]

Models could support the instruction of mathematics, not in the least by engaging Fancy.

Beddoes put forward a number of arguments supporting his claim that the teaching of geometry would benefit from the use of models. He was not advocating the truth of specific epistemological positions per se, but instead wanted to develop "the art of instruction." This practical aim may explain Beddoes' respectful tone of voice towards opposing positions, as in the *Extract of a Letter on early Instruction*. Referring to Kant's claim about the possibility of knowledge a priori, he remarked that it was unfair to judge an author on a single passage, since "a great work may doubtless contain much truth blended with a good deal of error."[20]

According to Beddoes, mathematics was based on experiment and observation.[21] Starting out from Locke's *Essay on Human Understanding*, Beddoes argued that the basis for the intuitive knowledge of mathematics lay in the senses. The power of abstraction built on the firm and distinct impressions of sense. If this was realized in youth, men could benefit from it in adulthood. He derived additional arguments for the pivotal role of demonstrative evidence from etymology and the study of language. For the former, the Dutch school of etymologists, including Jan Jacob Schultens, Tiberius Hemsterhuis, Lambert ten Kate, and Johannes Daniël van Lennep, served as a source of inspiration. The Leiden school examined the primary, *sensible* significance of words. Yet where the Dutch advocated the centrality of verbs, Beddoes followed the radical John Horne Tooke's claims that language started with nouns, and thus things. In the 1780s and 1790s, Tooke's *Diversions of Purley*[22] was a highly popular and exciting book. It

> offered more than a linguistic theory; it held out the promise of a simple and comprehensive system of knowledge which could lead to social reform. Tooke's lesson was that if all words could be understood in their primary meaning (his modest definition of etymology is "the real significance of the word"), semantic confusion would be no more and metaphysical fraud would be exposed.[23]

Tooke was not only a favourite of Beddoes, but of his friend Coleridge too.[24]

In sum, Beddoes was emphatic about the necessity of engaging the senses in teaching mathematics:

> Mathematics, in fine, teach either to measure or to count. The simplest and the shortest way we can acquire either of these arts, the better, I believe, in all respects. We cannot possibly set about to learn either of them otherwise, than by the use of the senses.[25]

Here too, the sense of touch was indispensable:

> If you lay a very small number of similar objects before a child, that has been learning to count for some time, and ask how many there are, he will be obliged to stretch out his hand, tell them over slowly. The eye, which takes its other lessons so admirably from the touch, never attains much readiness in the discrimination of number. No person, I apprehend, can distinguish eleven from twelve similar objects from a glance.[26]

In mathematics, Beddoes argued, it was almost as necessary as in mechanics to exhibit the objects whose qualities were taught, "and to call in the joint assistance of the hands and eyes."[27]

Beddoes elucidated Euclid's abstract notions with the help of tangible examples. He contrasted, for instance, Euclid's abstract definition of a point ("that which hath no parts, or which hath no magnitude") with observations built on demonstrative evidence: "a point is first the end of any thing sharp . . . then, by an easy derivation, any mark made by that sharp thing; and this is the meaning of a point in geometry."[28]

Reflecting upon the relationship between the different senses was common in eighteenth-century philosophy. In 1688, in a letter to John Locke, the Irish philosopher William Molyneux had asked whether a man born blind, having recovered his sight, would be able to distinguish between different shapes, relying only on sight. This led to a lively controversy among eighteenth-century British philosophers.[29] A related issue was addressed by George Berkeley, who highlighted the necessary interplay between the sense of touch and vision in estimating the height and distance of objects. Treatises reflecting upon the lack of one or more senses were common in the Enlightenment.[30]

Beddoes gave the philosophical controversy about the relationship between the senses of vision and touch a new turn by focusing on the utility of both senses in teaching and learning. This did not help, however, to make his approach less controversial. Richard Olson[31] situated Beddoes' radical empiricist position at one end of the Scottish spectrum of opinions about the nature and certainty of mathematics. James Beattie and John Playfair[32] were at the other, divorcing mathematics from sensory evidence. The relationship between tangible extension and shape (which one knows by touch alone) and visible extension and shape (which one knows by sight alone) was at the heart of that discussion.[33] John Leslie, a student in Dugald Stewart's moral philosophy class in Edinburgh and, later, from 1805 to 1819, holder of the chair of mathematics at Edinburgh, criticized Beddoes' approach:

> I read Dr. Beddoes book on its first appearance. I admired its ingenuity and approved of several of its leading principles. But I was convinced that the author proceeded much too far: . . . I admit, contrary to the adherents of the old school, that all of the primary notions and truths of geometry are derived from experience and observation. Here, however, I stop. The capital properties of figure and magnitude are only the *evolution* of those elements – and this is the province of reason, of mental exercise. To give instruction in geometry by the help of models seems as hollow as the quackish mode of teaching natural philosophy by the display of showy apparatus. It is surely easy to distinguish between proof and illustration.[34]

In Benjamin Donne, Beddoes was to hit upon a mathematician much more sympathetic to his views.

In touch with Benjamin Donne

Beddoes made contact with the remarkable Benjamin Donne in Bristol.[35] Donne had arrived in that city in 1764; there, from 1766, he ran an Academy, where he taught navigation and mathematics. He also gave many lectures as an itinerant lecturer throughout the West Country, and in London. When Beddoes arrived in Bristol, Donne was 64 and nearing the end of his life. Despite this, in February 1796 Beddoes

> engaged, with his usual activity, in promoting the sale of an Essay, accompanied by, and chiefly explanatory of, a set of schemes and models of the principal theorems both in plane and solid geometry; invented by a teacher of mathematics named Donne; a man well known in Bristol, and respected for unassuming talents and ingenuity. These models were, of course, particularly acceptable to [Beddoes] as they practically realized the plan of instruction in that science, which he had sketched out in his Essay on Demonstrative Evidence [in 1792].[36]

Beddoes was so convinced of the utility of Donne's models that he helped contribute to the printed Prospectus and wrote a letter included in Donne's *Essay on Mechanical Geometry*, published in Bristol in 1796. Beddoes also helped to secure many of his own friends and contacts as subscribers to this project and, presumably, to sales of Donne's models. This is all the more intriguing as Beddoes' old friend Davies Giddy had earlier been an unenthusiastic pupil at Donne's Bristol Academy:

> November 1782. I about this time began to attend the Academy or school of Mr. Benjamin Donne and I paid as a regular scholar but came

> perhaps 2 or 3 afternoons in a week. I learnt very little of Mr. Donne and he contrived to throw away my time. . . . Mr. Donne published several very good books on Arithmetic and Geometry . . . distinguished by their long Prefaces, although Mr. Donne seemed to me almost incapable of writing an ordinary letter.[37]

Beddoes' own thoughts regarding Donne are first revealed in February 1796, in correspondence with Tom Wedgwood.

> I had mentioned to you a plan for a publication on tangible geometry and you wished you could see it. I wish so too, as the invention has in my eyes extraordinary merit and deserves to be rescued from oblivion. It will have two excellent effects.
>
> 1 Prevent the liberally educated from conceiving a disgust to the exact sciences and
> 2 Put any students of geometry far on their way with little trouble and great satisfaction to themselves.
>
> It will furnish the demonstration of the principal proportions in plane and solid geometry. A child who has any power of apprehension may be convinced in a minute that the square of the hypotenuse of a [right] angled triangle = the sq[uare] of the other sides. If you wish for answers to particular questions, I will give you them with great pleasure. I will generally pledge my judgment that the invention will be of great service to all who desire to educate their children with clear ideas, and that it will obtain the approbation of all who wish well to useful knowledge. I am sorry the box with the book will amount to 2 guineas, as that sum will deprive this work of many purchasers. I wish to secure the poor old author [Donne] an indemnity and, for profit, he may take the chance of a general sale. He finds 40 subscribers necessary. I have procured near 30, and hope to fill up his number. If you can assist us, I should be glad. I would not even wish you to engage above five, if by any exertions you could do much more.[38]

Donne's printed Prospectus is dated 1796, and must have been distributed by March 28, the date of Beddoes' long letter written on the copy he soon sent to James Watt senior. The letter was mostly about Beddoes' treatment of consumptive patients, and only his final words related to the Prospectus: "If you know any person interested in education, shew him this. I do not mean to ask you to subscribe. I should think doing it directly would be more handsome. The scheme is afloat."[39]

Donne's Prospectus duly discussed the usefulness of geometry and noted that he had "invented mechanical or palpable demonstrations of geometrical proportions" about thirty years earlier, and had since exhibited them in his lectures in London, Bristol, and elsewhere. He had also demonstrated similar examples for architecture. The apparatus now offered was to consist of more than fifty models in card, paper, wood, and metal. The forty subscribers, as of March 1796, included Beddoes (no. 1),[40] William Clayfield (no. 3), Erasmus Darwin, Edward Long Fox, Benjamin Hobhouse (no. 4), William Reynolds (for six copies) and his brother Joseph, and three Wedgwoods, John (no. 7), Josiah and Thomas (for three copies). The Prospectus noted that Beddoes would add some preliminary remarks to Donne's forthcoming explanatory pamphlet, "who, without any previous knowledge of Mr. Donne's invention, had insisted upon the necessity of teaching the Elements of Geometry, on this Plan, in a work entitled, 'Observations on demonstrative Evidence.' "[41]

The scheme has been well described by Eric Robinson.[42] The promised pamphlet was announced in 1796, soon after Donne had been appointed "master of mechanics to King George III".[43] Donne's *Essay on Mechanical Geometry, chiefly explanatory of a set of Schemes and Models*[44] was published later that year, with the apparatus now selling for an increased "Two Guineas and a Half." An apparently unique set of Donne's complete apparatus survives in the London Science Museum,[45] along with the explanatory cards and the instruction book, in a mahogany box labelled "Donne's Mechanical Geometry". Closer inspection of this set has revealed that it relates to the later attempt to promote such Donne's models,[46] then made by Donne's daughter, Mary Ann Demedes Donne, then Clayton (1773–1836).[47]

One who was highly critical of Donne's scheme was the Scottish mathematician and natural philosopher John Leslie. In 1797 he wrote to his former pupil Tom Wedgwood, whom he had tutored at Etruria Hall, the Wedgwoods' family home Staffordshire, from 1790 to 1792. In his letter, Leslie claimed that "to give instructions in geometry by help of models seems as hollow as the quackish mode of teaching natural philosophy, by the display of showy apparatus", so he is clearly not in tune with Beddoes' or Wedgwood's ideas. His letter continued

> with regard to Donne, I shall do as you direct, but I cannot promise much success. His mathematical character is known in this part of the country, and does not, I believe, stand high. He has not done justice, I think, to Beddoes' [own] plan. His models are scanty and frivolous. Some of them might provoke a smile, – for instance the puzzle-pin and the mannikin taking the height of a steeple etc. Indeed he seems himself to doubt of his success, and advises those who desire a thorough knowledge of the subject to study Euclid or his Geometry. I might say a great deal, but room fails me.[48]

Donne's published *Essay* included a longer, and fascinating, list of ninety-one subscribers.[49] These now included numbers 1, 3, 4, 5 and 7 of Beddoes' future Toys Committee.[50] Donne's methods, using models, were similar to those which Beddoes had in mind in his *Observations on the Nature of Demonstrative Evidence*. Donne's main concern, apparent in the title of his *Essay*, was teaching the utility and use of geometry, "even to youth of an early age". In the preface of the *Essay on Mechanical Geometry* Donne endorsed the usefulness of "*mechanical*, or *palpable* demonstrations of the most important proportions in geometry." He indicated that he had already worked with them for a long time: "From 1766 to the present time, he has repeatedly exhibited them in London, Bristol, and other places." He too underlined that working with these models would be fun: "To acquire mathematical information will be rendered by this contrivance an amusement instead of a task."[51]

Donne's *Essay* was accompanied by Beddoes' promised letter recommending Donne's apparatus. Beddoes stated that he had not been active "in bringing out your mechanical demonstrations"; the propositions in his *Observations on the Nature of Demonstrative Evidence* "must stand by their own strength." Yet Beddoes supported Donne's project: "as all ideas are derived from sense, all argument must consist of a statement of facts or perceptions." Beddoes repeated his views related to the relevance of touch, while denouncing the available materials for the instruction of reading: "your tangible proofs of the properties of figures will be eminently serviceable to the intellectual faculties of young people. The effect will be exactly the reverse of that produced by *Reading-Made-Easy* and by the *Grammars* in use." Beddoes was referring to his own plans: "I have in view, not merely information in mechanics, chemistry and technology, but the improvement of the senses by presenting, in a certain order and upon principle, objects of touch along with objects of sight."[52]

Donne's and Beddoes' advocacy of the use of models and demonstrative evidence fitted the contemporary sensationalist philosophies and educational ideologies. Donne referred to "*palpable* demonstrations" of geometrical propositions;[53] Beddoes announced Donne's plan for "tangible proofs of the properties of figures".[54] Mathematics, however, represented a specific form of knowledge. In spite of the vogue for empiricism and the role of the senses, invoking the help of touch in teaching mathematics was highly controversial. That may explain Leslie's harsh criticism of Donne's scheme. Yet, unlike Donne's, Beddoes' ambitions for rational toys were not restricted to mathematics.

Beddoes' plan for a manufactory of rational toys

At the end of the eighteenth century, both "models" and "toys" were words with ambiguous meanings. Models could serve different purposes; instruction was only one of them. Manufacturers used models as prototypes; architects as conversation pieces with potential clients. Retailers used miniature models of their shops to show what they had on offer. The Victoria and Albert Museum of Youth and Childhood (Bethnal Green), for instance, displays a butcher's shop with meat hanging at the front side. Here the miniature shop shows the different kinds of meat for sale. The meaning of toys was even more ambiguous. There was a transitional phase, in which Beddoes was a main actor.

Coming from London to Bristol, Beddoes did not have a local network of relatives and friends. James Keir, whom Beddoes had met in 1791, introduced him to Richard Lovell Edgeworth, inventor, landholder, and, between 1766 and 1782, member of the Birmingham Lunar Society.[55] The Lunar Society included several individuals whom we have already encountered, and others, such as Edgeworth's intimate friend the educationalist Thomas Day.[56] Beddoes and Keir agreed with Edgeworth's high opinion of Day, who had died in 1789. In 1791, Keir published a biography of Day.[57] In 1791, Edgeworth was in Clifton while his young son Lovell was being treated for consumption at the Hotwells. Both Edgeworth and Beddoes were equally passionate about education and the need to reform it.[58]

In the letter recommending Donne's apparatus, Beddoes had included news of his own future manufacturing plans.

> For these several years, I have been corresponding and conversing with different friends about [another] project, much allied to that which you have now executed, and which will come in very well after yours. It is to establish a manufacture of RATIONAL TOYS. I believe parents are become sufficiently attentive to education to give such a scheme support; and fortunately it cannot alarm any prejudice. The design is to construct models, at first of the most simple, and afterwards of more complicated machines. The models are not to be very small, and they are to be so constructed, that a child may be able to take them to pieces, and to put them together again. The particulars of the design are too numerous to be given here. It comprehends engravings and a good deal of letter-press. I have in view, not merely information in mechanics, chemistry, and technology, but the improvement of the senses by presenting, in a certain order and upon principle, objects of touch along with objects of sight. In this important business we have hitherto trusted to chance. But there is every reason to suppose, that INTELLIGENT ART will produce a much quicker and greater effect. Should instruction addressed to sense be made in any country the principle of education, should the best methods of cultivating the senses be studied, and should proper exercises be devised for reproducing ideas (originally well-defined) sometimes with rapidity, at others in diversified trains, the consequence is to me obvious. . . . In the course of the ensuing winter I hope to get a few sets of these *rational toys*, with the engravings and the explanation, executed.[59]

When Donne died in 1798,[60] Beddoes was then alone in pursuing more extensive plans to manufacture rational toys. He commented on Donne's scheme in October 1799, noting the current "defective knowledge . . ., which evil still subsists in full force", in this field. He instead urged the adoption of "mechanical contrivances", or models, and added

> Some years ago, finding that the late Mr. B. Donne had mechanical demonstrations of the principal propositions of Euclid . . . which were plain and striking, I procured him a number of subscribers; and sets of demonstrations were made, and an explanatory book printed. By help of these, young persons soon, and easily, learn many properties of plane, and some of solid figures. Somewhat more than an hundred sets are dispersed through the kingdom; and it were to be wished, that more were in readiness. But as the project was, I believe, never advertised in any newspaper, or noticed in any literary journal, the knowledge of it was confined to a few hundred families. Those who have used these models, have been satisfied with them; though possibly they could be greatly improved.[61]

Beddoes clearly planned to do just that. Much of what we know directly of Beddoes' later plans for rational toys comes from Stock, who noted that Beddoes, sometime earlier in the 1790s, had "circulated among his friends, proposals for rational toys, to be constructed according to the principles contained in that chapter of Practical Education, but they were not made public till a future opportunity".[62] This came when the undated flyer announcing their intended manufacture appeared, which was later reproduced by Stock. Stock stated this was published in June 1800,[63] and it was soon reproduced in *Nicholson's Journal*.[64] Beddoes' Organizing Committee comprised eight men: Beddoes; John Billingsley; William Clayfield; Benjamin Hobhouse, M.P.; John Hodder Moggridge; James Stephens; John Wedgwood; and William Wynch.[65]

The final breakthrough that allowed the toys scheme to make progress was the recruitment of a "conductor of the planned Manufactory." This was "Mr Robert Weldon, c/o Pneumatic Institution, Bristol." Weldon is a sad figure.

The Weldons were an inventive family; Robert's elder brother, engineer James, stayed in Lichfield, taking out a patent in 1798 for a bark mill for tanning. In 1792, Robert, while still in Lichfield, had taken out a patent for his invention for "Raising and lowering weights from one level to another on canals", called a "caisson".[66] The only full-scale caisson built was erected between 1796 and 1800 on the Somerset Coal Canal. Sadly, this could not be made to function properly because of flaws in the masonry of the stone cistern in which it was to operate.[67] After Weldon died in Clifton in March 1804, the *Bath Journal* paid him this tribute:

> Lately died Mr. Richard [sic] Weldon, a young man of the most extraordinary natural parts, as well as of acquired abilities. He was the inventor of the Cassoon [Caissòn] Lock, the failure of the masonry of which,

> and not the failure of the invention, had very much injured his health and spirits. He was also the inventor of that extraordinary spinning wheel for which the Queen [Charlotte] so handsomely rewarded him with a medallion of merit. Thus fell in the prime of life a most noble engineer and, what is better, a very honest and reasonable man.[68]

Work on the actual caisson had begun about July 1796,[69] and a successful trial was made in December 1797[70] – but in February 1798 it failed when parts gave way.[71] Another trial in June 1798 was successful,[72] but by April 1799 the masonry "cistern was so leaky that it is impossible to continue any trials".[73] So, in May 1799, rival advertisements to either rebuild the caisson cistern or replace caissons with inclined planes appeared.[74] The Canal Committee and the Proprietors were at odds as to what to do next. By 11 February 1800 the lost Canal Minute Book showed it had been agreed to abandon the use of any caissons.[75] Weldon's canal career was over and he was, in theory, now free to superintend a toy factory. Since Billingsley and Stephens were both still closely involved with the Coal Canal, it seems likely they first suggested this act of philanthropy towards Weldon.

However, there was to be no rational toys manufacture. Stock stated that the plan had not been sufficiently encouraged.[76] Eliza Meteyard wrote only that "subscriptions were raised, but the matter, after some little agitation fell to the ground,"[77] In 1931 Ladbroke Black rightly noted that in this project, Beddoes had proved "a Georgian of the Fifth variety born in the time of the Third".[78] James Kendall, on the contrary, claimed that after 1799, Beddoes had simply "frittered away the rest of his life advocating drastic reforms in education, diet, dress and children's toys".[79] Yet in effect, Beddoes exercised a lasting influence on educational practice, especially through Maria and Richard Lovell Edgeworth's *Practical Education*, which went through two more editions in Beddoes' lifetime.[80] *Practical Education* became a classic pedagogic text. It won immediate acclaim on the Continent. In England, as we know, the political tide was strongly reactionary. As Maria Butler has argued, 1798 "was not a year for welcoming progressive books".[81] Yet *Practical Education* is still acknowledged as the most important work on general pedagogy between John Locke's *Some Thoughts Concerning Education* (1693) and Herbert Spencer's *Essay on Education* (1861).[82] Toys were the subject of the first chapter of *Practical Education*, and the authors acknowledged that "The first hint of the chapter on Toys was received from Dr. Beddoes." He was also credited as being the first to propose models that could be disassembled and reassembled, with the names of the different parts stamped upon them.[83]

Educational reform: toys

The Edgeworth family residence served almost as a home for Beddoes.[84] In 1794 he married Anna Edgeworth. He also had an important influence on

Anna's sister, Maria, who would become a prolific and extremely popular author. A lasting connection to Beddoes was established via *Practical Education* (1798), written by Maria and her father, Richard Lovell Edgeworth. The book was first announced immediately after Donne's death, in May or June 1798,[85] as "to be published in a few days", in two quarto volumes. Brian Simon has commented how

> it is impossible to do justice to the extraordinary richness of ideas concerning pedagogical methodology outlined in certain chapters of the book, perhaps particularly the chapter entitled "Toys" . . . and that on mechanics, the latter written by R.L. Edgeworth.[86]

In the preface of *Practical Education*, the authors explain that

> [a]ll that relates to the art of teaching to read in the chapter on Tasks, the chapters on Grammar and Classical Literature, Geography, Chronology, Arithmetic, Geometry, and Mechanics, were written by Mr. Edgeworth, and the rest of the book by Miss Edgeworth.[87]

However, Susan Manly has argued that Maria had a larger role than this suggests:

> [A]s family archives make clear, it was she who revived the registers of child observation, originated by her first stepmother, Honora Sneyd Edgeworth, which are the foundation of the work. Throughout the period of actual writing, moreover, as Mitzi Myers points out, *Practical Education* was regarded in the family as "Maria's great work". She habitually referred to it as my "toys and tasks" until late in its genesis, when other family members were enlisted for help with topics such as chemistry and classics. Only in later years, well after its first publication, was the credit for its pioneering programme transferred to R.L.E.[88]

Having lived in Lichfield, Richard Lovell Edgeworth was well acquainted with Birmingham's industry. In the second part of the eighteenth century Birmingham was the centre of England's toy production. Edmund Burke called Birmingham "the grand toy-shop of Europe".[89] Jay gives a vivid description: "A thousand small workshops and production lines churned out toys and buckles, cutlery, buttons and snuffboxes, flooding the country, and increasingly the world trinkets at rock-bottom prices."[90] However, as we have already seen, the meaning of toys was ambiguous in the second part of the eighteenth century. The toys Birmingham was famous for were miniatures, manufactured by skilled artisans. A lot of them were showpieces, for the mantelpiece, catering to a wealthy clientele. "Toyware" was connected to an emerging culture of consumption.[91]

Sketchley's Birmingham Directory of 1767 sums up the variety of toys and toy makers:

> An infinite variety of Articles that come under this denomination are made here; and it would be endless to attempt to give a list of the whole, but for the information of Strangers we shall here observe that these Artists are divided into several branches as the Gold and Silver Toy Makers, who make Trinkets, Seals, Tweezer and Tooth Pick cases, Smelling Bottles, Snuff Boxes, and Filligree Work, such as Toilets, Tea Chests, Inkstands &c &c. The Tortoishell Toy maker, makes a beautiful variety of the above and other Articles; as does also the Steel; who make Cork screws, Buckles, Draw and other Boxes: Snuffers, Watch Chains, Stay Hooks, Sugar knippers &c. and almost all these are likewise made in various metals.[92]

Sketchley's reveals here that toys were, to a large extent, precious goods, meant as well for adults. The historian John Brewer, however, looked up the word "toy" in Samuel Johnson's *A Dictionary of the English Language* where he found that:

> Dr Johnson defined "toy" in his famous Dictionary as "a pretty commodity; a trifle; a thing of no value; a plaything or bauble." There was absolutely no mention of children. The term "toy" meant any small inexpensive object or trinket sold to young and old alike. The travelling pedlar or chapman, the town's "toyman" offered cheap jewelry, buckles, bangles and hairpins.[93]

But as Liliane Hilaire-Peréz has challenged the narrow view of baubles, one may conclude that in the eighteenth century toys were not exclusively made for children – on the contrary. They could be cheap or dear; the word might have stood for a worthless trifle but could also have referred to expensive luxury goods.[94]

In *Practical Education*, Maria and Richard Lovell Edgeworth were promoting alternatives to the precious type of toys, highlighting the value of educational toys for youngsters. Maria Edgeworth and her father observed that children needed practical things to do. The first chapter, which was written principally by Maria Edgeworth, was dedicated to toys; as we have seen, Thomas Beddoes was credited as her first source of inspiration. In Edgeworth's opinion, children, "require to have things which exercise their senses of their imagination, their imitative and inventive powers".[95] The exercising of the senses should be taken literally. Pleasing the eyes was insufficient; toys should stimulate children to use and play with them: "The glaring colours, or the gilding of toys, may catch the eye, and please for a few minutes, but unless some use can be made of them, they will, and ought to be soon discarded."[96] Toys were considered as outstanding educational tools: "It is surprising how much children may learn from their playthings when they are judiciously chosen, and when the habit of reflexion and observation is associated with the ideas of amusement and happiness."[97] Playing with toys amounted to educating the senses, and to those who had acquired habits of observation, everything that is to be seen or heard could become a source of amusement. Edgeworth further linked the worlds of play and experiment:

> We think that a taste for science may early be given by making it entertaining, and by exciting young people to exercise their reasoning and inventive faculties upon every object which surrounds them. We may point out that great discoveries have often been made by attention to slight circumstances.[98]

Maria Edgeworth contrasted cheap educational toys with the older luxurious ones, the latter requiring particular care on the part of children, because they cost a great deal of money and had to be admired as miniatures of some of the fine things on which fine people prided themselves.[99] Edgeworth promoted children's toys not as precious objects, to be merely looked at; children's toys should invite play and discovery. If a toy was destroyed in play, children could gain insight into its inner mechanisms and its functioning, which would be a positive outcome.

Although *Practical Education* did not intend to provide a system of education,[100] some guiding principles were evident. Toys should be offered in a judicious order, attuned to the child's age:

> The first toys for infants should be merely such things as may be grasped without danger, and which might, by the difference of their sizes, invite comparison: round ivory or wooden sticks should be put into their little hands; by degrees they will learn to lift them to their mouths, and they will distinguish their sizes: square and circular bits of wood, balls, cubes, and triangles, with holes of different sizes made in them, to admit the sticks, should be their playthings.[101]

Furthermore, as we have seen, toys should not only please the sense of sight, but should also stimulate other senses, and should encourage activity. Therefore the nursery should "never have any furniture in it which they [children] are not to touch".[102] As children grew older, their toys should become more challenging. When scientific experiments were presented to children, the history of scientific inventions suggested a natural sequence:

> The history of the experiments which have been tried in the progress of any science, and of the manner in which observations of minute facts have led to great discoveries, will be useful to the understanding, and

> will gradually make the mind expert in that mental algebra, on which both reasoning and invention (which is perhaps only a rapid species of reasoning) depend.[103]

Maria Edgeworth gave an overview of the stock of a rational toy-shop, which should comprise all manner of garden and carpenter's tools and accessories, pencils, card, pasteboard, "substantial, but not sharp pointed, scissars [sic]", wire, gum, and wax – and, of course, work-benches. There should also be a wide variety of models, which children should be able to take to pieces, of gradually increasing complexity and abstraction: furniture, architecture, simple machines ("choosing at first such as can be immediately useful to children in their own amusements, such as wheelbarrows, carts"), complicated machinery, chemical toys, fossils for their cabinets, inexpensive microscopes, and experiments in optics (building on Dr. Priestley's *History of Vision*).[104] Donne's models were mentioned in the book's chapter XVI, dedicated to geometry. Once again, their *palpability* was highlighted. The Edgeworths did not succeed in procuring "a set of these models for our own pupils, but we have no doubt of their entire utility."[105] At the end of the first chapter, Maria Edgeworth stated that she hoped in the future to be able to offer a more detailed list of rational toys:

> We intended to have added to this chapter an inventory of the present most fashionable articles in our toy-shops, and *a list of the new assortment*, to speak in the true style of an advertisement; but we are to defer this for the present: upon a future occasion we shall submit it to the judgment of the public.[106]

In her novel, *The Good French Governess* (1801), Maria Edgeworth described what she believed the new rational toy-shop should be:

> "Have you any objection," said Mad. de Rosier, "to my buying for him some new toys?" [. . .] she took [. . . her pupils] with her one morning to a large toy-shop, or rather warehouse for toys, which had been lately opened, under the direction of an ingenious gentleman, who had employed proper workmen to execute rational toys for the rising generation.
>
> When Herbert entered "the rational toy-shop," he looked all around, and, with an air of disappointment, exclaimed, "Why, I see neither whips nor horses! nor phaetons, nor coaches!" – "Nor dressed dolls!" said Favoretta, in a reproachable tone – "nor baby houses!" – "Nor soldiers – nor a drum!" continued Herbert. – "I am sure I never saw such a toyshop," said Favoretta; "I expected the finest things that ever were seen, because it was such a new *great* shop, and here are nothing but vulgar-looking things – great carts and wheelbarrows, and things fit for orange-women's daughters, I think."[107]

On one hand, the educational reform the Edgeworths had in mind was revolutionary. On the other, it situated itself within established practices. For example, the Edgeworths did not envisage public education. The educational toys were thought to function in a domestic context.[108] Partly this tied in with Maria's own situation. She took care of the many children born to her father's successive wives. Another aspect that stood out in *Practical Education* was the repeated warning that contacts between children and the servants should be avoided at any time.[109] Hence, "The language and manners, the awkward and vulgar tricks which children learn in the society of servants, are immediately perceived, and disgust and shock well bred parents."[110]

Beddoes' and the Edgeworths' interest in toys tied in with changing views of education in the second part of the eighteenth century, which manifested itself on different levels. For one thing, the genre of books for children to read emerged.[111] John Newbery was the publisher who, in the mid-eighteenth century, first made children's books a popular and profitable part of his enterprise.[112] In the 1780s, books for children "suddenly became what they have been ever since, a substantial part of the bookseller's trade".[113] Beddoes disapproved of works with an overly moral tone, while promoting, for instance in his *Letter on Early Instruction*, books that tied in with children's daily surroundings, such as Thomas Day's much-praised *The History of Sandford and Merton* (1783–1789), Mrs. Laetitia Barbauld's *Evenings at Home* (1792–1796), and Mrs. Trimmer's *Fabulous Histories*, better known as *The History of the Robins* (1786). In addition to this genre directly addressing children, books on the principles of education were popular. John Locke's *Some Thoughts Concerning Education* was a seminal book that served as point of reference throughout the eighteenth century. Following in the footsteps of Locke, Rousseau, in *Emile ou de l'éducation*, proposed alternative educational principles, departing from nature and devising a program of experimental education. Julia V. Douthwaite claims that between 1762 and the turn of the century "[n]o fewer than two hundred treatises on pedagogy were published in English [. . .] the majority of which show the influence of Rousseau."[114] Richard Lovell Edgeworth even dared to experiment with educating his firstborn son Richard according to Rousseauian principles. The outcome was thoroughly discouraging.[115] Thomas Day had engaged before in a similar experiment in education, by bringing up two adopted young girls according to Rousseau's principles, which was also a complete failure.[116]

Combining guided instruction, while relying on the spontaneous, innate driving forces of children, furnished an interesting combination of Lockean and Rousseauian principles. So too did the Edgeworth's doctrine of rewards and punishments.[117] Well-chosen educational toys could provide both amusement and instruction. Acknowledging the fun factor fitted into the Enlightenment sensationalist context, as Roy Porter has pointed out:

> This sensationalist' psychology – man viewed as an ensemble of stimuli and responses, mediated through the senses – sanctioned a new hedonism. "Pleasure is now, and ought to be your business", Lord Chesterfield instructed his teenage son. The well-tempered pursuit of happiness

in the here and now – indeed, the right to happiness – became a leading theme of moral essayists.[118]

Beddoes was not the first to propose the use of toys for education. The most popular educational toy of the day – dissected (jigsaw puzzle) maps for the teaching of geography – had been introduced in the early 1760s.[119] Mme. Jeanne-Marie le Prince de Beaumont has been identified "as the first person known to have been selling [them]".[120] In *Practical Education*, Maria Edgeworth included observations of children playing with a dissected map:

> Whoever has watched children putting together a dissected map, must have been amused with the trial between Wit and Judgement. The child who quickly perceives resemblances catches instantly at the first bit of the wooden map, that has a single hook or hollow that seems likely to answer his purpose; he makes perhaps twenty different trials before he hits upon the right; whilst the wary youth, who has been accustomed to observe differences, cautiously examines with his eye the whole outline before his hand begins to move; and, having exactly compared the two indentures, he joins them with sober confidence, more proud of never disgracing his judgment by a fruitless attempt, than ambitious of rapid success. He is slow, but sure, and wins the day.[121]

The itinerant lecturer and mathematical instrument maker to George III, George Adams the younger, had made (probably in the 1770s) a box of geometrical models, of which a unique set survives in the London Science Museum.[122] In 1785 Lady Ellenor Fenn had published *The art of teaching in sport, designed as a prelude to a set of Toys, for enabling ladies to instill the rudiments of spelling, reading, grammar and arithmetic*. The accompanying set of toys was produced by John Marshall.[123] In 1785 Thomas Barnes, Unitarian minister and educational reformer then in Manchester, published his

> Plan for promoting and extending manufactures, by encouraging those Arts, on which manufactures principally depend. My first object was to provide a Public Repository among us for chemical and mechanic knowledge, [in which] I could wish models to be procured, of all such machines, in the various arts, as seem to bear the most distant relation to our own manufactures.[124]

We do not know how far Barnes succeeded.

Beddoes' plans for a manufactory of rational toys, and the Edgeworths' starting their book on *Practical Education* with a chapter on toys, are indicative of the interest in educational toys at the turn of the nineteenth century.

Beddoes' and the Edgeworths' lasting influence

There are different ways of evaluating what Beddoes' struggles for educational reform achieved. One way is to consider whether pupils were actually educated according to the principles that guided him, and which were also laid out in *Practical Education*.

Consider the two eldest sons of William Henry Lambton, M.P. for Durham, and Lady Anne Barbara Frances Villiers. Lambton became a consumptive patient of Beddoes; his health took a turn for the worse in 1796. He then went with his family first to Naples, in search of better health in a warmer climate; he moved to Rome in 1797, and finally to Pisa, where he died of consumption on 30 November 1797, leaving four young sons and a daughter.[125] Lambton wrote, just before his death, that he "would give the world Beddoes was with me!"[126] Beddoes described him as "the best man that I ever knew, or, that in his sphere of life [politics] I ever shall know."[127] Beddoes became the fatherless children's doctor, and in 1798 he took the two eldest boys, John George and Thomas Henry William, into his house for about four years; they became his pupils.[128] Because of their youth, they had barely known their father; John later wrote that "I have never felt the blessing of a father's care or advice, and, I fear, I have suffered much from it."[129]

Thomas Beddoes now determined to test his educational theories on these two Lambton boys, undertaking their moral, literary, and scientific formation.

> Mr. Sadler, son of the person who first accompanied Doctor Beddoes to Bristol, taught them chemistry and pyrotechny; and they had the advantage of receiving anatomical knowledge from Mr. King.[130] . . . The success of this experiment amply justified the departure from the common methods of instruction.[131]

When the boys went on to Eton, Stock went on to state that they were at least on a par with their fellow students.[132] As so often with Beddoes' studies, the surviving record is simply insufficient to for us to know all that Beddoes attempted towards the improvement of his fellow men (in this case, children). What we do know is that in spite of losing his father, John George Lambton went on to enter politics, became known as Radical Jack, first Earl of Durham, and served as Governor General of Canada.

Another way of evaluating Beddoes' achievements is by studying institutional connections and the circulation of ideas. When taking stock of Beddoes' life, Mike Jay concluded that,

> [I]n many ways, the spirit of the 1820s had become far more sympathetic to Beddoes than to Davy. The Royal Institution was a harbinger of a new generation of projects that aimed to harness science for the public good. As Beddoes had urged Sir Joseph Banks, the state began to take centralised measures to quantify society's ills: infant mortality statistics were collected. . .. Francis Bacon became a touchstone for the new Utilitarian movement, for which science and education were the key to practical advancement. . .. Many of the causes and projects that

> drove this new spirit were those for which Beddoes, in the 1790s, had been a voice in the wilderness.[133]

Based on a qualitative analysis of the presence of "toys" in London's newspaper *The Morning Chronicle* in the period 1800–1827, Wachelder has concluded that the market for educational toys took off in the second decade of the nineteenth century.[134] Combining instruction and amusement then became a trope in the marketing of such toys. In the first two decades of the nineteenth century, advertisements of educational books and toys subordinated moral education to amusement and instruction. One may conjecture that Beddoes' and (to a lesser extent) Richard Lovell Edgeworth's known political radicalism might have impeded a rapid dissemination of their educational theories and the rational toys they endorsed. Edgeworth continued his efforts to produce rational toys. In 1807, in a letter to a friend, he discussed a proposal to open a "Rational Toyshop" in London:

> I have never abandoned this scheme of a Rational Toyshop. . .. I think that three hundred copies of each toy or model should be completed before we open shop; and that a pamphlet should be written unfolding the general plan, besides particular descriptions and proper references to explain each particular.[135]

In this letter, Edgeworth highlighted, once again, the relevance of the bodily appropriation of toys, by distinguishing between philosophical apparatus and instructive toys:

> The difference between instructive toys and philosophical apparatus consists in this, the latter are exhibited by masters and carefully preserved from the hands of learners; the former are intended for the rude hands of unpractised children, and should be constructed so as to bear hardship and to excite the attention of those who use them, to the effects of mechanical powers and mechanical resistances, to the means of communicating motion in various manners and in various directions. To fix the attention is then the great object and this can never be done without the toy or instrument necessarily requires the bodily action of the child who uses it.[136]

In this context, Edgeworth wrote to Mr. Tabart,[137] who ran a Juvenile Library at No. 157, New Bond Street, London, about rational toys. In an advertisement in *The Morning Chronicle*, dated 31 December 1803, Tabart announced the re-opening of his "well-known and much approved Establishment for the sale of Juvenile and School Books . . . he has laid in a new and valuable assortment of the best books of amusement and instruction which have appeared in the English language".[138] Immediately, Tabart added that "no work is permitted to be sold in it which is in the slightest degree inimical to the interests of religion, virtue and morality." We do not know if Tabart ever sold rational or educational toys.

What can be documented, however, is that, from the 1820s onwards, both Beddoes and Maria Edgeworth were celebrated for their pioneering work on such toys. A reference to both Dr. Beddoes and Maria Edgeworth can, for instance, be found in Francis West's 1836 catalogue of toys "as recommended by Dr. Beddoes, Miss Edgeworth, &c".[139] In the introduction to this elaborate sales catalogue, we can recognize some of the arguments brought up before by Beddoes and Maria Edgeworth: "The common toys of childhood, are not only expensive and quickly broken, but useless and senseless in themselves, as giving no information to the inquisitive mind of youth, and having no interest but a short-lived novelty."[140] The optical, mathematical, and philosophical instrument maker Francis West offered some thirty educational toys, adapted from those employed by public lecturers, to cater to young persons. In 1831 he had offered a much more limited number of intellectual toys. Maria Edgeworth was the dedicatee of John Ayrton Paris's book *Philosophy in sport made science in earnest; Being an attempt to illustrate the first principles of natural philosophy by the aid of popular toys and sports* (1827). Paris is best known as the first biographer of Humphry Davy, yet his *Philosophy in Sport* had a larger readership in the nineteenth century. The book would go through many editions, and it marked the establishment of the idea that toys were beneficial to children's upbringing, especially in raising their interest in science.[141] There are several references to Maria Edgeworth throughout *Philosophy in Sport*, which endorses her plea for rational toys.

Meanwhile, the centre of commercial activities had moved to London. Paris was also the alleged, yet contested, inventor of an optical toy, the thaumatrope,

> a scientific toy illustrating the persistence of visual impressions, consisting of a card or disk with two different figures drawn upon the two sides, which are apparently combined into one when the disk is rotated rapidly; [the word is] also applied to a disk or cylinder bearing a series of figures which, on being rapidly rotated and viewed through a slit, produce the impression of a moving object.[142]

In *Philosophy in Sport* Paris mentioned that the thaumatrope "lately made its appearance in the scientific circles of London".[143] Whether or not he was the inventor of the toy, the *Oxford English Dictionary* attributes the first use of the word "thaumatrope" to him. David Brewster stated in his *Letters on Natural Magic*:

> In virtue of this property of the eye an object may be seen in many places at once; and we may even exhibit at the same instant the two opposite sides of the same object, or two pictures painted on the opposite sides

> of as piece of card. It was found by a French philosopher M. D'Arcet [sic.],[144] that the impression of light continued on the retina about the eighth part of a second after the luminous body was withdrawn, and upon this principle Dr. Paris has constructed the pretty little instrument, called the *Thaumatrope*, or *Wonder-turner*.[145]

The thaumatrope was sold at the Royal Institution at a price of more than seven shillings. It was definitively not a bargain at a time when the average wage of a competent craftsman was about thirty shillings per week. Even so, the audience did not always include artisans.[146] That price, and the status of its outlet, the Royal Institution, made the thaumatrope a serious present, though other manufacturers offered cheap imitations, about which Paris issued a warning: "We have since learned that Mr. Beacham obtained this toy at Mr. William Phillips's, George Yard, Lombard Street, the publisher. We mention this circumstance to guard the reader against those inferior imitations which are vended in the shops of London."[147] William Phillips, printer and geologist, was born in London in 1773. His father was a Quaker, originally from Cornwall, who had settled in London in 1768, acquiring the major Quaker printing and publishing house of Mary Hinde at George Yard, Lombard Street in 1775. In 1814 Phillips gave free lectures on geology and mineralogy to young people in his home village, Tottenham. These were printed and published as an *Outline of Mineralogy and Geology . . . for . . . Young Persons* in 1815 (4th edn., 1826). In 1816 he published *An Elementary Introduction to . . . Mineralogy . . . for the Student*. His crystallographic work in this, using W.H. Wollaston's reflecting goniometer of 1809, drew high praise from its inventor. His *Eight Familiar Lectures on Astronomy*, another interest, followed in 1817.[148]

In Peter Mark Roget[149] we find a direct connection between the Bristol Pneumatic Institute in the 1790s and the London-based scientists, instrument makers and traders with diverse backgrounds who invented and sold educational toys in the 1820s. In 1798, the young Edinburgh graduate Peter Mark Roget had landed in Bristol, in search of spas and resorts to cure his persistent chest complaint.[150] For that reason, he joined the circle surrounding Beddoes and Davy, experimenting with nitrous oxide. Roget volunteered as a medical assistant. In 1799, somewhat disappointed, he left Beddoes and Bristol.[151] Perhaps Beddoes had not been without influence, however. In 1809, Roget obtained a position in London, where he engaged in many projects to improve public hygiene and health. Between 1827 and 1848 he served as Secretary to the Royal Society. He became holder of the Fullerian chair in physiology at the Royal Institution. In 1824, he contributed an overview of Beddoes' work to the Supplement of the 4th, 5th, and 6th editions of the *Encyclopaedia Britannica* (vol. 2). Stansfield characterizes the piece as "a warm account, especially when we remember that Roget had not been at ease during his year at Clifton".[152] Roget became involved in a controversy regarding the invention of another philosophical toy, called the Phantasmascope or Phenakistiscope. This consisted of

> a disk with figures upon it arranged radially, representing a moving object in successive positions; on turning it round rapidly, and viewing the figures through a fixed slit (or their reflexions in a mirror through radial slits in the disk itself), the persistence of the successive visual images produces the impression of actual motion.[153]

For this, Michael Faraday and Joseph Plateau were potential competitors. Roget claimed to have developed a similar contrivance, and even demonstrated it to friends, but did not consider the gadget worth much attention. In 1831, Roget reflected that Faraday's article[154]

> again directed my attention to the subject, and led me to the invention of the instrument which has since been introduced into notice under the name Phantasmascope or Phenakisticope. I constructed several of these at that period (in the spring of 1831) which I showed to many of my friends; but in consequence of occupations and cares of a more serious kind, I did not publish any account of this invention, which was last year reproduced on the continent.[155]

By the 1820s, the idea of rational, intellectual, or philosophical toys had struck root in London circles around the Royal Institution. The toys from this era were clearly more spectacular and often more controversial than Donne's models or Beddoes' rational toys, which were particularly didactic. Yet the toys shared the original goal of combining instruction and amusement of the young by stimulating the senses.

Excursus note

Beddoes' Organising Committee comprised eight men:[156]

1 Thomas Beddoes, M.D.
2 John Billingsley (1747–1811), a wealthy landowner, came from a well-known Presbyterian (and later Unitarian) family.[157] He was active as an agricultural improver, and an equally active exponent of Canal improvements. He wrote the Board of Agriculture's account of agriculture in Somerset,[158] in which Robert Weldon first described his ill-fated caisson. By March 1800, he was a patient at Beddoes' Pneumatic Institution. Since he had not been a subscriber to Donne's 1796 *Essay*, this was probably how he became involved with Beddoes and the toys project. His obituary was published by his fellow Rational Toys Committee member Hobhouse (no. 4).[159]

3 William Clayfield (c.1772–1837) came from a Gloucestershire family that had settled in Bristol as merchants by 1775. He became a close friend of Beddoes as soon as Beddoes arrived there; they shared interests in botany, chemistry, and geology. Clayfield had trained at James Watt's Soho works and was in charge of the production of gases at the Pneumatic Institution.[160] He was also an active experimenter, balloonist, patentee,[161] and was the inventor of the mercurial airholder,[162] which allowed his portrait to be identified in 1988.[163] He was elected a founding Honorary Member of the Geological Society of London in 1807.[164] Beddoes described him as "the most ingenious philosopher in this part of the world [Bristol] . . . whom nothing but money-getting prevents from being among the most successful explorers of nature anywhere".[165] Clayfield died in Clifton on 3 March 1837.[166]

4 Benjamin Hobhouse (1757–1831), M.P., was a champion of civil liberties in Parliament. He was born in Bristol and trained as a barrister; through his second wife, he was drawn into dissenting circles. He was M.P. for Bletchingley in 1797 and for Gamound in 1802. He had subscribed to Beddoes' Pneumatic Institution by 1795 and his Bath Bank was the bank of the Somerset Coal Canal,[167] one of the five banks to which subscriptions towards the Rational Toys scheme were to be sent. He was an Original Proprietor of the Royal Institution in London from April 1799. He was also a member of the Board of Agriculture.[168]

5 John H(odder) Moggridge (1771–1834) came from a Devon maritime family that had moved into cloth making at Bradford-on-Avon. Here they became Unitarian dissenters.[169] In 1786, he published a note on cross-bedded rocks at Bath in the *Monthly Magazine*, to which he became a frequent "improving contributor".[170] In 1798 he and his father were involved in the over-optimistic Hill House Colliery, near Newent, which involved both coal and lime.[171] This produced a new need for a canal, which was opened in March 1798.[172] In 1809 he was Sherriff of Gloucestershire.

6 James Stephens (1748–1816) was a son of Philip Stephens (1718–1780) a lawyer of Chancery Lane, London, who inherited property at Camerton circa 1750. James was admitted at Lincoln's Inn in 1765, before immediately moving to Oxford University, where he graduated M.A. in 1774; he inherited Camerton, with its many coal workings, in 1780. Here the pioneer geologist William Smith in 1796 "first put in practice his ideas of draining derived from a knowledge of the strata", while working for Stephens and the nearby Somerset Coal Canal, of which Stephens was a founder and first chairman.[173] Stephens had taken a real interest in such subjects before meeting Smith, and was also a competent botanist.[174] He married in 1781 and again in 1804.[175] Since he, along with Billingsley, had not subscribed to Donne's *Essay*, it seems

likely that it was their joint involvement with Weldon and his caisson on the Somerset Coal Canal that had drawn them into Beddoes' circles.

7 John Wedgwood (1766–1844),[176] like his two younger brothers, had subscribed to Donne's *Essay*. John's membership in the Committee on behalf of the newly rich, Unitarian dissenting, family of Staffordshire potters made sense because of his proximity to Bristol.[177] Beddoes' patient, Tom Wedgwood, had been an early and principal supporter of Beddoes' Pneumatic Institution. Tom's educational plans[178] also involved the Rational Toys project.[179] John had real interests in botany (he was a founder of the Royal Horticultural Society in 1804, and its Treasurer 1804–1806),[180] and an early interest in mineralogy,[181] giving him a common scientific interest with Beddoes. His connection with this Rational Toys scheme was enhanced by his having become a junior partner, in 1792, in the London bank of Alexander Davison and Co.

8 William Wynch (1750–1819) came from a family with a very long connection with the East India Company. He started as a factor in India, first at Masulipatnam (now Machilipatnam) in Andhra Pradesh in 1774, then in Trichinopoly in Tamil Nadu in 1776. In 1788 he returned to England,[182] married, and lived in London. His connection with Beddoes very probably came through the Royal Institution. On 5 May 1800, he, when "of Grosvenor Place, London, was proposed as a candidate for election as Proprietor by Count Rumford."[183] In 1803 he was a major subscriber to the Royal Institution's New Library and Reference Collection.[184] By 1814 he was a director of the United Company of Merchants of England trading to the East Indies. He lived in London until his death in 1819.[185]

Notes

1 M. O. Tremayne, *The Value of a Maimed Life* (London: Daniel, 1912), 71–72.

2 Roy Porter, review of Stansfield, *BJHS*, **19** (1986), 121–122.

3 See D. Stansfield, "Thomas Beddoes and Education", *History of Education Society Bulletin*, **23** (1970), 7–14.

4 Dorothy A. Stansfield and Ronald G. Stansfield, "Dr. Thomas Beddoes and James Watt: Preparatory Work 1794–1796 for the Bristol Pneumatic Institute", *Medical History*, **30** (1986), 276–302; Roy Porter, "Taking Histories, Medical Lives: Thomas Beddoes and Biography", in *Telling Lives in Science. Essays on Scientific Biography*, eds. M. Shortland and Richard Yeo (Cambridge: Cambridge University Press, 1996), 215–240.

5 Maria Edgeworth and Richard Lovell Edgeworth, *Practical Education*, 2 vols. (London: Printed for J. Johnson, 1798), reprinted in S. Manly, ed., *The Pickering Masters: The Novels and Selected Works of Maria Edgeworth* (London: Pickering & Chatto, 2003). We refer to this edition. A second edition appeared in 1801, followed by a third in 1811. The foundation of this work had come during the years 1777–1780. But this project had been abandoned, in 1780, after the death, from consumption, of Richard Edgeworth's second wife Honora.

6 David Hartley, *Observations on Man, His Frame, His Duty, and His Expectations* in two parts (London: S. Richardson for James Leake and Wm. Frederick

booksellers in Bath and sold by Charles Hitch and Stephen Austen, Booksellers in London, 1749); Stansfield, 25; Andrew O'Malley, *The Making of the Modern Child: Children's Literature and Childhood in the Late Eighteenth Century* (New York & London: Routledge, 2003), 66–69; Jay, 55.

7 John Locke, *Some Thoughts Concerning Education: By John Locke, Esq.; to Which Are Added, New Thoughts Concerning Education, by Mr. Rollin* (Dublin: Printed by R. Reilly . . . for G. Risk, E. Ewing, and W. Smith, 1738).

8 J. J. Rousseau, *Émile, ou de l'éducation*, 4 vols. (A la Haye: chez Jean Néaulme, 1762); also (Francfort [false: London, at least in part]: 1762); Jean Bloch, *Rousseauism and Education in Eighteenth Century France* (Oxford: Voltaire Foundation, 1995); John Lawson and Harold Silver, *A Social History of Education in England* (New York: Methuen, 1981).

9 Trevor H. Levere, "Dr. Thomas Beddoes (1750–1808): Science and Medicine in Politics and Society", *BJHS*, **17** (1984), 187–204, esp. 189.

10 Beddoes, *Extract of a Letter on Early Instruction, Particularly That of the Poor* (1792), 21–22.

11 Ibid., 3.

12 [Anna L. Aikin, afterwards Barbauld], *Lessons for Children from Three to Four Years Old* (London: J. Johnson, 1788).

13 Beddoes, *Extract*, 5.

14 Ibid., 4.

15 Stansfield, 83.

16 Beddoes, *Extract*, 4.

17 Ibid., 9.

18 Trevor Levere, "Dr. Thomas Beddoes (1760–1808): Chemistry, Medicine, and Books in the French and Chemical Revolutions", in *New Narratives in Eighteenth-Century Chemistry: Contributions from the First Francis Bacon Workshop, 21–23 April 2005, California Institute of Technology, Pasadena, California*, ed. L. Principe (Dordrecht: Springer, 2007), 157–176; 166.

19 Thomas Beddoes, *Observations on the Nature of Demonstrative Evidence; with an Explanation of Certain Difficulties Occurring in the Elements of Geometry: And Reflections on Language* (London: J. Johnson, 1793), vii–viii.

20 Ibid., 91.

21 Ibid., 15.

22 John Horne Tooke, Εϖεα Πτεροεντα, *or the Diversions of Purley* (London: J. Johnson, 1786). On Tooke, see Chapter 3, this volume.

23 H. J. Jackson, "Coleridge, Etymology and Etymologic", *Journal of the History of Ideas*, **44** (1) (1983), 75–88; 76.

24 Ibid., 79.

25 Beddoes, *Observations*, 114.

26 Ibid., 114.

27 Ibid., 113.

28 Ibid., 31–32.

29 Thomas Reid, *An Inquiry into the Human Mind, on the Principle of Common Sense* (Dublin: A. Ewing, 1764); J. B. Mérian, "Sur le problème de Molyneux. Premier – Huitième mémoire", *Nouveaux Mémoires de l'Académie Royale des Sciences et Belles Lettres*, 1770–1782, in *Sur le Problème de Molyneux*, ed. F. Markovits (Paris: Flammarion, 1984), 5–191; J. W. Davis, "The Molyneux Problem", *Journal of the History of Ideas*, **21** (1960), 392–408; M. J. Morgan, *Molyneux's Question: Vision, Touch and the Philosophy of Perception* (Cambridge: Cambridge University Press, 1977); M. Degenaar, *Het probleem van Molyneux* (Dissertation, University of Rotterdam, 1992).

30 E. Von Erhardt-Siebold, "Harmony of the Senses in English, German and French Romanticism", *Publications of the Modern Language Association of America*, **47** (1932), 577–592; K. Maclean, *John Locke and English Literature of*

the Eighteenth Century (New Haven: Yale University Press, 1936); M. Hope-Nicolson, *Newton Demands the Muse: Newton's Opticks and the Eighteenth Century Poets* (Princeton: Princeton University Press, 1946); W. R. Paulson, *Enlightenment, Romanticism, and the Blind in France* (Princeton: Princeton University Press, 1987); Jessica Riskin, *Science in the Age of Sensibility: The Sentimental Empiricists of the French Enlightenment* (Chicago and London: University of Chicago Press, 2002).

31 Richard Olson, "Scottish Philosophy and Mathematics 1750–1830", *Journal of the History of Ideas*, **32** (1971), 29–44.

32 John Playfair, *Elements of Geometry, Containing the First Six Books of Euclid, with Two Books on Geometry of Solids* (Edinburgh: Bell and Bradfute, 1795). See Amy Ackerberg-Hastings, "Analysis and Synthesis in John Playfair's *Elements of Geometry*", *BJHS*, **35** (2002), 43–72; James Beattie, *An Essay on the Immutability of Truth, in Opposition to Sophistry and Skepticism*, 2nd edn. (Edinburgh: Kincaid and Bell, 1771); there were several editions.

33 Olson, "Scottish Philosophy and Mathematics 1750–1830", 36.

34 John Leslie to Thomas Wedgwood, 14 May 1797, WMB, MS "Eturia 256" quoted in Olson (1971), 39.

35 E. Robinson, "Benjamin Donne (1729–1798): Teacher of Mathematics and Navigation", *Annals of Science*, **19** (1963), 27–36; W. L. D. Ravenhill, "Benjamin Donne (1729–1798): Map-Maker and Master of Mechanics", *Reports and Transactions of the Devonshire Association for the Advancement of Science*, **97** (1965), 179–183.

36 Stock, 128.

37 Davies Giddy diary, CRO MS DD DG 12, 4–10 November 1782, note added in 1835.

38 Beddoes to Tom Wedgwood, 22 February 1796. WMB MS W/M 80.

39 Beddoes to James Watt, 28 March 1796. BCL JWP MS 3219/4/29 no.54 c1/30.

40 These numbers refer to the 1800 Rational Toys Organizing Committee discussed in the Excursus note at the end of this chapter.

41 Benjamin Donne, *Prospectus of an Essay on Mechanical Geometry*, 2.

42 E. Robinson, "Dr. Beddoes and Benjamin Donne on the Teaching of Mathematics", *Mathematical Gazette*, **40** (1956), 133–135. See also "John Sadler", in ed. Torrens, "James Sadler", *ODNB*.

43 *Monthly Magazine*, **2** (1796), no. 8, September, 651.

44 Benjamin Donne, *An Essay on Mechanical Geometry, Chiefly Explanatory of a Set of Schemes and Models, by Which the Knowledge of the Most Useful Propositions of Euclid, and Other Celebrated Geometricians, May be Clearly and Expeditiously Conveyed, Even to Youth of an Early Age: By Benjamin Donne, Master of the Mechanics in Ordinary to His Majesty* (Bristol: Printed for and sold by the author. Sold in London by Messrs. Vernor and Hood, Booksellers, Birchin-Lane, Cornhill; and by the following mathematical instrument makers: Mr. Adams and Mr. Troughton, Fleet-Street, and Mr. Jones, Holborn, 1796.)

45 Registered 1951–653.

46 See *Bristol Mercury*, 5 December 1825 and 10 July 1826.

47 See *Bristol Mercury*, 7 January 1837.

48 Leslie to Tom Wedgwood, Largo 14 May 1797, WMB MS Eturia 256.

49 Donne, *Essay*, xiii–xvi and 91.

50 Brief biographies of the members of the Toys Committee are given in the Excursus note at the end of this chapter.

51 Donne, *Essay*, ix–x.

52 T. Beddoes, "Letter from Dr. Beddoes to Mr. Donne", in *Essay*, ed. Donne, v–vii.

53 Donne, *Essay*, ix.
54 Beddoes, "Letter from Dr. Beddoes to Mr. Donne", vii.
55 Stansfield, 62–64.
56 Robert E. Schofield, *The Lunar Society of Birmingham: A Social History of Provincial Science and Industry in Eighteenth-Century England* (Oxford: Clarendon, 1963); Christina Colvin, "Introduction", in *Maria Edgeworth, Letters from England, 1813–1844*, ed. Colvin (Oxford: Clarendon Press, 1971), xiii-xlii; xviii.
57 James Keir, *An Account of the Life and Writings of Thomas Day Esq.* (London: Printed for John Stockdale, 1791).
58 Jay, 80.
59 Beddoes in Donne, *Essay*, vii–viii.
60 *Bristol Gazette*, 31 May 1798, and *Monthly Magazine*, 5 (June, 1798), 470 and **6** (July, 1798), 76–77.
61 "On Teaching Arithmetic and Mathematics", *Monthly Magazine*, 8 (1) (1803), 677–678.
62 Stock, 92.
63 Stock, Appendix 8, xlix–li.
64 "Rational Toys", *A Journal of Natural Philosophy, Chemistry and the Arts [Nicholson's Journal]*, **4** (1800), 287–288.
65 See Excursus note at the end of this chapter.
66 For a survey, see D. H. Tew, *Canal Inclines and Lifts* (Gloucester: Sutton, 1984).
67 See H. S. Torrens, "The Somersetshire Coal Canal Caisson Lock", *Bristol Industrial Archaeological Society Journal*, **8** (1975), 4–10; D. K. Brown, "The Combe Hay Caisson Lock: An Early Submarine", *Five Arches*, **9** (1989), 5–56; M. Bates, "Simply Brilliant & Brilliantly Simple, Robert Weldon's Patent Hydrostatic or Caisson Lock", *Bristol Industrial Archaeological Society Journal*, **35** (2002), 2–19 and H. S. Torrens, "The Water-related Work of William Smith (1769–1839)", in *200 years of British Hydrogeology*, ed. J. D. Mather, Geological Society of London, Special Publication, **225** (2004), 17–21.
68 *Bath Journal*, 9 April 1804.
69 *Bath Herald*, 27 August 1796.
70 *Bath Chronicle*, 12 December 1797.
71 *Sun* (London), 14 February 1798.
72 *Bath Herald*, 9 June 1798 and *Observer* (London), 19 June 1798.
73 *Bath Chronicle*, 13 April 1799 and *Observer* (London), 12 April 1799.
74 *Bath Chronicle*, 23 and 30 May 1799.
75 W. S. Mitchell in *Bath Chronicle*, 8 August 1872, 8.
76 Stock, 152
77 Eliza Meteyard, *A Group of Englishmen (1795–1815) Being Records of the Younger Wedgwoods and their Friends . . .* (London: Longmans, Green, 1871), 92.
78 Ladbroke Black, *Some Queer People* (London: Sampson Low, 1931), 196.
79 James Kendall, "The First Chemical Society, the First Chemical Journal, and the Chemical Revolution", *Proceedings of the Royal Society of Edinburgh*, **63A** (1949–1952), 392.
80 See also Brian Simon, *Studies in the History of Education, 1780–1870* (London: Lawrence and Wishart, 1960); Teresa Michaels, "Experiments before Breakfast: Toys, Education and Middle-Class Childhood", in *The Nineteenth-Century Child and Consumer Culture*, ed. Dennis Denisoff (Aldershot: Ashgate, 2008), 29–42. C. E. Colvin and C. Morgenstern, "The Edgeworths: Some Early Educational Books", *Book Collector*, **26** (1977), 39–43.
81 Butler, *Maria Edgeworth*, 172.
82 M. and R. L. Edgeworth, *Practical Education* (1798), ix. Simon, *Studies in the History of Education*, 89; Alice Paterson, *The Edgeworths: A Study of later*

Eighteenth-Century Education (London: W. B. Clive, 1914), v–vi; Julia V. Douthwaite, *The Wild Girl, Natural Man, and the Monster: Dangerous Experiments in the Age of Enlightenment* (Chicago and London: The University of Chicago Press, 2002), 152.
83 M. and R. L. Edgeworth, *Practical Education*, ix, 23.
84 Butler, *Maria Edgeworth*, 109.
85 *Star*, 31 May 1798 and *Morning Chronicle*, 1 June 1798.
86 Simon, *Studies in the History of Education*, 53.
87 M. and R. L. Edgeworth, *Practical Education*, ix–x.
88 Manly, "Introductory Note", vii.
89 Quoted in Jay, 13.
90 Jay, 13.
91 Liliane Hilaire-Peréz, "Technology as a Public Culture in the Eighteenth Century: The Artisans' Legacy", *History of Science*, **45** (2007), 135–153; cf. Hilaire-Peréz, *La pièce et le geste: Artisans, marchands et savoir technique à Londres au XVIIIe siècle* (Paris: Albin Michel, 2013).
92 Quoted in J. Uglow, *The Lunar Men: Five Friends Who Made the Future 1730–1810* (New York: Farrar, Strauss and Giroux, 2002), 17.
93 J. Brewer, "Childhood Revisited: The Genesis of the Modern Toy", in *Educational Toys in America: 1800 to the Present*, eds. K. Hewitt and L. Roomet (Burlington, VT: The Robert Hull Fleming Museum, 1979), 3–10. See also J. Brewer, "The Genesis of the Modern Toy", *History Today*, **30** (1980), 32–39.
94 For the ambiguities and range of the word "toys" around 1800, see K. D. Brown, *The British Toy Business: A History since 1700* (London and Rio Grande: The Hambledon Press, 1996), 26.
95 M. and R. L. Edgeworth, *Practical Education*, 11 [(1798), 2].
96 Ibid., 11 [(1798), 1].
97 Ibid., 20 [(1798), 18].
98 Ibid., 27–28 [(1798), 31].
99 Ibid., 12 [(1798), 3].
100 Ibid., 5 [(1798), v].
101 Ibid., 13 [(1798), 5–6].
102 Ibid., 16 [(1798), 11].
103 Ibid., 28 [(1789), 31/2].
104 Joseph Priestley, *The History and Present State of Discoveries Relating to Vision, Light, and Colours* (London: J. Johnson, 1772).
105 M. and R. L. Edgeworth, *Practical Education*, 257.
106 Ibid., 30.
107 Maria Edgeworth, "The Good French Governess", in *Moral Tales for Young People* (London: J. Johnson, 1801), 291.
108 A. Paterson, *The Edgeworths: A Study of later Eighteenth-Century Education* (London: W. B. Clive, 1914), 29; Stansfield, 95.
109 Paterson, *The Edgeworths*, 30.
110 M. and R. L. Edgeworth, *Practical Education*, 77 [(1798), 122].
111 A. O'Malley, *The Making of the Modern Child* (New York and London: Routledge, 2003).
112 Paterson, *The Edgeworths*, 2. J. R. Townsend, ed. *Trade and Plumb-Cake for ever, Huzza! The Life and Work of John Newbery, 1713–67* (Cambridge: Colt Books, 1994).
113 Butler, *Maria Edgeworth*, 156.
114 Douthwaite, *The Wild Girl, Natural Man, and the Monster*, 134.
115 Butler, *Maria Edgeworth*, 37/8; 58/9; Stansfield, 65; Douthwaite, *The Wild Girl, Natural Man, and the Monster*, 136–138.
116 Douthwaite, *The Wild Girl, Natural Man, and the Monster*, 138–141. Day adopted two girls from foundling hospitals, took them to France, considered that one of them was promising, and brought her back to England in hopes of

making her his wife; he decided that she was unsuitable. See Wendy Moore, *How to Create the Perfect Wife: Britain's Most Ineligible Bachelor and His Enlightened Quest to Train the Ideal Mate* (London: Weidenfeld and Nicolson, 2013).

117 Paterson, *The Edgeworths*, 67.

118 Roy Porter, "Enlightenment and Pleasure", in *Pleasure in the Eighteenth Century*, eds. Roy Porter and Marie Mulvey Roberts (Houndmills, Basingstoke, Hampshire and London: MacMillan, 1996), 1–18; 15.

119 Dissected maps were maps mounted on a board and cut into different shapes, i.e. jigsaw puzzles. Linda Hannas, *The English Jigsaw Puzzle 1760–1890: With a Descriptive Check-List of Puzzles in the Museums of Great Britain* (London: Wayland Publishers, 1972); Jill Shefrin, *Neatly Dissected for the Instruction of Young Ladies and Gentlemen in the Knowledge of Geography: John Spilsbury and Early Dissected Puzzles* (Los Angeles: Cotsen Occasional Press, 1999).

120 Jill Shefrin, *Such Constant Affectionate Care: Lady Charlotte Finch – Royal Governess & The Children of George III* (Los Angeles: The Cotsen Occasional Press, 2003), 6.

121 M. and R. L. Edgeworth, *Practical Education*, 21 [(1798), 19].

122 Science Museum, London: http://www.sciencemuseum.org.uk/about_us/press_and_media/press_releases/2005/05/321.aspx. For another set of geometric shapes and models, made by George Adams, see Alan Q. Morton and Jane A. Wess, *Public & Private Science: The King George III Collection* (Oxford: Oxford University Press in association with the Science Museum, 1993), 202–206.

123 Shefrin, *Such Constant Affectionate Care*, 58.

124 *Memoirs of the Literary and Philosophical Society of Manchester*, **1** (1785), 72–89.

125 R. G. Thorne, *The House of Commons 1790–1820: 4, Members G-P* (London: Secker & Warburg, 1986), 371–373.

126 Quoted in Stuart J. Reid, *Life and Letters of the First Earl of Durham, 1792–1840*, 2 vols. (London: Longmans, Green and Co., 1906), I, 26–28.

127 Reid, I, 36.

128 Reid, I, 38–53, and Stock, 150–151.

129 Ged Martin, "Lambton, John George, First Earl of Durham (1792–1840)", ODNB.

130 John King was the Anglicized name of the Swiss refugee Johann Koenig, who married Anna Beddoes' sister Emmeline in 1802.

131 Stock, 151–153.

132 Ibid.

133 Jay, 255–256.

134 J. Wachelder, "Toys, Christmas Gifts and Consumption Culture in London's *Morning Chronicle*, 1800–1827", *ICON: Journal of the International Committee for the History of Technology*, **19** (2013), 13–32.

135 R. L. Edgeworth, Letter to a Friend, 6 December, 1807, in the Cotsen Children's Library, Princeton, Shelf mark CTSN MSS Q 26758, quoted in Shefrin, *Neatly Dissected for the Instruction*, 22.

136 Ibid.

137 Edgeworth MSS, 1724–1817 in the National Library of Ireland: Reel 16, MS 22471: Correspondence and associated papers of Richard Lovell Edgeworth regarding Education. Fifty-eight items. c. 1772–1817, with four pages of typescript listing and seven folders). c. 1809 – "Rational Toys in a letter to Mr Tabart". For more on Tabart, see "Benjamin Tabart", *Wikipedia*.

138 *The Morning Chronicle*, 31 December 1803.

139 Francis West, *A Descriptive Account of a Variety of Intellectual Toys made for the Instruction and Amusement of Youth of both Sexes, as Recommended by Dr. Beddoes, Miss Edgeworth, &c, &c.* Illustrated by 30 Engravings of

the Various Instruments (London: 83, Fleet Street, near St. Bride's Avenue, J. Barker printer, 1836). Educational toys were often then referred to as "intellectual toys".

140 West, *A Descriptive Account of a Variety of Intellectual Toys*, 7.

141 J. A. Paris, *Philosophy in Sport Made Science in Earnest; Being an Attempt to Illustrate the First Principles of Natural Philosophy by the Aid of Popular Toys and Sports*, 3 vols. with sketches by George Cruikshank (London: Longman, et al., 1827).

142 OED, which credits Paris with the invention. Others who have been suggested as inventors of the thaumatrope are John Herschel, William Henry Fitton, and William Hyde Wollaston.

143 Paris, *Philosophy in Sport Made Science in Earnest*, III, 2.

144 This should be Chevalier Patrice D'Arcy.

145 David Brewster, *Letters on Natural Magic: Addressed to Sir Walter Scott* (London: Murray, 1832), 26.

146 R. J. Leskosky, "Two-State Animation: The Thaumatrope and Its Spin-offs", *Animation Journal*, 2 (1993), 20–35; R. J. Leskosky, "Phenakiscope: 19th Century Science turned to Animation", *Film History*, 5 (1993), 176–189.

147 Paris, *Philosophy in Sport Made Science in Earnest*, III, 12.

148 H. S. Torrens, "Phillips, William (1773–1828)", ODNB.

149 D. L. Emblen, *Peter Mark Roget: The Word and the Man* (London: Longman, 1970).

150 Jay, 166.

151 Stansfield, 247.

152 Ibid., 248.

153 OED.

154 Michael Faraday, "On a Peculiar Class of Optical Deceptions", *Journal of the Royal Institution of Great Britain*, **1** (1830–1831), 205–223.

155 P. M. Roget, *Animal and Vegetable Physiology Considered with Reference to Natural Theology*: *The Bridgewater Treatises on the Power Wisdom and Goodness of God as Manifested in the Creation: Treatise V*, 2 vols. (London: William Pickering, 1834), II, 524.

156 It clearly helped for all those hoping to help promote Beddoes' plans for Rational Toys to be sufficiently wealthy. Until 12 January 1858, all English wills had to be proved by church and other courts. The PCC was the most important of these, dealing only with relatively wealthy individuals who lived mainly in the south of England, as did all the eight listed here.

157 J. Murch, *A History of the Presbyterian and General Baptist Churches in the West of England* (London: Hunter Ashwick, 1835), 156–164. R. Atthill, *Old Mendip*, 2nd edn. (Newton Abbot: David and Charles, 1971), Chapter 4.

158 *General View of the Agriculture of the County of Somerset* (1794); 2nd edn. (Bath: Printed for the author, 1798). Weldon's description of his caisson ("now building and nearly completed") is on 317–310, and plate 15.

159 B. Hobhouse, "Eulogy on the Late John Billingsley Esq", *Letters and Papers of the Bath and West Society*, **13** (1813), 90–99.

160 F. F. Cartwright, *The English Pioneers of Anaesthesia* (Bristol: Wright, 1952), 104–105.

161 P. C. Rushen, *Old Time Invention in the Four Shires* (Evesham: Journal Press, 1916), 60–61.

162 See Chapter 1 in this volume, Figure 1.12.

163 W. D. A. Smith, "William Clayfield's Mercurial Airholder", in *Essays on the History of Anaesthesia* (London: Royal Society of Medicine Press, 1996), 13–15.

164 H. B. Woodward, *The History of the Geological Society of London* (London: The Society, 1907), 269.
165 Letter to Erasmus Darwin, 18 April (1799 or 1800), Wood Library & Museum of Anesthesiology, Park Ridge, Illinois, in Clayfield Correspondence MS WZ 100 C579 1805–1825 RB oversized.
166 *Felix Farley's Bristol Journal*, 11 March 1837.
167 K. R. Clew, *The Somersetshire Coal Canal and Railways* (Newton Abbot: David and Charles, 1970), 24.
168 Mr. Hobhouse, *Public Characters of 1807* (London: Phillips), 101–135. An obituary appeared in *Gentleman's Magazine*, **101** (1831), 371–372 and 653.
169 J. E. Gethyn-Jones, *Dymock Down the Ages* (Gloucester: Smith, 1966), 30, 98 and 152. Boyce Court is illustrated on plate 6.
170 Torrens, Variorum, III, 242–243.
171 D. E. Bick, *The Mines of Newent and Ross* (Newent: Pound House, 1987).
172 *Monthly Magazine*, 5 April 1798, 313 and D. E. Bick, *The Hereford and Gloucester Canal* (Newent: Pound House, 1979).
173 Torrens, Variorum, III, 230–231 and 241–242.
174 See Copy of a letter from Mr. Stephens on diseases of wheat, dated 22 August, 1800, *Proceedings of the Bath Natural History and Field Club*, 3 (1877), 12–16.
175 *Bath Chronicle*, 19 July, 1781 and *Bath Journal*, 8 July 1804.
176 B. Wedgwood, "John Wedgwood (1766–1844)", in *Wedgwood; Its Competitors and Imitators 1800–1830, the 22nd Wedgwood International Seminar,* ed. Arthur R. Luedders (Ann Arbor: Ars Ceramica Ltd, 1977), 251–265.
177 E. Meteyard, *A Group of Englishmen*; B & H. Wedgwood, *The Wedgwood Circle* (London: Studio Vista, 1980).
178 David Erdman, "Coleridge, Wordsworth and the Wedgwood Fund", *Bulletin of the New York Public Library*, **60** (1956), 425–443 and 487–507.
179 The family were enthusiastic proponents of educational improvement. F. Doherty, "The Wedgwood System of Education", *American Wedgwoodian*, 6 (8) (1983), 182–187, describes the system of education which Josiah junior (1769–1843) developed in 1797–1799.
180 H. R. Fletcher, *The Story of the Royal Horticultural Society 1894–1968* (Oxford: Oxford University Press for the Society, 1969).
181 F. W. Steer, "John Wedgwood as a Mineralogist", *Proceedings of the Wedgwood Society*, 5 (1963), 47–59.
182 *Whitehall Evening Post*, 8 July 1788.
183 *Royal Institution Minutes*, **2** (1800), 77.
184 *Morning Post*, 16 April 1803.
185 *Gentleman's Magazine*, **89** (1) (1819), 657.

Appendix 1. The mystery of Dr. John Edmonds Stock, Beddoes' first biographer

Hugh Torrens

Introduction

Stock's 1811 biography of Beddoes has acquired something of a reputation. John Barrell called it "probably the most boring biography ever written",[1] while Roy Porter, a fellow Beddoes biographer, considered it "one of the least informative biographies of all time".[2] But Dorothy Stansfield, whose biography appeared in 1984, had rightly noted that Beddoes' widow's choice of "Dr Stock may have been for reasons other than his ability to edit Beddoes' medical work. Stock could be trusted to refer discreetly to Beddoes' political activities."[3] We shall see why.

Stock's ancestry

Part of Stock's previously shady ancestry was recorded in the pedigree given by Joseph Alfred Bradney in his *History of Monmouthshire*.[4] This was established by his Edmonds family's inheritance of Welch estates at Gwaelod-y-Wlad, Trealy and Church-farm, in Monmouthshire, which were sold in 1865 by Stock's only son, John Shapland Edmonds Stock.

Dr. Stock's father was here recorded as "John Stock, of Chipping Sodbury", Gloucestershire. He had first been apprenticed as an apothecary[5], who then became a Bristol paper-maker and stationer with his warehouse opposite the Bridgewater Slip, on the Back, in Bristol. Here he was active from 1744 to 1771[6] and prospered enough to have retired from business just before he married Ann or Anne Lowle, daughter of the late Anglican Rev. Robert Lowle, at St. Michael's, Bristol, on 22 October 1772.[7] John Stock's will[8] showed he had died in the second quarter of 1788. His will was first drawn up in June 1782, when he lived in the out-parish of St. James, Bristol. It made careful provision for the best "liberal education" of his two very young sons. This was to be under the care of his executors, named as "John Norris of Stokes Croft, Bristol, Esquire", "Harry Canby, of Bristol, Accomptant", and his friend, the attorney George Rolph senior,[9] of nearby Thornbury.

Stock's widow, mother of our biographer, Ann, née Lowle, remarried at Chipping Sodbury on 25 October 1789 to Thomas Higgs,[10] also of Chipping Sodbury. This was the second marriage for both. Thomas had been born there on 28 July 1732, son of Daniel Higgs and Mary Hulbert. His first marriage was to Elizabeth Richardson in 1757, and they had four children. Thomas Higgs died on 3 October 1802 at Chipping Sodbury. He may have been connected with the two Josiah Higgs here, listed as brewer and maltster in the *Universal British Directory* of the 1790s.[11] This is of some interest as it would connect them with the Shapland family at Marshfield, into which John Edmonds Stock was to marry in 1803. Ann, now Higgs, died at Wickwar, "at the house there of her second son Thomas on 4 March 1819, [and named as] mother of Dr. Stock of Clifton."[12] There were only two children of her first marriage, our John Edmonds and Thomas Edmonds of Wickwar, Gloucestershire,[13] both of whom were born in Bristol. The only signs yet noted of any nonconformity in the Stock family are the facts that Stock had been sent by his father's executors to a Unitarian school in Bristol, and then to the Hackney Academy, before his matriculation at Oxford.

John Edmonds Stock in Britain

John Edmonds Stock, eldest son, had first been entered at Lincoln's Inn, London, clearly intended for a legal career, on 9 November 1790.[14] But soon, on 10 March 1791, "aged 16, son of John of Bristol, gentleman", he matriculated at Exeter College, Oxford.[15] Here, having acquired "conscientious scruples as to subscription", to the 39 articles of the Church of England, having become a Unitarian, "he left the university without taking any degree."[16] The next we hear of him is on 14 November 1793 when, following in Beddoes' footsteps, he had become a medical student at Edinburgh, and was then elected a member of the University Student's Natural History Society. To this Society he read a paper on "Instinct" in the 1793–1794 session.[17] In the same year he also joined the student Royal Medical Society at Edinburgh, again following Beddoes' lead.[18]

In Edinburgh, Stock became a fervent reformer (whom Pitt would have considered a revolutionary), and joined a Reform Society. In this he was active by February 1794, since, on about 1 March 1794, he was sending out copies of an inflammatory circular letter from the secretary of the London Corresponding Society (LCS) to the Reform Society in Edinburgh. This was written by the radical London bootmaker, and Scottish-born founder of the LCS, Thomas Hardy.[19] This circular letter was of considerable significance as it proved that the LCS was now seemingly acting as leader of all Reform Societies in both England and Scotland.

In the event, the British Government panicked and, on 15 August 1794, indicted three local men for High Treason to a grand jury at Edinburgh. These were Robert Watt, David Downie, and Stock.[20] All three had been part of a seven-man secret Edinburgh Committee of Ways and Means, formed there in March 1794, following the suppression of the famous convention of British radicals held in Edinburgh the previous December. Stock was probably responsible for writing the handbill, dated from Dundee, 12 April 1794, which was distributed to the Hopetoun Fencibles in Dalkeith, then quartered there prior to marching into England, and which had encouraged them to mutiny by playing on their dissatisfaction at being transferred illegally to Liverpool, leaving "foreign mercenaries [to] violate the chastity of your wives, your sisters and your daughters."[21] Watt was sentenced to be "hung drawn and quartered", but when executed on 15 October 1794 he was merely beheaded. Downie was similarly sentenced, but was later pardoned and banished to America.[22]

Stock's escape to America

Stock, meanwhile, had been alerted to the very real danger he was now in. In May he managed to escape from Edinburgh by ship, with the Edinburgh Committee's secret answer to Thomas Hardy's LCS circular letter of March. He probably arrived in London after Hardy had been arrested on 12 May 1794, thus leaving no evidence of the delivery of this answer. Stock went into hiding in London, where he was visited by his close Bristol school friend John Edye (see below) in June while government agents were hunting for him, as reported on 21 August 1794.[23] Some time before, or perhaps very soon after, Stock must have escaped from London by ship to America, finally settling in Philadelphia.[24] Here he returned to his medical studies, graduating at the University of Pennsylvania at Philadelphia on 12 May 1797, with the degree of M.D. for his *Essay on the Effects of Cold upon the Human Body*.[25] He also here made translations from German.[26] He remained in exile for at least eight years.

His wish to return to England

All that was known previous to this study was that he had been permitted to re-enter England from France during the truce of 1803.[27] But a letter of 17 April [1802] from John Edye to Charles Bragge, M.P. for Bristol from 1796 to 1812, tells us much more:

> The interest of a much loved and early friend impels me to obtrude myself almost meekly upon you, unknown and unintroduced, a circumstance which I hope you will have the goodness to excuse, as I have not time to procure such a recommendation as might entitle me to your notice. The name of the gentleman I allude to is Stock, a native of the city which you so respectably represent [Bristol]. My friend is possessed [torn – illegible] of a very high and distinguished rank, and his moral character is unimpeachable, but unhappily he became acquainted in the early part of his life while a student in the University of Edinburgh, with Watt and Downie and others of the same designation – By the self-persuasion of these men, he became,

in the eye of Government, implicated in plans respecting which it is difficult to say whether wickedness or folly were the predominant feature.

At the early age of nineteen, a bill of high Treason was found against him, and a reward of £200 offered for his apprehension. He had the good fortune to make his escape to America, but, though the vengeance of the law has not reached his person, it has robbed him of his happiness – not only does he feel himself exiled in a Country where the views and pursuits of the inhabitants differ so widely from his own, but England is the repository of everything that he admires and loves – He wishes most earnestly for permission to return to his offended, but I trust not irreconcilably offended, Country. As his earliest and confidential friend, I have taken upon myself to make an effort in his favor and most happy shall I esteem myself if my efforts, or those of his other friends, are ultimately attended with success – Lord Westmorland[28] has been so obliging as to put a memorial in my friend's favor into the hands of L[ord] Pelham.[29] The noble Secretary wrote immediately to the Lord Advocate of Scotland,[30] whose answer was that the circumstances of Mr. Stock's case were such, as not to warrant him in recommending his return at present – I am by no means surprised at the Lord Advocate's answer. In his situation I have little doubt that I should myself have returned the same. I know how difficult, nay how impossible it is, that men high in office should descend to the minute shades of difference which distinguish individual cases and degrees of guilt – In this predicament, my friend's case may rest for ever. No representations in his favor from the South of England can be expected to reach the conviction of the Lord Advocate of Scotland. I am confident however that could it be on[ly – torn] stated, such is the equity and moderation [of] Ministers, that no impediment would be thrown in the way of his return – I take the liberty therefore of appealing to you who am so nearly connected with Administration, and requesting the favor of a few moments of your time in my friend's behalf. His case is such as to court enquiry, and you as the representative of Bristol have the most unequivocal means of informing yourself respecting it. If I do not convince you that the object of my solicitations is a man of most respectable character, that his political sentiments have undergone an undisputable change, and that his conduct in America has been uniformly peaceable and moderate; in short if I do not find the means of satisfying you that he is likely to become an ornament to his Country instead of endangering its repose, I will not ask the [oper]ation of your very powerful interest with [torn] in his favor. I propose to myself [the fav]or of calling on you Monday morning next – Should I not have the good fortune of meeting you at home, allow me to request that you will leave word with your servant, at what time and on what day I may see you.

In the meantime, I remain Sir
Your most obt Humble Servant

John Edye

The letter had been correctly endorsed "Mr Edye" (not Edge) and was dated 17 April [1802], from No. 5, Robert Street, Adelphi (London).[31]

It was accompanied by a "Copy of a Memorial in favour of Dr. Stock, addressed to Lord Pelham, His Majesty's principal Secretary of State for the Home Dept. The Memorial of John Edye of Lyme Regis in the County of Dorset, sent 28 April 1802".[32] This added that

> The Memorialist was bound by the most sincere and early friendship to John Edmonds Stock, late of Little Sodbury House [Gloucestershire, to which Stock's father had moved between 1782 and 1788], but now of Philadelphia. [Edye added that Stock had] attended [Estlin's] Grammar School in Bristol and was then placed by his three guardians (all since dead [and listed in his father's 1788 will]) at New College, Hackney,[33] and admitted [in November 1790] a Member of the Honorable Society of Lincoln's Inn. Then, after two years, they removed him to Oxford University where he remained nearly three years [to 1793]. Having changed from Law to Physic, they now placed him in Edinburgh at 19 years old. Here he became political and left Edinburgh in May 1794, determined to break off all connection with political Societies, as your memorialist found him [in London] in June 1794. Some of his friends were now in custody and he needed to withdraw to America for his personal security (a reward of £200 was offered for his apprehension then) where he has now resided 7 years.

Edye next quoted from a letter Stock had written from America, on 26 December 1800, to the antiquary and author Rev. Edward Davies. He was then curate of Olveston, near Bristol. This letter concerned the changes in Stock's political attitudes. Edye also noted that Stock's "aged mother [Ann, née Lowle, 1742–1819] during 7 years towards the close of life, had been severed from his embrace."

These documents were in the archives of the then M.P. for Bristol, Charles Bragge Bathurst, which city he had represented from 1796 to 1812. Other documents, included those sent in further support.[34] These included a group from the three "gentleman who had successively the care of [Stock's] Education." These comprised:

1 A letter of 25 April 1802 from John Prior Estlin of Bristol, who had been his former Grammar School master there. Estlin had studied at the dissenters' Warrington Academy from 1764, where he became a Unitarian, and after 1771 he ran his own school in Bristol, which produced many distinguished alumni.[35]
2 A letter of 22 April 1802 from the encyclopaedist and dissenter Abraham Rees who taught Stock at New College, Hackney.
3 A letter of 24 April 1802 from Henry Richards (1747–1807), then in London, and Stock's former tutor at Exeter College, Oxford, who would later become Vice-Chancellor of that University from 1806 to 1807.[36]

The memorialist, John Edye junior (1773–1818), was Stock's contemporary and school friend under Estlin. Edye was a member of an another Bristol Unitarian family and his initiative to get Stock pardoned had clearly been inspired by this friendship, after the start of the Peace of Amiens, in March 1802. Edye's birth had been registered at Lewin's Mead Unitarian chapel, Bristol on 27 January 1774, when he was ten weeks old.[37] His father, John Edye (1732–1794) senior, was a Bristol banker.[38] His mother, Hannah (née King), who had married the first John Edye senior in April 1759. She died in July 1800 at Lyme Regis,[39] whence the family had moved, after her own husband John Edye junior died in 1794.

The family's acquisition of the Pinhay, or Pinney, estate near Lyme Regis on the Devon/Dorset border had come as a result of John Edye junior's marriage at Wootton Fitzpaine, Dorset on 2 April 1797 to Frances Walrond Oke, the heiress of Walter Oke of Combpyne, Devon.[40] It must have been here that Stock and his blue-stocking wife (introduced next) subsequently stayed with the Edyes on their many visits to Lyme.

Stock returns to England, 1803–1835

The first we hear of Stock after his return to England is his marriage at Marshfield, Gloucestershire, on 3 July 1803 to Charlotte Shapland of that place.[41] The Shaplands had long been based there. The family had arrived with Angel Shapland (c.1691–1748) who had come, from Crediton in Devon, to become minister of the Unitarian chapel there,[42] probably in the 1720s. One of Angel's sons was John Shapland of Marshfield near Bath,[43] who had died by September 1789.[44] Charlotte (c.1779–1861) and her brother Anglican Rev. Joseph (c.1774–1837) were two of John's children by his wife Martha. The later *Universal British Directory* of the 1790s records this Martha Shapland was a maltster at Marshfield,[45] which must indicate how the Shapland family fortune had been made. Charlotte's brother Joseph was an Oxford graduate,[46] who then became the vicar of St. Peter's, Worcester.[47] Charlotte had clear aspirations as a blue stocking. She, along with Beddoes and Estlin, subscribed to the poet and orientalist Charles Fox's *A Series of Poems . . . of Achmed Ardebeili*, published in 1797.

Married life

Meanwhile, on 5 December 1803, Stock had been admitted an Extra-Licentiate of the College of Physicians of London.[48] Newly re-qualified here, with his American degree, Stock could start as medical assistant to Beddoes, from the spring of 1804.[49] On 30 July 1805, the publication of Stock's *Medical Collections on the effects of Cold*, work he had started in America, was announced in London.[50] It was well reviewed. His later medical career in Bristol is well covered by Munro Smith.[51]

By 13 February 1808 Samuel Taylor Coleridge reported to Robert Southey on Stock and his now "tyrannical wife":

> I could tell you a good story of Dr. Stock, whom I had one evening cock-pecked into an opinion of Wordsworth's merit as a Poet, and who came next morning . . . complaining of my conduct . . . the poor man had been hen-pecked out of it again. . . . I heard from all quarters of the insolence & overbearing Self-conceit of Mrs Stock – and the poor Doctor, who seems by nature good & kind, she treats openly, as a mere Insignificant.[52]

Shortly after this unflattering portrait, Stock's intended *Life of Beddoes* was announced in February 1809.[53] It was ready, and "in the Press", by April 1810.[54] Stock soon stood for election as a physician to the Bristol Royal Infirmary, from September 1810, but he was successfully elected from March 1811.[55] By this time he and his wife were clearly able to spend much time at Lyme Regis with the Edyes. This had been much facilitated by the will of their Bristol Baptist friend, Joseph Holland,[56] who, upon his death in 1812, left

> Charlotte Stock, the wife of John Edmonds Stock, of the City of Bristol, Doctor of Physic, the sum of five thousand pounds, as a token of my gratitude for her kind attentions and service to me during the ill state of my health, and of my friendships for her.[57]

Stock crises, 1813–1816

Late in 1816, Dr. Stock seceded from the Unitarian Church. This caused a major sensation among dissenters, both here and abroad.[58] Stock's wife remained active in Lyme, and got duly pilloried in *The Lymiad*, the blank verse poem written there in 1818 by Charlotte Jane Skinner (1785–1872), which has recently been published.[59] Here she – and only she, since *The Lymiad*'s female author made no reference to Mrs Stock's husband – becomes "Mrs Manly, a masculine, whist enthusiast, a blue-stocking and learned craniologist".[60] Mrs. Stock also appears in the MSS diaries 1831–1833 of Anna Maria Pinney. Here Mrs. Stock was reported, in October 1831, to have earlier "employed Mary Anning to run errands and, in about 1813, gave her the first book on Geology she ever read",[61] and which diaries give further evidence of "Mrs Stock's singularities". Anna Maria Pinney's diary added this, on 23 January 1832, about

> the way [Mary Anning had since] unfolded to me the character of Mrs Stock. What with her craniology, her metaphysics and her literature, she amused me extremely, and I should have been more intimate with her, had it not been for Mary's showing me, that, if in her higher

> flights, she rose higher than common minds, her real character was grovelling, and [that] vanity, passion, and covetousness, so reigned in her mind, that it would be hard to say which governed her most.[62]

Also about this time "Mrs. Stock" seems likely to be the same person as recalled by the later painter and traveller, Marianne North. She recorded of her childhood,

> another old lady [sometime between 1835–1845] used to come and stay with us [in Hastings, Sussex], a Mrs Stock, who impressed me greatly, as she wore stick-up collars, played splendidly on the piano, and had a mania for phrenology [craniology]. Whenever she came victims were collected, with their back hair let down, to have their bumps felt and registered, and the drawing-room looked like a hair-dresser's shop; under her influence my mother became quite a believer.[63]

It seems likely there was only one such female phrenologist called Mrs Stock.

Latterly John Stock became an active member of the Bristol Institution for the Advancement of Science, Literature and the Arts, which opened in 1823.[64] He also made the Institution donations of fossils (e.g. of "Two fine Dudley trilobites and seven Dudley slabs")[65] in 1824. He was also, certainly, the "Stock. – Esq., M.D." who duly subscribed to George Roberts, *History and Antiquities of Lyme Regis* (1834).

John Edmonds Stock died, "after a long and most suffering illness", at the Lodge, near Tewkesbury, home of his brother-in-law, now Mr. Joseph Shapland, on 4 October 1835. Despite his having renounced Unitarianism in 1816 as untenable, and left the church, Stock was still buried in the Unitarian Lewin's Mead Burial ground, in Brunswick Square, Bristol, where his tomb still survives.[66]

Stock's tyrannical, and clearly opinionated, wife long outlived him. She continued to live at 14 Royal York Crescent, Clifton, until 1847, when she announced the auction sale of its house contents. She then moved to 7 Upper Brook Street, London, where she died on 15 October 1861. Their only son, John Shapland Edmonds Stock (1804–1867), became a noted judge and author.

Notes

1 *London Review of Books*, **31** (19 November, 2009), 21.
2 Roy Porter, "Review of Stansfield, *Thomas Beddoes MD, 1760–1809*", *British Journal for the History of Science*, **19** (1986), 121.
3 Stansfield, 3.
4 Joseph Alfred Bradney, *A History of Monmouthshire* (London: 1904–1932), II (1913), part 2, 186.
5 G. Munro Smith, *A History of the Bristol Royal Infirmary* (Bristol: Arrowsmith, 1917), 180–182.

6 A. H. Shorter, *Paper Mills and Paper Makers in England 1495–1800* (Hilversum: Paper Publications Society, 1957), 59, 66 and 167–168.
7 *General Evening Post*, 24–27 October 1772.
8 NA, Prerogative Court of Canterbury (PCC), PROB 11/1167/142, proved 17 June 1788.
9 NA, PCC will, PROB 11/1228/86, proved 31 January 1793 and see http://sms.thornburyroots.co.uk/CS%209%20owners%20the%20Rolph%20Family.htm.
10 *Felix Farley's Bristol Journal*, 31 October 1789.
11 *Universal British Directory of Trade, Commerce, and Manufacture, 1791–1798* (1795–1798), IV, part 1, 429–430.
12 *Bristol Mercury*, 15 March 1819.
13 Obituary in *Berrow's Worcester Journal*, 3 March 1825.
14 *The Records of the Honourable Society of Lincoln's Inn: Volume I, Admissions Register A.D. 1420-A.D. 1799* (London: Lincoln's Inn, 1896), 537.
15 J. Foster, *Alumni Oxoniensis*, part 2, 1715–1886 (Oxford: Parker, 1888), IV, 1356.
16 W. Munk, 1878, *The Roll of the Royal College of Physicians of London*, 2nd edn. (London: The Royal College of Physicians: London, 1878), III, 12.
17 Anon., *Laws of the Society Instituted at Edinburgh 1782 for the Investigation of Natural History* (Edinburgh: 1803), 30 and 57 (copy in Edinburgh Central Library). His essay survives, in MS, in EUL, in the Society's volumes of *Essays*, XII, for 1793–1794.
18 J. Gray, *History of the Royal Medical Society* (Edinburgh: Edinburgh University Press, 1952), 79.
19 M. Thale, *Selections from the Papers of the London Corresponding Society* (Cambridge: Cambridge University Press, 1983), 119–120.
20 For Robert Watt see ODNB. *Lloyd's Evening Post*, 18–21 August 1794, and see *The Trial of Robert Watt, Late Wine Merchant in Edinburgh, for High Treason* (Edinburgh: Printed for William Brown, 1795), 11.
21 *The Trial of Robert Watt*, 136.
22 On Downie, see John Kay, *Kay's Original Portraits* (Edinburgh: Black, 1877), I, 352–355. Downie's pardon was on condition he be imprisoned for one year and afterwards banished for life from Britain (*Annual Register*, 1795, 17). For America, see M. Durey, *Transatlantic Radicals and the Early American Republic* (Lawrence, KS: Kansas University Press, 1997), 78–79 and 137–138. For the most recent discussion of these Watt-Downie trials, see J. Barrell, *Imagining the King's Death* (Oxford: Oxford University Press, 2000), Chapter 9, 252–284.
23 M. Durey, "William Wickham, the Christ Church Connection and the Rise and Fall of the Security Service in Britain, 1793–1801", *English Historical Review*, **121** (2006), 735.
24 See M. Durey, "Transatlantic Patriotism: Political Exiles and America in the Age of Revolutions", in *Artisans, Peasants & Proletarians 1760–1860*, eds. C. Emsley and J. Walvin (London: Croom Helm, 1985), 13.
25 Stock, *Essay on the Effects of Cold upon the Human Body* (Philadelphia: Printed by Joseph Gales, 1797); see also John Andrews, *Address to the Graduates in Medicine . . . Held May 12, 1797* (Philadelphia: Printed by Ormrod and Conrad, 1797).
26 See W. J. Morris, "John Quincy Adams and Alexander Hill Everett: Pathfinders of German Studies in America", in *Anglo-German and American-German Crosscurrents*, ed. A. O. Lewis (1990), IV, 28–29.
27 The truce was the Peace of Amiens, which lasted from 25 March 1802 until 18 May 1803. N. Vickers, *Coleridge and the Doctors* (Oxford: Clarendon Press, 2004), 37.
28 John Fane, tenth Earl of Westmorland (1759–1841) and politician of Lyme Regis, his pocket borough – see ODNB.

29 Thomas Pelham, second Earl of Chichester (1756–1826) and politician – see ODNB.
30 Charles Hope, Lord Granton (1763–1851), judge, see ODNB.
31 John Edye to Charles Bragge, in the latter's Lydney Park papers, GRO, D421/X10/7/13.
32 GRO, D421/X9/20.
33 Then a major dissenting Academy between 1786 and 1796, see H. McLachlan, *English Education under the Test Acts* (Manchester: Manchester University Press, 1931), 246–255.
34 GRO, D421/X10/8–10.
35 GRO, D421/X10/10 and see ODNB.
36 GRO, D421/X10/9 and, for Richards, see *Jackson's Oxford Journal*, 26 December 1807.
37 Lewin's Mead register, Bristol Record Office.
38 C. H. Cave, *A History of Banking in Bristol* (Bristol: Hemmons, 1899), 47, and *Felix Farley's Bristol Journal*, 7 June 1794, 3 (but there wrongly recorded as Joseph).
39 *Felix Farley's Bristol Journal*, 5 July 1800, 3.
40 A. J. P. Skinner, "Marriage of John Edye", *Devon and Cornwall Notes and Queries*, **9** (2) (1916), 47–48.
41 *Gloucester Journal*, Monday 11 July 1803, 3.
42 J. Murch, *A History of the Presbyterian and General Baptist Churches in the West of England* (London: R. Hunter, 1835), 36–41.
43 See his PCC will, NA, PROB 11/764/426, proved 17 September 1748.
44 *Bath Chronicle*, 24 September 1789, when his daughter Martha's marriage was announced.
45 *Universal British Directory* (1795–1798), III, part 2, 908.
46 J. Foster, *Alumni Oxoniensis*, part 2, 1715–1886 (Oxford: Parker, 1888), IV, 1279.
47 *Gentleman's Magazine*, **68** (2), Supplement (1798), 1157.
48 Munk, *Roll*, III, 12.
49 Stock, 331.
50 *Morning Post*. The book was published by Longmans, London.
51 G. Munro Smith, *A History of the Bristol Royal Infirmary*.
52 E. L. Griggs, *Collected Letters of Samuel Taylor Coleridge* (Oxford: Oxford University Press, 1959), **III**, 68–72 and see also K. Coburn, *The Notebooks of Samuel Taylor Coleridge*, 1804–1808 (London: Routledge & Kegan Paul, 1957), **II**, 3136, where Coleridge again refers to "Dr. Stock and his uxor Tyranna".
53 *Monthly Magazine*, **27**, February, 1809, 68 and *Nicholson's Journal*, **22**, February, 1809, 158 (here recorded as to be by "Dr. Hook").
54 *Gentleman's Magazine*, **80** (1810), 336.
55 A. B. Beaven (1899), *Bristol Lists* (1899), "Bristol, Times & Mirror", 257.
56 NA, PCC will, PROB 11/1537/467.
57 *Felix Farley's Bristol Journal*, 12 September 1812, 3.
58 See "A Layman", *Unitarianism Tried by Scripture and Experience* (London: Hamilton Adam, 1840, 103–104, which reprints a "Narrative of his renunciation", and 105–113; also see Stock's letter, dated 8 November 1816, adopting Trinitarianism, 105–113). Murch (note 42, 117) records how Stock "long a zealous member of the Unitarian congregation, became, in the course of a few weeks, a Calvinistic Baptist."
59 *The Lymiad – A Poem in the form of Letters from Lyme to a Friend in Bath, Written during the Autumn of 1818*, eds. John Fowles and John Constable (Lyme Regis: Philpot Museum, 2011). This poem was certainly written by a female visitor to Lyme. The original MSS even carries her name, as "Charlotte Wm Nth Skinner" on page 1, of whom the second editor could only note: "no trace of a Charlotte Skinner has been found in local records or in online sources"

(8). But this name had recorded, in only slightly misleading form, the poem's actual author. She was born Charlotte Jane Parslow (1785–1872) in Manchester, who then married attorney William North. The friend to whom she was writing her poem, away in Bath, was Maria Sarah Wallis (or Wallace) later Ogle (1792–1844), her first cousin.

60 *The Lymiad*, lines 904–934. Craniology, generally called "phrenology", was all the rage at the time.

61 W. D. Lang, 1956, Mary Anning and Anna Maria Pinney, *Proceedings of the Dorset Natural History and Archaeological Society*, **76** (1956), 146–152.

62 Bristol University archives, diary for 1831–1832, Pinney papers, Box 9.

63 *Some Further Recollections of a Happy Life, Selected from the Journals of Marianne North, Chiefly between the Years 1859 and* 1869, ed. Janet Catherine (North) Symonds (Mrs John Addington Symonds), 2 vols. (London: Macmillan & Co, 1893), I, 10.

64 Michael A. Taylor, "The Plesiosaur's Birthplace: The Bristol Institution and Its Contribution to Vertebrate Palaeontology", *Zoological Journal of the Linnean Society*, **112** (1964), 179–196.

65 *Proceedings of the Annual Meeting. Bristol Institution for the Advancement of Science and Literature and the Arts*, **2** (1824), 22.

66 Beaver H. Blacker, ed., "Munk's 'Roll of Physicans', Gloucestershire Names", *Gloucester Notes and Queries*, **4** (1890), 490.

Appendix 2. Beddoes' borrowings from the Bristol Library Society

Trevor Levere

Bristol Library Society, borrowing register, in Bristol Public Library. 1798

1 3 Apr. – 10 Apr., *Phil. Trans.* vol. 73
2 3 Apr. – 18 Apr., *Phil. Trans.* vol. (77 or 87, illegible)
3 9 Apr. – 10 Apr., Newton's *Optics* (sic), London 1730, 8vo
4 10 Apr. – 12 Apr., *R. Irish Academy*, vol. 5
5 12 Apr. – 18 Apr., *Phil. Trans.* vol. 82
6 18 Apr. – 22 May, *Manchester Mem.*, vol. 4, pt. 2
7 ditto, pt. 1
8 1 June – 13 June, (Edward) Bancroft on Colours, vol. 1(*Philosophy of Permanent Colours, &c.*, London 1794, 8vo)
9 13 June – 28 June, Ingenhousz, *Expériences sur les Végétaux*, vol. 1 (2 vols., Paris, 1787–9, 8vo.)
10 4 July – 9 July, Massinger's *Works*, vol. 2 (*Dramatic Works, with notes by J. M. Mason*, 4 vols., London, 1799, 8vo.)
11 4 July – 10 July, Search(e)'s *Light of Nature* vol. 1
12 9 July – 11 July, " " " vol. 2
13 10 July – 11 July, " " " vol. 3
14 11 July – 13 July, " " " vol. 4
15 13 July – 17 July, *Manchester Memoirs*, vol. 5
16 13 July – 17 July, *Edinburgh Trans.*, vol. 4
17 17 July – 21 Aug., Search's *Light of Nature* vols. 1 and 2
18 21 Aug. – 27 Sept., " " " vol. 1
19 21 Aug. – 29 Aug., " " " vol. 2
20 29 Aug. – 10 Sept., *Phil. Trans.*, vol. 79
21 17 Sept. – 8 Oct., *Arabian Nights*, vol. 4 (*Arabian Nights Entertainment*, 4 vols., London, 1783, 12mo.)
22 27 Sept. – 8 Oct., Plutarch, Morals, by Wittenbach (Plutarch's Morals., Gr. and Lat. a Wyttenbach, Tom. 1, Oxonii 1795, 4to.) [*Ploutarchou tou chaironeos ta ethika. Plutarchi chæronensis moralia, id est opera, exceptis vitis, reliqua. Græca emendavit, . . . animadversiones explicandis rebus ac verbis, item indices copiosos, adjecit, Daniel Wyttenbach, . . .*], vol. 5 (Oxford, Clarendon Press, 1795)

23 8 Oct.–12 Oct., *Forster's Journey*, vol. 1 (probably Geo. Forster's *Voyage round the world, with Capt. Cook, 1772, &c.*, 2 vols., London, 1777, 4to.)
24 12 Oct. – 20 Nov., Strutt's Dresses (*Complete View of the Dress and Habits of the People of England*, London 1796, vol. 2, pt. 1.2.)
25 12 Oct. – 16 Oct., Foster's (Forster's) *travels*, vol. 2
26 Dirom's Narrative (*of the campaign in India, in 1792*, 4to., 1794)
27 2 Nov. – 20 Nov., *Edin. Trans.*, vol. 4
28 20 Nov. – 10 Dec., Strutt's *Dresses*
29 20 Nov. – 10 Dec., *Phil. Trans.* for 1792, vol. 82
30 10 Dec. – 10 Jan. 1799, Strutt's *Dresses*
31 10 Dec. – 27 Dec., (David) Collins, *Account of the English Colony in New South Wales*, 2 vols., 4to. 1798–1802, vol. 1
32 [27 Dec., no return date], *Medical Trans., by the College of Physicians in London*, 4 vol. 1772–85, 8vo., vol. 3

The record shows that Humphry Davy, working for Beddoes, also borrowed volumes of the *Phil. Trans.*, along with books by Priestley, Kirwan, and Lavoisier; an entry on 22 August 1799 is for Davy, but in Beddoes' hand, borrowing William Woodville's "Medical Botany" (*Medical Botany* **2** [1790], **3** [1792], Supp. [1794]). For the borrowings by Coleridge, who was soon a friend of Beddoes and of Davy, see G. Whalley, "The Bristol Library borrowings of Southey and Coleridge, 1793–8", *Library*, **4** (1949), 114–131.

Index

Note: Square brackets indicate uncertain identification or information. Dates are given for Beddoes' contemporaries or sources, but not for authors of secondary works. *Italics* indicate that the individual has an entry in the *ODNB*.